COMBINATION THERAPY AGAINST MULTIDRUG RESISTANCE

COMBINATION THERAPY AGAINST MULTIDRUG RESISTANCE

Edited by

Mohmmad Younus Wani

Assistant Professor, Department of Chemistry, College of Science, University of Jeddah, Jeddah, Saudi Arabia

Aijaz Ahmad

Lecturer, Department of Clinical Microbiology and Infectious Diseases, School of Pathology, Health Sciences, University of the Witwatersrand, Johannesburg, South Africa and Medical Scientist, Division of Infection Control, Charlotte Maxeke Johannesburg Academic Hospital, National Health Laboratory Service (NHLS), Johannesburg, South Africa

ELSEVIER

ACADEMIC PRESS

An imprint of Elsevier

Academic Press is an imprint of Elsevier
125 London Wall, London EC2Y 5AS, United Kingdom
525 B Street, Suite 1650, San Diego, CA 92101, United States
50 Hampshire Street, 5th Floor, Cambridge, MA 02139, United States
The Boulevard, Langford Lane, Kidlington, Oxford OX5 1GB, United Kingdom

Notices

Knowledge and best practice in this field are constantly changing. As new research and experience broaden our understanding, changes in research methods, professional practices, or medical treatment may become necessary.

Practitioners and researchers must always rely on their own experience and knowledge in evaluating and using any information, methods, compounds, or experiments described herein. In using such information or methods they should be mindful of their own safety and the safety of others, including parties for whom they have a professional responsibility.

To the fullest extent of the law, neither the Publisher nor the authors, contributors, or editors, assume any liability for any injury and/or damage to persons or property as a matter of products liability, negligence or otherwise, or from any use or operation of any methods, products, instructions, or ideas contained in the material herein.

Library of Congress Cataloging-in-Publication Data
A catalog record for this book is available from the Library of Congress

British Library Cataloguing-in-Publication Data
A catalogue record for this book is available from the British Library

ISBN: 978-0-12-820576-1

For information on all Academic Press publications
visit our website at https://www.elsevier.com/books-and-journals

Publisher: Andre Gerhard Wolff
Acquisitions Editor: Erin Hill-Parks
Editorial Project Manager: Barbara Makinster
Production Project Manager: Sreejith Viswanathan
Cover Designer: Matthew Limbert

Typeset by SPi Global, India

Contents

4 Combination therapy against human infections caused by *Candida* species

Indresh Kumar Maurya, Ruchi Badoni Semwal, and Deepak Kumar Semwal

5 Metallodrug-driven combination chemotherapy in cancer treatment

Afzal Hussain, Mohamed Fahad AlAjmi, Prince F. Iqbal, and Waseem A. Wani

6 Combination antituberculosis therapy: Opportunities and challenges to combat drug-resistant tuberculosis

Sudarkodi Sukumar, Md. Zafaryab, Md. Khurshid Alam Khan, Krishnan Hajela, and Mohammad Nasiruddin,

7 Synergistic effect of drugs against multiple drug-resistant swine pathogen *Streptococcus suis*

Shama Khan

Contributors

Aijaz Ahmad Department of Clinical Microbiology and Infectious Diseases, School of Pathology, Health Sciences, University of the Witwatersrand; Division of Infection Control, Charlotte Maxeke Johannesburg Academic Hospital, National Health Laboratory Service (NHLS), Johannesburg, South Africa

Mohamed Fahad AlAjmi Department of Pharmacognosy, College of Pharmacy, King Saud University, Riyadh, Saudi Arabia

Mahmood A. Alam Lord Kelvin/Adam Smith (LKAS) fellow, Wellcome Centre for Integrative Parasitology, Institute of Infection, Immunity and Inflammation, University of Glasgow, Glasgow, Scotland

Othman A. Alghamdi Department of Biological Sciences, Faculty of Science, University of Jeddah, Jeddah, Kingdom of Saudi Arabia

N. Bakthavatchala Reddy Ural Federal University, Chemical Engineering Institute, Yekaterinburg, Russian Federation

A. Balakrishna Department of Chemistry, Rajeev Gandhi Memorial College of Engineering and Technology (Autonomous), Nandyal, Andhra Pradesh, India

Adriano Duse Department of Clinical Microbiology and Infectious Diseases, School of Pathology, Health Sciences, University of the Witwatersrand; Division of Infection Control, Charlotte Maxeke Johannesburg Academic Hospital, National Health Laboratory Service (NHLS), Johannesburg, South Africa

Abdul Hafiz Faculty of Medicine, Umm Al-Qura University, Makkah, Kingdom of Saudi Arabia

Krishnan Hajela School of Life Sciences, Devi Ahilya Vishwavidyalaya, Indore, Madhya Pradesh, India

Athar Adil Hashmi Bioinorganic Lab., Department of Chemistry, Jamia Millia Islamia (Central University), New Delhi, India

Syed Masood Husain Molecular Synthesis and Drug Discovery Unit, Centre of Biomedical Research, Lucknow, India

Afzal Hussain Department of Pharmacognosy, College of Pharmacy, King Saud University, Riyadh, Saudi Arabia

Prince F. Iqbal Department of Chemistry, Government Degree College Boys, Pulwama, Jammu and Kashmir, India

Shama Khan Department of Clinical Microbiology and Infectious Diseases, School of Pathology, University of Witwatersrand, Johannesburg, South Africa

Md. Khurshid Alam Khan School of Life Sciences, BS Abdur Rahman Crescent Institute of Science and Technology, Chennai, Tamil Nadu, India

Manzoor Ahmad Malik Bioinorganic Lab., Department of Chemistry, Jamia Millia Islamia (Central University), New Delhi, India

Musa Marimani Department of Clinical Microbiology and Infectious Diseases, School of Pathology, Health Sciences, University of the Witwatersrand, Johannesburg, South Africa

Indresh Kumar Maurya Department of Microbial Biotechnology, Panjab University, Chandigarh, India

Arif Mohammed Department of Biological Sciences, Faculty of Science, University of Jeddah, Jeddah, Kingdom of Saudi Arabia

Rifat Munir Department of Clinical Microbiology and Infectious Diseases; Department of Biosystems, University of Manitoba, Winnipeg, MB, Canada; Department of Immunology, University of Witwatersrand, Johannesburg, Gauteng, South Africa

Mohammad Nasiruddin Clinical Genomics and Molecular Diagnostics, Neuberg Anand Reference Laboratory; Molecular Oncology and Transplant Immunology, MedGenome Labs Pvt Ltd., Bangalore, Karnataka, India

Deepak Kumar Semwal Department of Phytochemistry, Faculty of Biomedical Sciences, Uttarakhand Ayurved University, Dehradun, India

Ruchi Badoni Semwal Department of Chemistry, Pt. Lalit Mohan Sharma Government Postgraduate College, Rishikesh, Uttarakhand, India

Shailesh Kumar Singh Molecular Synthesis and Drug Discovery Unit, Centre of Biomedical Research, Lucknow, India

G. Sravya Ural Federal University, Chemical Engineering Institute, Yekaterinburg, Russian Federation

Sudarkodi Sukumar School of Life Sciences, BS Abdur Rahman Crescent Institute of Science and Technology, Chennai, Tamil Nadu, India

T.V. Surendra Department of Chemistry, Rajeev Gandhi Memorial College of Engineering and Technology (Autonomous), Nandyal, Andhra Pradesh, India

C. Suresh Reddy Department of Organic Chemistry, Sri Venkateswara University, Tirupati, Andhra Pradesh, India

Mohmmad Younus Wani Department of Chemistry, College of Science, University of Jeddah, Jeddah, Saudi Arabia

Waseem A. Wani Department of Chemistry, Government Degree College, Tral, Jammu and Kashmir, India

Md. Zafaryab Department of Zoology, Kirori Mal College, University of Delhi, New Delhi, Delhi, India; School of Life Sciences, Devi Ahilya Vishwavidyalaya, Indore, Madhya Pradesh, India

Grigory V. Zyryanov Ural Federal University, Chemical Engineering Institute; I. Ya. Postovskiy Institute of Organic Synthesis, Ural Division of the Russian Academy of Sciences, Yekaterinburg, Russian Federation

Preface

Humans have been plagued by dreadful diseases for centuries, but during the last few centuries tremendous advances have been made that have either completely cured or minimized the impact of many diseases. Today, hundreds of thousands of powerful medicines are used to treat and often cure conditions that were thought to be incurable or untreatable a couple of decades ago. But, sadly enough, the misuse, overuse, and abuse of these drugs have resulted in the development of multidrug resistance (MDR), which is a global problem of great concern that needs immediate attention and action. New resistance mechanisms are developing and spreading globally, more rapidly than the development of new drugs, therefore threatening our ability to treat any multidrug resistant pathogen or disease. In 2016, 490,000 people globally developed multidrug-resistant TB, and drug resistance has also started to complicate the fight against other diseases like malaria, HIV, and cancer. In recent decades, bacterial resistance to antibiotics has developed faster than the production of new antibiotics, making bacterial infections increasingly difficult to treat. In addition, pharmaceutical companies are showing less interest in developing new antibiotics. Scientists worry that a particularly virulent and deadly "superbug" could one day join the ranks of existing untreatable bacteria, causing a public health catastrophe.

Among the various strategies to combat MDR, combination therapy shows great promise, as it offers potential benefits such as a broad spectrum of efficacy, greater potency than the drugs used in monotherapy, improved safety and tolerability, and reduction in the number of resistant organisms. Combination therapy (or polytherapy) is the use of a combination of drugs to treat a drug-resistant infection or disease, on the theory that if one drug can do something, two or three could accomplish more. While it typically denotes the use of two or more drugs, it can also include immunotherapy and nonmedical therapies, including psychological therapy and other means of therapy or treatment. Combination therapy has been a standard treatment for infections of human immunodeficiency virus (HIV), *Plasmodium falciparum*, *Mycobacterium tuberculosis*, and *Pseudomonas aeruginosa*, and in cystic fibrosis (CF) patients. The growing clinical studies and the recent FDA approval of different combination drugs and regimes for the treatment of different diseases clearly argues that combination therapy affords great opportunities for the discovery and development of novel medicines in the 21st century.

The time is right to provide a state-of-the-art study of MDR and therapeutic strategies against it. Although a plethora of material is currently available in the form of research articles and reviews covering the multidrug resistance problem, there is no comprehensive coverage of the potential of combination therapy as an efficient strategy to combat multidrug resistance in the form of a book. Previous works on combination therapies are either a chapter in a book or a section in a book chapter. Therefore a comprehensive book devoted entirely to combination therapy against multidrug resistance will be an important addition to the literature.

This book is intended to bring to its audience crucial information on multidrug resistance and the potential of combination therapy as an efficient strategy to combat it. The book will enlighten readers on the views of experts, from different countries and having a wide spectrum of backgrounds, who have made significant contributions in understanding drug resistance and handling it with different combination therapies.

This book will also serve as a comprehensive literature guide for beginning researchers and as reference material for the academic and research community involved in tackling the multidrug resistance problem. This book is especially aimed at focusing the attention of researchers in the pharmaceutical industry toward making use of combination therapy as a treatment strategy to tackle the drug-resistance problem. This strategy has already been quite successful in combating drug-resistance in many cases, but work is still needed, both from the research community and industrial sectors, to streamline efforts and understand the need to develop treatment strategies to augment the therapies currently being used. We hope the readers of this book will find in its pages valuable information and views on MDR and combination therapies, and will use it to develop new treatment strategies.

Mohmmad Younus Wani
Aijaz Ahmad

CHAPTER

1

Combination therapy: Current status and future perspectives

Manzoor Ahmad Malik[a], *Mohmmad Younus Wani*[b], *Athar Adil Hashmi*[a]

[a]Bioinorganic Lab., Department of Chemistry, Jamia Millia Islamia (Central University), New Delhi, India, [b]Department of Chemistry, College of Science, University of Jeddah, Jeddah, Saudi Arabia

Abbreviations

AIDS	acquired immunodeficiency syndrome
AMP	antimicrobial peptides
ART	antiretroviral therapy
ARVs	antiretroviral drugs
CDC	Centers for Disease Control
DNA	deoxyribose nucleic acid
FDA	Food and Drug Administration
FDC	fixed dose combination
HIV	human immunodeficiency virus
MBL	metallo-beta-lactamase
MDR	multidrug resistance
MRSA	methicillin-resistant *Staphylococcus aureus*
WHO	World Health Organization

1.1 Introduction

Our bodies are equipped with natural (specific and nonspecific) defense mechanisms to fight off invading microbes or unwanted changes that may cause disease. An infection or a disease occurs when the body's defense system gives up or loses the battle and that is when drugs come to our rescue. Possibly the earliest written accounts of medical therapeutics used by humans are found in the famous Ebers papyrus, which is an almost 20-m long, 110-page medical scroll named after the German

https://doi.org/10.1016/B978-0-12-820576-1.00001-1

Egyptologist Georg Ebers, who acquired it in 1872. Hundreds of treatment options are described in the Ebers papyrus for diseases afflicting ancient Egyptians in ~ 1500 BC, and these treatments involved mixing various herbs, leaves, shrubs, minerals, and animal excreta—forming a combination (Jones, 2011).

The isolation and characterization of the active principles in medicinal plants signified a major challenge, which was later met by the development of synthetic drugs. Today we have 10,000 + ever-more targeted and increasingly powerful high-tech medicines that can treat and often cure conditions that have confounded healers for thousands of years. These drugs, particularly antibiotics, have moved us from being helpless victims of epidemics to being able to fight off potentially deadly diseases. However, sadly enough, with these new medicines has come an increase in the misuse, overuse, and abuse of some of them, which has led to the emergence of multidrug resistance (MDR), a global problem of great concern.

The upsurge in microbial resistance has become a significant threat worldwide. Development of resistance to the available drug candidates has resulted in considerable patient mortality and morbidity (Tamma, Cosgrove, & Maragakis, 2012). Although researchers are involved in the development of new drug candidates for combating the serious problems created by rising multidrug resistance (MDR), the situation appears to be far more complex. During the past two decades, only two new classes of antibiotics have been introduced into clinical use, but none of them is assuredly active against gram-negative bacteria. Daptomycin, which was offered clinically in 2003, lost ground a year later due to the development of resistance in patients with *Enterococcus faecium* and MRSA infections (Worthington & Melander, 2013). Multidrug resistance against other life-threatening diseases such as malaria, tuberculosis, cancer, and viral diseases such as HIV/AIDS is already ringing alarm bells; it is feared that we may lose the battle if new treatment modalities or strategies are not discovered. Among the various strategies that could augment the current treatment regimen and add to the armament against multidrug resistance is combination therapy. It is imperative to examine the influence of drugs beyond what they can accomplish alone. A combination of drug candidates can act as a multiplier and thereby increase the sum of their benefits.

Drug combinations have been discussed for the treatment of diagnosed conditions for some time, such as aspolol (a combination of atenolol and aspirin) for patients diagnosed with cardiovascular disease. For preventive use, Wald and Law proposed the use of a combination of well-known and inexpensive medications in one pill (called a polypill) for defense against cardiovascular disease (Wald & Law, 2003). Based on the emerging and ongoing clinical studies and research outcomes it is now clear that the use of drug combinations is an effective approach for the treatment of complicated and refractory diseases. The Drug Combination Database (DCDB) has a collection of 1363 drug combinations (330 approved and 1033 investigational, including 237 unsuccessful usages), involving 904 individual drugs and 805 targets; the purpose of the database is to facilitate in-depth analyses and to provide a basis for theoretical modeling and simulation of such drug combinations (Yanbin et al., 2014). Almost 10,000 clinical trials are presently registered in the United States alone, exploring combination therapies for the successful eradication of cancer, infectious diseases, and cardiovascular, neurological, autoimmune, and metabolic disorders. Numerous research articles contain details on drug combinations. Yet, these numbers are quite modest relative to all possible and potent combinations that could be tested (Rationalizing combination therapies, 2017). The numbers of US Food and Drug Administration (FDA) approvals of different classes of combination drugs have, however, increased since the approval of the first combination drugs in the 1940s (Fig. 1.1). Between the 1940s and 1950s the highest growth rate (37.5%) occurred, while the lowest growth rate occurred between the 1960s and 1970s (9%). This initial surge followed by the slowed approval rate of combination drugs was due to the FDA's new strict criteria and support guidelines announced in 1971. Since 1971, a strong indication has been required to show that a combination drug offers a therapeutic benefit in comparison to each individual drug entity (Finland, 1974). In 1943, the combination drug hycodan (homatropine + hydrocodone), which has been discontinued, was

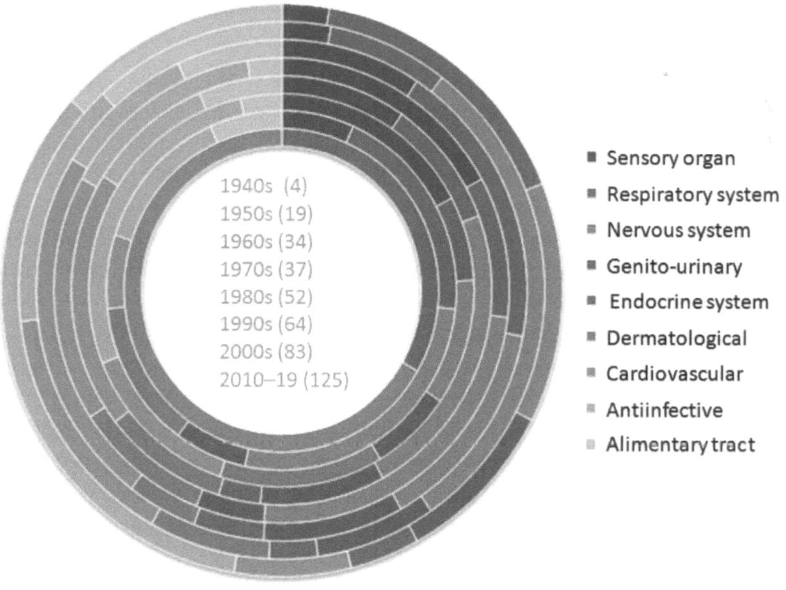

FIG. 1.1 FDA approval of combination drugs by decade (1940s–2019).

the earliest combination to be approved. Cafergot, a combination of ergotamine and caffeine, was approved in 1948, and since then more than 400 new drug combinations have been approved. The FDA approval of drug combinations for different classes of diseases, including HIV, cancer, and infectious diseases, is now on the rise, with the majority of drug combinations approved for treating infectious diseases (Fig. 1.1).

Drug combinations can have synergistic or even additive effects. However, it is imperative to note that not all seemingly rational combinations of drugs will yield better efficacy or safety. For example, a 2004 trial disclosed that torcetrapib could raise HDL and lower LDL both with and without an added statin (Brousseau et al., 2004). However, its combination with atorvastatin (Lipitor) led to a 60% increase in deaths (Nissen et al., 2007). Therefore, in-depth investigations of the known cases of successful and unsuccessful drug combinations are needed to recognize the patterns of valuable drug interactions and to deliver the bases for rational design of efficient drug combinations.

1.2 Combination rule

A combination drug is a fixed-dose combination (FDC) of two or more pharmaceutically active ingredients combined in a single dosage form. As one or both drugs are typically already FDA approved, there might appear to be no need for approval of many drug combinations. However, since the FDA approval is for the drugs alone and not for the combination, FDA approval for the combination is therefore required. For example, to develop an aztreonam-avibactum combination drug, which involves the FDA-approved beta-lactamase resistant aztreonam and a non-beta-lactam-beta-lactamase inhibitor, avibactum, which also restores aztreonam's activity against isolates expressing multiple beta-lactamases, phase II and phase III clinical studies were required for FDA approval. Another example is the combination of naltraxone HCl and bupropion HCl for weight management, for which similar clinical trials were required for approval. In many cases, additional studies beyond those required for the original approval of the two component drugs are required because the FDA must apply what is commonly referred to as the "combination rule." This rule states that two or more drugs may be combined in a single dosage form when each component makes a contribution to the claimed effects and the dosage of each component (amount, frequency, duration) is such that the combination is safe and effective for a significant patient population requiring such concurrent therapy as defined in the labeling for the drug.

In order to fulfill the requirements of the combination rule, a combination must demonstrate that each component contributes to the safety or efficacy of the product. This typically requires a multifactorial study in which each component is compared separately and in combination with a placebo control. For example, if the product aims to combine Drug A and Drug B, clinical studies with the following treatment arms may be required to demonstrate that each component contributes to the overall effect of the drug:

Placebo
Drug A
Drug B
Drug A and Drug B combined.

Developing a triple combination (three-drug combination) may require eight treatment arms. However, the complexity increases even more when one considers that the dose of each drug must be justified in terms of dosing frequency, amount of each drug, and duration of each effect. If the dosage requires clinical justification, the required studies may be so numerous or so large that they may not be logistically or financially possible.

1.3 Combination therapy: Current status

In combination therapy, two or more drug candidates are used together to achieve better results than the routinely used monotherapy. Typically, combination therapy has one or more of the following aims: (i) decreasing the rate at which acquired resistance arises by combining drugs with minimal cross-resistance, such that emergence of resistance requires attainment of multiple mutations in rapid succession—an unlikely event; (ii) lowering the doses of drugs with nonoverlapping toxicity and similar therapeutic profile so as to attain efficiency with fewer side effects; (iii) sensitizing cells to the action of a drug through the use of another drug (chemosensitization) or radiation (radiosensitization), often by altering cell-cycle stage or growth properties (cytokinetic optimization); and (iv) achieving enhanced potency by exploiting additivity, or better yet, greater-than-additive effects, in the biochemical activities of two drugs. The objectives of combination therapy are not mutually exclusive, and good combinations like ABV (doxorubicin, bleomycin, vinblastine) or BEP (bleomycin, etoposide, cisplatin) accomplish several objectives, including positive cytokinetic and biological interaction (with and without surgery), and reduced toxicity (Fitzgerald, Schoeberl, Nielsen, & Sorger, 2006). Combination therapy has been a standard treatment for infections of human immunodeficiency virus (HIV) and in cystic fibrosis (CF), *Plasmodium falciparum*, *Mycobacterium tuberculosis*, and *Pseudomonas aeruginosa* patients (Vestergaard et al., 2016). Previous studies have endorsed that the advantage of combination therapy against MDR *P. aeruginosa* infections was mainly due to an increased possibility of selecting an effective agent rather than an in vitro synergy among antimicrobials or prevention of resistance (Garnacho-Montero, Sa-Borges, Sole-Violan, et al., 2007). Against other MDR gram-negative bacteria, however, quite a few studies have revealed higher efficiency and lesser levels of resistance for combination treatment in comparison to monotherapy (Batirel, Balkan, Karabay, et al., 2014; Daikos, Petrikkos, Psichogiou, et al., 2009; Zarkotou, Pournaras, Tselioti, et al., 2011). An upsurge in the survival rate was detected when colistin was administered as part of combination treatments with tigecycline (an aminoglycoside) or meropenem (particularly for carbapenem, minimal inhibitory concentration (MIC) below 4mg/L). Higher elimination rates were also confirmed for the combination of colistin with rifampin compared to colistin monotherapy against *A. baumannii* (Durante-Mangoni, Signoriello, Andini, et al., 2013). Similarly, synergistic effects of fosfomycin have been established with carbapenems (along with a decrease in the appearance of resistance), aminoglycosides, and quinolones (Samonis, Maraki, Karageorgopoulos, et al., 2012). Lower mortality rates were observed in *Klebsiella pneumoniae* carbapenemase (KPC)-associated infections on using a triple combination therapy containing a carbapenem, tigecycline, and colistin. A recent metaanalysis, including studies carried out in the United States, Greece, and Italy, has related the clinical results of combination therapy versus monotherapy for treating carbapenemase-producing *Enterobacteriaceae* infections, primarily KPC bloodstream infections (BSIs) (Tzouvelekis, Markogiannakis, Piperaki, et al., 2014). A significant difference in the mortality rates for carbapenem-containing regimens (18.8%) compared to the noncarbapenem-containing regimens (30.7) signifies that the presence of a carbapenem in the combination may afford a better survival benefit. Overall, current data backs the usage of combination therapy involving colistin and/or tigecycline along with a carbapenem in the therapy of invasive infections caused due to carbapenem-resistant *K. pneumoniae*, specifically in severe infections.

1.3.1 Bacterial infections

The overuse and misuse of antibiotics in the healthcare and agricultural industries has resulted in the spread of bacterial resistance worldwide. The recent rise of multidrug resistant (MDR) bacteria, particularly the strains of earlier susceptible bacteria (notably *staphylococci*) and other bacterial species (like those of *aerobacter*, *proteus*, and *pseudomonas*) resistant to the commonly used antibacterial agents has resulted in a call for the adoption and development of new strategies to tackle this global issue. The rise of extended-spectrum beta-lactamase (ESBL)-producing bacteria, carbapenem-hydrolyzing beta-lactamases, or carbapenemases including metallo-beta-lactamases (MBLs; e.g., New Delhi metallo-β-lactamases, plasmid-mediated imipenem-type carbapenemases, Verona integron-encoded metallo-β-lactamases) and the oxacillinase group of β-lactamases (OXA)-type carbapenemases (e.g., OXA-23 in *A. baumannii* and OXA-48 in *K. pneumoniae*) has significantly increased the need for new antibiotics. The delay in the evolution of new antibiotics led to the successful use of combination drugs to treat multidrug-resistant bacterial infections, which began with the use of a combination of tetracycline and oleandomycin. Several new combinations or some mixtures of antibiotics with several sulfonamides and with other drugs exhibiting pharmacologic properties have been introduced. Since the FDA approval of third-generation cephalosporin and the β-lactamase inhibitor combination ceftazidime/avibactam in 2015, the combination of meropenem with the β-lactamase inhibitor vaborbactam in 2017 represented a paradigm shift from the development of new antibiotics to the use of already available drugs in combination (Table 1.1). As of July 2018, a total of about 48 antibiotics (along with some combinations) and 10 biologicals that target the WHO priority pathogens *Mycobacterium tuberculosis* and *Clostridium difficile* were in the pipeline (Table 1.2) (WHO, 2018). Quite recently the FDA approved the antibacterial injection drug Recarbrio, a combination of imipenem, cilastatin, and relebactum, to treat adults suffering from complicated urinary tract infections (cUTIs) and complicated intraabdominal infections (cIAIs) (FDA, 2019).

TABLE 1.1 FDA-approved new drug combination for treating bacterial infections.

Trade name	Combination	Treatment
Recarbrio	Imipenem, cilastatin, and relebactam	Treating adults with complicated urinary tract infections and complicated intraabdominal infections
Dalbavancin	Dalvance, Xydalba	For treatment options for methicillin-resistant *S. aureus* (MRSA) infections
Zerbaxa	Ceftolozane-tazobactam	For the treatment of complicated intraabdominal and urinary tract infections and bacterial pneumonia
Avycaz	Ceftazidime-avibactam	For the treatment of complicated intraabdominal infections in combination with metronidazole and urinary tract infections
Ceftobiprole	Zevtera, Mabelio	For the treatment of community-acquired pneumonia and hospital-acquired pneumonia (HAP) in adults
Coartem	Artemether/lumefantrine	For the treatment of malaria infections due to *Plasmodium falciparum*

TABLE 1.2 Antimalarial combinations deployed for clinical use.

Trade name	Combination	Treatment
Coartem	Artemether and lumefantrine	For treating nonsevere malaria
Malarone	Atovaquone and proguanil	Prevent malaria by interfering with the growth of parasites in the red blood cells of the human body
Fansidar	Sulfadoxine and pyrimethamine	For the treatment of acute, uncomplicated *P. falciparum* malaria
Fansimef	Mefloquine-suladoxine and pyrimethamine	For prophylaxis of falciparum malaria
Coarsucam or ASAQ	Artesunate/amodiaquine	Recommended by the WHO for uncomplicated falciparum malaria
Artequin or ASMQ	Artesunate and mefloquine	Recommended by the WHO for uncomplicated falciparum malaria
Ariplus or Amalar plus	Artesunate and sulfadoxine/pyrimethamine	Recommended by the WHO for uncomplicated falciparum malaria
Pyramax	Pyronaridine and artesunate	For the treatment of *P. falciparum* and *P. vivax*

Combination therapy has largely been used to fight MDR bacterial infections and the current focus on their applications using antimicrobial peptides (AMPs) may permit antibiotics to be potent against resistant bacterial strains. Research on a wide range of AMPs displayed promising activity against some resistant strains. Increased understanding of the mode of action of AMPs has revealed similarity and complementarity to conventional antibiotics and the combination of both has led to synergistic effects in some cases. By joining an antibiotic with an AMP, a new compound may be obtained with possibly superior activity and reduced side-effects and toxicity. Such antimicrobial peptide conjugates can act as unique adjuvants for the antibiotic by disturbing the resistance mechanisms of bacteria, thereby permitting the antibiotic to once again be effective (Sheard, O'Brien-Simpson, Wade, & Separovic, 2019).

The combination of fosfomycin with other antimicrobial agents has displayed good efficiency against multidrug-resistant (MDR) bacteria in both clinical and in vitro studies. Further studies have been carried out to investigate the synergism and also the optimal intravenous dosing regimens of fosfomycin with meropenem against MDR and MBL-producing *P. aeruginosa* strains. Rifampicin acts as an in vivo as well as in vitro bactericidal agent but, due to the rapid development of resistance, a combination of this antibiotic along with another agent is always used to avoid resistance for treating serious infections. Hamilton-Miller noticed synergy between fosfomycin and rifampicin against methicillin-resistant *Staphylococcus aureus* (MRSA) and coagulase-negative *staphylococci* (Katharina, Manfred, Kristian, Miglioli, & Allerberger, 2001). In a study carried out by Nakazawa et al., it was revealed that the combination of fosfomycin with various antibiotics enhanced their antimicrobial potential against MRSA (Nakazawa, Kikuchi, Honda, Isago, & Nozaki, 2003). Yu et al. studied the in vitro antibacterial efficiency of a combination of fosfomycin and other antimicrobial agents against KPC-producing *Klebsiella pneumoniae*. They reported that the selected five drug combinations (fosfomycin combined with imipenem, ertapenem, tigecycline, colistin, or amikacin) displayed noteworthy additive effect against 136 KPC-Kp strains in an in vitro checkerboard assay. Furthermore, the time-kill assays exposed the fact that fosfomycin improved the bactericidal activity of the five other antimicrobial agents (Yu et al., 2017).

The high affinity of avibactam (NXL104) (with class A and class C beta-lactamases) was studied to observe the resistances caused by extended-spectrum β-lactamases (ESBLs), *K. pneumoniae* carbapenemases (KPCs), the oxacillinase group of b-lactamases (OXA), and ampicillin type b-lactamases (AmpC) generating organisms. Mostly the combination of avibactam with ceftazidime has been explored and it is also in clinical development in combination with aztreonam and ceftaroline. Among these, the aztreonam/avibactam combination could conceivably be employed in NDM-1-producing bacteria, signifying optimistic results in treating KPC and ESBL-producing strains (Crandon & Nicolau, 2013). In another report, the ceftaroline/avibactam combination has displayed in vitro and in vivo synergistic activity against ESBL and KPC-producing *Klebsiella* species, and against ceftazidime-resistant *Enterobacter cloacae* strains, but not against *P. aeruginosa* and *A. baumannii* (Sader, Fritsche, Kaniga, Ge, & Jones, 2005).

A recent report on the inhibition of different targets within the same pathway involves the targeting of wall teichoic acid synthesis by the natural product tunicamycin, which prevents the first enzyme in the biosynthesis of wall teichoic acid, the *N*-acetylglucosamine-1-phosphate transferase TarO, in *S. aureus*; it also shows a dramatic synergy with β-lactam antibiotics, reducing the minimum inhibitory concentration (MIC) of oxacillin against one MRSA clinical isolate from 50 to 0.4 μg/mL at a concentration of just 0.08 μg/mL (Campbell et al., 2011). Another TarO inhibitor was later recognized for the ability to enhance the activity of the cephalosporin cefuroxime against MRSA. The drug ticlopidine (antiplatelet drug) did not show antibiotic activity alone, but was strongly synergistic with cefuroxime against various MRSA strains, dropping MICs by 64-fold. TarO was identified as the molecular target of ticlopidine and the activity was shown to be dependent on the presence of the tarO gene (Farha et al., 2013).

Furthermore, there are a number of known inhibitors of cell wall synthesis (fosfomycin, bacitracin, teicoplanin, β-chloro-D-alanine, D-cycloserine, and vancomycin) at subinhibitory concentrations ($0.25 \times$ MIC), which display notable synergy with β-lactam antibiotics and have the capability to bring a decrease in the methicillin resistant *S. aureus* strain, causing MIC reductions of 2-fold (for β-chloro-D-alanine) to 128-fold (for fosfomycin) (Sieradzki & Tomasz, 1997).

1.3.2 Fungal infections

Candida albicans, a leading opportunistic fungal pathogen, has resulted in increased mortality rates, particularly in immunocompromised and high-risk surgical patients. Systemic infections triggered by *C. albicans* have become a serious public health threat and are accountable for more than 50% of human candidiasis infections. Moreover, the development of antifungal drugs is relatively slower than that of antibacterial antibiotics, and the increasing evidence for antifungal drug resistance reduces the efficacy of known antifungals. Synergistic drug combinations have been demonstrated to be a valid and pragmatic approach for seeking drugs that have novel mode of actions. Targeting virulence factors is an important alternate approach that will open a new pipeline for the discovery of antifungal drugs. Cui et al., studied the synergistic combinations, including antifungals and antivirulence agents, against *C. albicans*. It was established that the synergistic combination of antifungal drugs and antivirulence agents proved to be a promising method to combat the drug-resistant strains of *C. albicans*, by targeting both the pathogenic process and the cell growth (Cui, Ren, Tong, Dai, & Zhang, 2015).

A combination of echinocandin and an azole has proved to be another well-known antifungal combination for the treatment of invasive candida infections. No clinical trials on this combination have been reported, but the combinations of posaconazole with caspofungin or micafungin have been examined in vitro and in animals (Cui et al., 2015). Chen et al. confirmed that posaconazole displays in vitro and in vivo synergy with caspofungin against *C. albicans*, including echinocandin-resistant isolates (Chen, Lehman, Averette, Perfect, & Heitman, 2013).

Barchiesi, Falconi Di Francesco, and Scalise (1997) studied the combination interaction of terbinafine-fluconazole and terbinafine-itraconazole, and in more than 40% of the combinations synergistic activity was noted. Comparable conclusions were described when a terbinafine-azole combination was scrutinized in vitro against *C. neoformans* (Guerra, Ishida, Nucci, & Rozenta, 2012) and dermatophytes (Ashley & Johnson, 2011).

Many reports suggest the successful combination of flucytosine and amphotericin (AmB) for targeting cryptococcal meningitis (Tobias, Wrigley, & Shaw, 1976). Bennett et al., showed that such a combination

TABLE 1.3 Some commonly used combination antifungal (CAF) therapies.

Disease	CAF therapy
Invasive candidiasis	AmB + flucytosine AmB + azole Echinocandin + azole
Invasive aspergillosis	Azole or AmB + echinocandin Voriconazole + anidulafungin
Mucormycosis	AmB + posaconazole or caspofungin

AmB, amphotericin B.

regimen displayed an increased cure rate and fewer relapses than the monotherapy with AmB, and, more remarkably, the combination regimen displayed lower nephrotoxicity than the monotherapy regimen (Bennett, Dismukes, et al., 1979). Some commonly used combination antifungal (CAF) therapies for the treatment of invasive candidiasis, invasive aspergillosis, and mucormycosis are given in Table 1.3.

1.3.3 Tuberculosis

Therapeutic options for tuberculosis (TB) are inadequate and notoriously ineffective, regardless of the wide variety of potent antibiotics accessible for the treatment of other bacterial infections. The WHO has estimated that there are about 9.4 million new TB cases and almost 1.7 million deaths occur annually, worldwide (WHO, 2010). The currently recommended TB regimen for drug-susceptible (DS) disease is prolonged (at least 6 months), with a successful cure rate of about 95% under optimal conditions (Koul, Arnoult, Lounis, Guillemont, & Andries, 2011). Moreover, the emergence of resistant strains and coinfection with HIV have confirmed TB as a rising global public health threat (WHO, 2010). Multidrug-resistant *Mycobacterium tuberculosis* (MDR-TB) strains are resistant to rifampin and isoniazid, the two first-line TB drugs; extensively drug-resistant *M. tuberculosis* (XDR-TB) strains have, in addition, acquired resistance to any fluoroquinolone and to any one of the three injectable second-line anti-TB drugs (amikacin, kanamycin, or capreomycin) (Jassal & Bishai, 2009). Effective MDR-TB therapy is more toxic to patients than conventional treatment for DS-TB, and is more costly, more prolonged (lasting up to 2 years), and more uncertain (cure rates typically range from 50% to 70%) (Koul et al., 2011). All these complications are even more severe for XDR-TB patients (Chiang, Centis, & Migliori, 2010). While more new compounds are under examination (Koul et al., 2011), alternate therapies are immediately required both to curtail the duration of the existing TB treatment and effectively treat MDR- and XDR-TB. One economical

TABLE 1.4 Combination antituberculosis drugs.

Drug name	Combination
IsonaRif	Isoniazid/rifampin
Rifamate (Pro)	Isoniazid/rifampin
Rifater (Pro)	Isoniazid/pyrazinamide/rifampin

solution is to find new usages for existing drugs ("repurposing") (Boguski, Mandl, & Sukhatme, 2009; Chong & Sullivan, 2007), either alone or in combination therapies. Antituberculosis combinations containing more than one drug are given simultaneously to treat TB, as listed in Table 1.4. These different drugs have different mechanisms of action and are directed in combination to evade the development of drug resistant *M. tuberculosis* strains. Employing drug combinations that have different mechanisms of action and that target the bacterial species in different ways can make the treatment options more effective.

1.3.4 Malaria

Malaria is a grave global health concern that accounts for substantial morbidity and mortality (Olliaro & Taylor, 2004). It is accepted that, out of the four human malarial species *Plasmodium malariae, Plasmodium ovale, Plasmodium vivax*, and *Plasmodium falciparum*, *P.falciparum* and *P.vivax* account for more than 95% of malarial infections in humans (Loy et al., 2017). Estimations of the global burden differ but a figure of 400–600 million cases per year appears to be reasonable (Pacorel et al., 2010). To tackle this rising global concern, several strategies are currently being used, including impregnated bed nets, integrated vector control, reducing anemia and parasitemia in pregnant women through recurrent presumptive treatment, and quick access to reliable diagnosis and effective treatment. Chemotherapy has been adopted as a sustainable and realistic approach by the WHO in the control of malaria. Indiscriminate use of available antimalaria drugs has resulted in the development of resistance, consequently rendering them ineffective. Resistance to the present antimalarial drugs has been an important factor in increased rates of morbidity and mortality. The key mechanism for the development of resistance in the microorganisms is naturally occurring genetic mutation in the malaria parasite, which confers a survival advantage. Such mutations result in the decline of drug sensitivity that depends on the class of antimalarial drug (White & Pongtavornpinyo, 2003). Inadequate treatment (like subtherapeutic dose or suboptimal drug) of a large biomass infection will seldom kill mutant parasites and thus becomes the main selective pressure for resistance. Transmission of such resistant parasites to other individuals occurs by mosquitoes. Furthermore, drugs with long half-lives have more chances to be select for resistance, because lower drug concentrations delay and are capable of killing sensitive parasites only (Nzila et al., 2000).

The idea of delaying drug resistance by combination of drugs with varied targets was first pioneered in treating tuberculosis and was subsequently implemented in cancer and HIV infection treatment. The usage of drug combinations in antimalarial treatment to delay antimicrobial drug resistance is a new strategy that will extend the life of existing drugs (Peters, 1990). However, the rationale behind most of the combinations is to exploit marked synergy (chlorproguanil plus sulfones, pyrimethamine plus sulfonamides, atovaquone plus proguanil). Some antimalarial drug combinations can be shown to be antagonistic in vitro, but the effects are generally small, and there is no evidence that this translates into a reduced in vivo effect. There is certainly no reason to reject a drug combination on the basis of mild antagonism between the compounds, particularly where one of the drugs is rapidly eliminated (White, 1998). In Thailand in 1984 a combination of three known antimalarial drugs, i.e., mefloquine, pyrimethamine, and sulfadoxine, was intended against the treatment of multidrug resistant *falciparum* explicitly to delay the onset of resistance to mefloquine, but it did not work because *P. falciparum* in Thailand was already highly resistant to pyrimethamine-sulfadoxine. Recently, artemisinin and its derivatives have been combined with other antimalarials, notably mefloquine, thus resulting in accelerated recoveries, increased cure rates, reduced transmissibility; this appears to have delayed the development of further resistance (Peters & Robinson, 2016; White, 1999).

In a study carried out by Nakornchai et al., synergistic effects were observed against parasite growth on using the combinations of chloroquine with azithromycin or erythromycin. Quinine-azithromycin and quinine-erythromycin combinations displayed potentiation. Additive effects were observed in mefloquine-azithromycin and mefloquine-erythromycin combinations. Similar results were also produced by pyronaridine in combination with azithromycin or erythromycin. However, artesunate-azithromycin and artesunate-erythromycin combinations had antagonistic effects. The in vitro data advocates that azithromycin and erythromycin will have clinical utility in combination with chloroquine and quinine (Nakornchai & Konthiang, 2006).

1.3.5 Cancer

The complexity and risks associated with cancer underscores the need for an assortment of treatment approaches, which is why a combination of one or more therapeutic interventions or treatment regimens is often used to battle cancer. Use of combination therapy is sought out as a superior approach to treat different cancers as compared to monotherapy. Combination therapy has been found to achieve higher efficacy with lower doses of less toxic drugs, or even lower doses of toxic drugs, with much better results. In combination therapy for cancer, two or more drugs are used to chemo-sensitize cells, sometimes making the additional compound more potent. The combination can have additive or synergistic effects, minimize drug resistance, or fight against expected resistance. Although monotherapy is still a common approach, it is often believed that even better results could potentially be obtained when these therapies become rationally combined with others. Combination therapy has been extended for treating drug-resistant tumors and has long been adopted as the first-line treatment of several malignancies to improve the clinical outcome. Furthermore, anticancer drug combinations have been shown to generally induce synergistic drug actions and deter the onset of drug resistance. Scientists are pioneering many different methods to discover and test novel drug combinations that may be able to overcome multiple mechanisms of resistance or delay their emergence. If these efforts are successful, it could potentially transform cancer for many patients.

One combination treatment approach is to coadminister drugs that work by different molecular mechanisms, thereby increasing tumor cell killing while reducing the likelihood of drug resistance and minimizing overlapping toxicity. Another approach is to treat patients with drugs that block the particular mechanism of resistance by which their tumors have developed, and then treat them again with the drug to which they grew resistant. The idea is that this combination approach may "resensitize" the patients to the original treatment.

Almost all the drugs used for treatment of different cancers face several transportation barriers on their tortuous journey to their sites of action. To overcome these barriers, an effective drug delivery system for cancer therapy is imperative. A self-delivery system, which actually involves a combination approach in which anticancer drugs could be delivered by themselves without any carriers, has been developed. In this drug delivery system, an amphiphilic drug-drug conjugate (ADDC) has been synthesized from the hydrophilic anticancer drug irinotecan (Ir) and the hydrophobic anticancer drug chlorambucil (Cb) via a hydrolyzable ester linkage (Fig. 1.2). This amphiphilic conjugate has the ability to self-assemble into nanoparticles in water and exhibit longer blood retention half-life compared to the free drugs, which facilitates its accumulation in the tumor tissues and promotes their cellular uptake. After internalization, the conjugate undergoes hydrolytic cleavage of the ester linkage due to different cleavable enzymes, thus releasing the active drugs at the tumor sites, resulting in enhanced and better antitumor activity compared to the drugs alone. This drug conjugate also has the potential to overcome multidrug resistance (Huang et al., 2014).

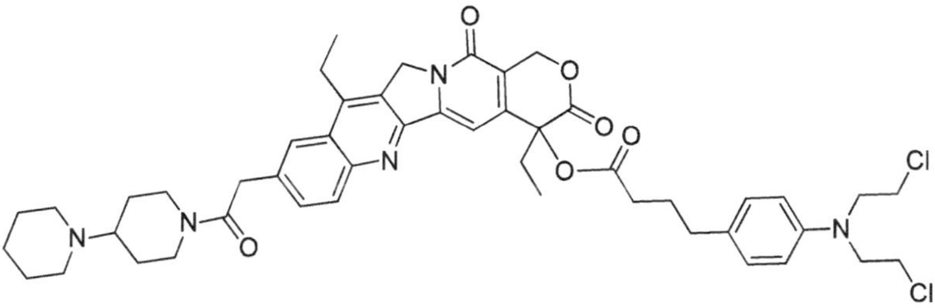

FIG. 1.2 Amphiphilic drug-drug conjugate (ADDC).

Trastuzumab, a humanized monoclonal antibody targeting the ErbB2 extracellular domain, was the first clinical breakthrough for the treatment of ErbB2-positive breast cancer, yet most of the metastatic breast cancer does not respond to trastuzumab therapy alone. A randomized study of lapatinib alone or in combination with trastuzumab in ErbB2-positive breast cancer patients has demonstrated a synergistic interaction between trastuzumab and lapatinib. This study showed that dual blockade by using two anticancer agents is more effective than a single agent alone. The combination of trastuzumab and lapatinib provided a statistically significant improvement in progression-free survival compared with single-agent lapatinib, with a hazard ratio of 0.73 (95% CI, 0.57–0.93; $P = .008$) (Blackwell et al., 2010).

The aromatase inhibitor anastrazole inhibits estrogen synthesis and fulvestrant, on the other hand, binds and accelerates degradation of estrogen receptors. Mehta and his coworkers hypothesized that the combination of these two agents could result in higher efficacy in patients with hormone-receptor (HR)-positive metastatic breast cancer. It was found that the combination of anastrazole and fulvestrant was superior to anastrazole alone or sequential anastrazole and fulvestrant for the treatment of hormone-receptor (HR)-positive metastatic breast cancer, with no significant interactions (Mehta, Barlow, et al., 2012).

The combination of standard anticancer chemotherapy drugs with histone deacetylases (HDACs) is a promising strategy to treat hard-to-treat cancers. In some recent in vitro and in vivo studies, the use of well-tolerated doses of HDACs in combination with platinum-based chemotherapeutics has resulted in significant improvements in many cancer types and stages. The use of HDAC inhibitors for chemosensitization to increase the efficacy of Pt-based chemotherapeutics has been explored in several studies and it was found that the HDACs were beneficial for even cross-resistant strains of malignancies at very low doses. Preclinical studies in which standard drugs and standard-of-care Pt-based agents have been used in combination with HDACs have shown synergistic improvement in cytotoxicity to cancer cells (Diyabalanage, Granda, & Hooker, 2013).

Cisplatin, a platinum-based chemotherapeutic agent has been effectively used for the treatment of cervical cancer. Despite its excellent antineoplastic activity, some severe side effects like nephrotoxicity and ototoxicity limit its use. In a study performed by Jin and coworkers, a combination treatment of cisplatin and suberoylanilide hydroxamic acid (SAHA) was tested on HeLa cells. In this study it was found that even a low concentration of SAHA worked synergistically with cisplatin to induce cytotoxicity to a greater extent than with either agent alone. The combination of cisplatin and SAHA induced a marked reduction in cell viability, as determined by Western blot analysis and DAPI staining, showing significantly increased caspase-3 activation and apoptosis in HeLa cells treated with the combination drugs compared to the treatment with cisplatin alone (Jin et al., 2010). Carboplatin, an analog of cisplatin with less nonhematologic toxicity than the parent molecule has been used with paclitaxel, a non-Pt-based mitotic inhibitor. A combination of carboplatin and paclitaxel has been successfully used to treat solid malignancies, resulting in less platelet toxicity compare to the treatment with carboplatin alone (Bali et al., 2005; Ozols, Bundy, Greer, et al., 2003). Further studies on the same combination showed that SAHA improved the efficacy of carboplatin and paclitaxel in patients with advanced nonsmall-cell lung cancer (NSCLC) (Ramalingam, Maitland, Frankel, et al., 2010).

To improve the efficiency of Pt-based chemotherapeutic agents, in a study by Poon and coworkers a novel nanoparticle platform was developed based on nanoscale coordination polymer-1 (NCP-1) for simultaneous delivery of two chemotherapeutics, oxaliplatin and gemcitabine monophosphate (GMP) at 30 wt% and 12 wt% drug loadings, respectively. A strong synergistic effect of oxaliplatin and GMP was observed in vitro against AsPc-1 and BxPc-3 pancreatic cancer cells. NCP-1 acted as a novel nanocarrier for the codelivery of two chemotherapeutic agents having distinctive mechanisms of action to simultaneously disrupt multiple cancer pathways with maximal therapeutic efficacy and minimal side effects (Poon, He, Liu, Lu, & Lin, 2015).

A combination of methotrexate, 6-mercaptopurine, vincristine, and prednisone (formally known as the POMP regimen) has been successfully used in reducing tumor burden with prolonged remission in pediatric patients with acute lymphocytic leukemia (Frei et al., 1965). The success of the POMP regimen stimulated further intensive research, which focused on the use of combination therapies that could target different pathways to deliver synergistic or an additive effect.

Quinn et al. in a study discovered that sabutoclax, a pan-Bcl-2 inhibitor, in combination with minocycline, an antibiotic that has previously shown anticancer activity, acted synergistically on the intrinsic apoptotic pathway. In addition, this combination exhibited selective toxicity and an attenuation in tumor growth in vitro and in vivo on pancreatic ductal adenocarcinoma (Quinn et al., 2015). Thus, by means of a drug combination that targets different pathways, a synergistic or potentiation effect could yield significant anticancer results. A new, early-phase clinical trial has established that the immunotherapy agent tremelimumab, combined with microwave ablation therapy, is safe for patients with advanced biliary tract cancer (BTC). The results, published December 22, 2018 in the journal *Hepatology*, mark an important stepping-stone toward pinpointing which patients may benefit most from a combination of a new class of immunotherapy drugs called immune checkpoint inhibitors with radiation therapy (Xie et al., 2019).

The combined use of ABRAXANE (paclitaxel protein-bound particles for injectable suspension, albumin-bound) and gemcitabine has been approved by the FDA for the treatment of metastatic adenocarcinoma of the pancreas. Each one of the drugs used in the combination suits a specific purpose: paclitaxel acts as a microtubule inhibitor, keeping cancer cells from growing and spreading, while gemcitabine replaces segments of cell DNA to force apoptosis (programmed cell death).

Over the past decades, more than 30 randomized phase III clinical trials have been conducted in patients with advanced pancreatic cancer, and ABRAXANE (paclitaxal protein-bound) in combination with gemcitabine is the first new treatment approved for metastatic adenocarcinoma of the pancreas. Some FDA-approved and in clinical trials drug combinations against different cancers are given in Table 1.5.

1.3.6 Viral diseases

1.3.6.1 HIV (human immunodeficiency virus)

The drugs used to treat HIV are called antiretroviral drugs (ARVs). There are different types of antiretroviral drugs and they work in different ways. HIV treatment is made up of three or more antiretroviral drugs taken together, sometimes combined into one pill. Since there are a lot of ARVs, they can be combined in different ways. The World Health Organization (WHO) has recommended a combination of antiretroviral drugs for people starting HIV treatment. People on ART (antiretroviral treatment) take a combination of HIV medicines (called an HIV treatment regimen) every day. A person's initial HIV regimen generally includes three HIV medicines from at least two different HIV drug classes. The different classes of HIV drugs are broadly described as follows and are listed in Table 1.6.

1. **Nucleoside reverse transcriptase inhibitors (NRTIs):** NRTIs block reverse transcriptase, an enzyme HIV needs to make copies of itself. These drugs also have other actions that prevent HIV from replicating in the body. They are also referred to as "nukes."
2. **Nonnucleoside reverse transcriptase inhibitors (NNRTIs):** NNRTIs bind to and later alter reverse transcriptase. They stop the virus from replicating.

TABLE 1.5 FDA-approved/in-clinical-trial drug combinations against different cancers.

Drug combination	Treatment
In clinical trials (phase I, II, and III)	
Everolimus (Afinitor) and exemestane	Breast cancer
Trastuzumab (Herceptin) and pertuzumab (Perjeta)	HER2-positive metastatic breast cancer
Olaparib and cediranib	Ovarian cancer
TAS 102 and ramucirumab	Gastric or gastroesophageal junction cancer
Temozolomide with trifluridine and tipiracil hydrochloride (TAS-102)	Neuroendocrine tumors
Trifluridine and tipiracil hydrochloride (TAS-102) and oxaliplatin	Stage IV colon or colorectal cancer
Venetoclax (Venclexta) with obinutuzumab (Gazyva)	Chronic lymphocytic leukemia (CLL) or small lymphocytic lymphoma (SLL)
Venetoclax (Venclexta) and rituximab (Rituxan)	Chronic lymphocytic leukemia (CLL)
Obinutuzumab plus chlorambucil	Untreated CLL
Afatinib (Gilotrif) and cetuximab (Erbitux)	Nonsmall cell lung cancer
Bortezomib (Velcade), lenalidomide (Revlimid) and dexamethasone	Multiple myeloma
Binimetinib and palbociclib	KRAS and NRAS mutation positive colorectal cancer
Vincristine and YK-4-279 (an experimental drug)	Ewing sarcoma tumors
Nivolumab (Opdivo) and ipilimumab (Yervoy)	Advanced kidney cancer
Carfilzomib, lenalidomide (Revlimid) and dexamethasone	Multiple myeloma
Trastuzumab (Herceptin), pertuzumab (Perjeta) and docetaxel	HER2-positive tumors
Carboplatin and gemcitabine plus bevacizumab	Ovarian cancer
Carboplatin and paclitaxel plus bevacizumab	Ovarian cancer
Nivolumab plus tetrahydrouridine-decitabine	Nonsmall cell lung cancer

TABLE 1.5 FDA-approved/in-clinical-trial drug combinations against different cancers—cont'd

Drug combination	Treatment
Rituximab plus Hu5F9-G4 (experimental drug)	Non-Hodgkin lymphoma
Metformin and bicalutamide	Prostate cancer
Bortezomib and MD5-1 (agonist antibody)	Lung metastases of kidney or breast cancer
Pembrolizumab plus LMB-100 (immunotoxin)	Malignant mesothelioma
Liposomal irinotecan and veliparib	Metastatic tumors
FDA-approved drug combinations (2014–19)	
Xpovio (selinexor) plus dexamethasone	Relapsed or refractory multiple myeloma (RRMM)
Darzalex, lenalidomide and dexamethasone	Multiple myeloma
Polatuzumab vedotin-piiq (Polivy), bendamustine, and rituximab	Relapsed or refractory diffuse large B-cell lymphoma (DLBCL)
Lenalidomide (Revlimid) and rituximab	Follicular lymphoma (FL), marginal zone lymphoma (MZL)
Avelumab (Bavencio) and axitinib pembrolizumab (Keytruda) plus axitinib	Advanced renal cell carcinoma (RCC)
Atezolizumab (Tecentriq), carboplatin, and etoposide	Extensive-stage small cell lung cancer (ES-SCLC)
Atezolizumab (Tecentriq), bevacizumab, paclitaxel, and carboplatin	Metastatic nonsquamous, nonsmall cell lung cancer (NSq NSCLC) with no EGFR or ALK genomic tumor aberrations
Venetoclax (Venclexta), azacytidine or decitabine	Newly diagnosed acute myeloid leukemia (AML)
Glasdegib (Daurismo) with low-dose cytarabine (LDAC)	Newly diagnosed acute myeloid leukemia (AML)
Pembrolizumab (Keytruda), carboplatin, and either paclitaxel or nab-paclitaxel	Metastatic squamous nonsmall cell lung cancer (NSCLC)
Pembrolizumab (Keytruda), pemetrexed and platinum	Metastatic, nonsquamous nonsmall cell lung cancer (NSqNSCLC)
Pertuzumab (Perjeta) and trastuzumab	HER2-positive early breast cancer
Abemaciclib (Verzenio) and fulvestrant	HR-positive, HER2-negative advanced or metastatic breast cancer
Cabazitaxel (20 mg/m^2 every 3 weeks) (Jevtana) and prednisone	Metastatic castration-resistant prostate cancer

Continued

TABLE 1.5 FDA-approved/in-clinical-trial drug combinations against different cancers—cont'd

Drug combination	Treatment
Liposome-encapsulated combination of daunorubicin and cytarabine (Vyxeos)	Newly diagnosed therapy-related AML (t-AML) or AML with myelodysplasia-related changes (AML-MRC)
Dabrafenib and trametinib	Metastatic nonsmall cell lung cancer (NSCLC)
Rituximab and hyaluronidase human (Rituxan Hycela)	Follicular lymphoma, diffuse large B-cell lymphoma, and chronic lymphocytic leukemia
Pembrolizumab (Keytruda), pemetrexed, and carboplatin	Metastatic nonsquamous nonsmall cell lung cancer (NSCLC)
Ribociclib (Kisqali) in combination with an aromatase inhibitor	Postmenopausal women with hormone receptor (HR)-positive, human epidermal growth factor receptor 2 (HER2)-negative advanced or metastatic breast cancer
Daratumumab (Darzalex) in combination with lenalidomide and dexamethasone, or bortezomib and dexamethasone	Multiple myeloma
Lenvatinib (Lenvima) and everolimus	Advanced renal cell carcinoma, locally recurrent or metastatic, progressive, radioactive iodine-refractory differentiated thyroid cancer
Obinutuzumab in combination with bendamustine followed by obinutuzumab monotherapy	Follicular lymphoma (FL)
Obinutuzumab and chlorambucil	Chronic lymphocytic leukemia
Palbociclib (Ibrance) and fulvestrant	Hormone receptor (HR)-positive, (HER2)-negative advanced or metastatic breast cancer
Elotuzumab (Empliciti), lenalidomide, and dexamethasone	Multiple myeloma
Necitumumab (Portrazza), gemcitabine, and cisplatin	Metastatic squamous nonsmall cell lung cancer (NSCLC)
Trametinib (Mekinist) plus dabrafenib (Tafinlar)	Unresectable or metastatic melanoma with BRAF V600E or V600K mutations
Ixazomib (Ninlaro), lenalidomide, and dexamethasone	Multiple myeloma
Carfilzomib (Kyprolis), lenalidomide, and dexamethasone	Relapsed multiple myeloma

TABLE 1.5 FDA-approved/in-clinical-trial drug combinations against different cancers—cont'd

Drug combination	Treatment
Ramucirumab (Cyramza) in combination with FOLFIRI (made up of folinic acid, fluorouracil and irinotecan)	Metastatic colorectal cancer (mCRC)
Dinutuximab (Unituxin) in combination with granulocyte-macrophage colony-stimulating factor (GM-CSF), interleukin-2 (IL-2), and 13-cis-retinoic acid (RA)	Pediatric patients with high-risk neuroblastoma
Panobinostat (Farydak), bortezomib, and dexamethasone	Multiple myeloma
Palbociclib (Ibrance) plus letrozole	Postmenopausal women with estrogen receptor (ER)-positive, human epidermal growth factor receptor 2 (HER2)-negative advanced breast cancer
Ramucirumab (Cyramza injection) in combination with docetaxel	Metastatic nonsmall cell lung cancer (NSCLC)
Bevacizumab solution for intravenous infusion in combination with paclitaxel, pegylated liposomal doxorubicin, or topotecan	Platinum-resistant, recurrent epithelial ovarian, fallopian tube, or primary peritoneal cancer
Ramucirumab (Cyramza) plus paclitaxel	Advanced gastric or gastroesophageal junction (GEJ) adenocarcinoma
Ofatumumab (Arzerra injection), for intravenous infusion in combination with chlorambucil	Chronic lymphocytic leukemia (CLL)
Trametinib and dabrafenib	Unresectable or metastatic melanoma with a BRAF V600E or V600K mutation as detected by an FDA-approved test

Data taken from National Cancer Institute (NCI) and FDA website.

TABLE 1.6 FDA approval of HIV medicines—Timeline.

1981	First AIDS cases reported in the United States
1987	Zidovudine (NTRI)
1991	Didanosine (NTRI)
1992	Zalcitabine (NTRI)
1994	Stavudine (NTRI)
1995	Lamivudine (NTRI), Saquinavir (PI)

Continued

TABLE 1.6 FDA approval of HIV medicines—Timeline—cont'd

1996	Indinavir (PI), Nevirapine (NNRTI), Ritonavir (PI)
1997	Combivir (FDC), Delavirdine (NNRTI), Nelfinavir (PI)
1998	Abacavir (NTRI), Efavirenz (NNRTI)
1999	Amprenavir (PI)
2000	Didanosine EC (NTRI), Kaletra (FDC), Trizivir (FDC)
2001	Tenofovir DF(NRTI)
2003	Atazanavir (PI), Emtricitabine (NTRI), Enfuvirtide (FI), Fosamprenavir (PI)
2004	Epzicom (FDC), Truvada (FDC)
2005	Tipranavir (PI)
2006	Atripla (FDC), Darunavir (PI)
2007	Maraviroc (CA), Raltegravir (INSTI)
2008	Etravirine (NNRTI)
2011	Complera (FDC), Nevirapine XR (NNRTI), Rilpivirine (NNRTI)
2012	Stribild (FDC)
2013	Dolutegravir (INSTI)
2014	Cobicistat (PE), Elvitegravir (INSTI), Triumeq (FDC)
2015	Evotaz (FDC), Genvoya (FDC), Prezcobix (FDC)
2016	Descovy (FDC), Odefsey (FDC)
2017	Juluca (FDC)
2018	Biktarvy (FDC), Cimduo (FDC), Delstrigo (FDC), Doravirine (NNRTI), Ibalizumab-uiyk (PAI), Symfi (FDC), Symfi Lo (FDC), Symtuza (FDC), Temixys (FDC)
2019	Dovato (FDC)

CA, CCR5 antagonist; *FDC*, fixed dose combination; *FI*, fusion inhibitors; *INSTI*, integrase inhibitor; *NNRTI*, nonnucleoside reverse transcriptase inhibitors; *NRTI*, nucleoside reverse transcriptase inhibitors; *PE*, pharmacokinetic enhancer; *PI*, protease inhibitor; *PAI*, postattachment inhibitor.
Data taken from the AIDSinfo website, US Department of Health and Human Services.

3. **Protease inhibitors (PIs):** PIs block HIV protease. This is a protein that HIV needs to replicate in the body. When protease can't do its job, the virus can't complete the process that makes new copies. This reduces the number of viruses that can infect more cells.
4. **Fusion inhibitors (FIs):** Also known as entry inhibitors, they block HIV from entering the CD4 cells of the immune system.
5. **CCR5 antagonists (CAs):** CCR5 antagonists block CCR5 coreceptors on the surface of certain immune cells that HIV needs to enter the cells.

6. **Integrase inhibitors (INSTIs):** Integrase inhibitors block HIV integrase, a viral enzyme that HIV uses to infect T-cells. Integrase inhibitors are usually among the first HIV drugs used in patients who have recently contracted HIV because they work well and have minimal side effects.
7. **Postattachment inhibitors (PAIs):** Postattachment inhibitors block CD4 receptors on the surface of certain immune cells that HIV needs to enter the cells.
8. **Pharmacokinetic enhancers (PEs):** Pharmacokinetic enhancers are used in HIV treatment to increase the effectiveness of an HIV medicine included in an HIV regimen (US Department, n.d.).

People with HIV need to take more than one drug, because attacking HIV from multiple directions reduces the viral load more quickly. It also helps prevent resistance to the drug being used. Combination drugs combine medications from different classes into one drug form. The drugs are combined to make a complete HIV regimen, as listed in Table 1.7. This type of regimen is usually used to treat people who have never taken HIV medication before.

1.3.6.2 Influenza

Influenza viruses belong to the *Orthomyxoviridae* family and are classified as A, B, or C. Influenza A viruses circulate in several species, including humans, horses and related animals, swine, and birds, while type B affects only humans. Influenza caused by type A and B is indistinguishable; in contrast, type C causes mild respiratory symptoms (Joosting, Head, Bynoe, & Tyrrell, 1968; Kauppila et al., 2014; Mosnier et al., 2015). The emergence of pandemic H1N1 influenza viruses in April 2009 and the continuous evolution of highly pathogenic H5N1 influenza viruses underscore the urgency of novel approaches to tackle this human influenza infection. There are currently six anti-influenza drugs available on the market but only four are FDA approved and recommended by CDC, which includes oseltamivir phosphate (available as a generic version or under the trade name Tamiflu), zanamivir (trade name Relenza), peramivir (trade name Rapivab), and baloxavirmarboxil (trade name Xofluza) (CDC, 2018). The first three drugs inhibit the virus by blocking a viral enzyme called neuraminidase, so they are also called neuraminidase inhibitors (NAIs). In contrast, baloxavir inhibits a subunit of the viral polymerase, the enzyme responsible for influenza virus replication. In simple terms, baloxavir interferes with the ability of the flu virus to multiply, while the other drugs interfere with the ability of the flu virus to spread within the body. The other two drugs, amantadine and rimantadine, are not recommended, because circulating viruses have a high rate of resistance to these compounds and are ineffective against type B and

TABLE 1.7 Combination HIV medicines.

Year approved	Trade name	Combination
1997	Combivir	Lamivudine and zidovudine
2000	Kaletra	Lopinavir and ritonavir
	Trizivir	Abacavir, lamivudine, and zidovudine
2004	Epzicom	Abacavir and lamivudine
	Truvada	Emtricitabine and tenofovir disoproxil fumarate
2006	Atripla	Efavirenz, emtricitabine, and tenofovir disoproxil fumarate
2011	Complera	Emtricitabine, rilpivirine, and tenofovir disoproxil fumarate
2012	Stribild	Elvitegravir, cobicistat, emtricitabine, and tenofovir disoproxil fumarate
2014	Triumeq	Abacavir, dolutegravir, and lamivudine
2015	Evotaz	Atazanavir and cobicistat
	Prezcobix	Darunavir and cobicistat
	Genvoya	Elvitegravir, cobicistat, emtricitabine, and tenofovir alafenamide
2016	Odefsey	Emtricitabine, rilpivirine, and tenofovir alafenamide
	Descovy	Emtricitabine and tenofovir alafenamide
2017	Juluca	Dolutegravir and rilpivirine
2018	Biktarvy	Bictegravir, emtricitabine, and tenofovir alafenamide
	Symtuza	Darunavir, cobicistat, emtricitabine, and tenofovir alafenamide
	Delstrigo	Doravirine, lamivudine, and tenofovir disoproxil fumarate
	Symfi	Efavirenz, lamivudine, and tenofovir disoproxil fumarate
	Symfi Lo	Efavirenz, lamivudine, and tenofovir disoproxil fumarate
	Cimduo	Lamivudine and tenofovir disoproxil fumarate
2019	Dovato	Dolutegravir and lamivudine

Data taken from the AIDSinfo website, US Department of Health and Human Services.

C influenza viruses (Hussain, Galvin, Haw, Nutsford, & Husain, 2017). However, some studies have found that the combination of amantadine or rimantadine with ribavirin and/or interferon showed additive or synergistic antiviral activity against a range of influenza A viruses, without increased toxicity (Hayden, 1986). Another combination that has shown potentially more efficacy in experimental murine influenza is parenterally administered rimantadine and aprotinin, a polypeptide antiprotease that inhibits cleavage activation of the virus hemagglutinin (Zhirnov, 1987). Drug combination studies that have assessed the effects of double (two) and triple (three) drug combinations on influenza virus infection in vitro and in vivo (Table 1.8) have shown the potential for synergistic or additive antiviral activity and for inhibiting the development of resistance; these qualities must now be demonstrated in clinical trials. All the available information supports the initiation of clinical trials on combination chemotherapy for influenza.

1.4 Conclusions

With limited treatment options and drugs to tackle multidrug resistance in different infections and diseases, combination therapy affords great opportunities in the discovery of novel medicines in the 21st century. The well-established use of appropriate antibiotic combinations to achieve broad-spectrum coverage in cases where the infective organism is unknown and where the need for rapid treatment is acute remains the best argument for empirical combination of antibiotics. Combination of antibiotics with nonantibiotic adjutants or potentiators offers a very promising and effective area for antibiotic discovery and development. In an era where discovery and development of new antibiotics is at a nadir, reviving our existing antibiotic drug classes provides an excellent opportunity to extend the life of well-established and clinically validated drugs. The only disadvantage in the combined use of antibiotics is the opportunity for unnecessary antibiotic exposure that fuels resistance in patients and in the healthcare settings. Therefore, combined antibiotics or antimicrobials should be given only after careful diagnosis has been made, and in doses that would be optimal for each drug if used alone. For example, for delaying the emergence of resistant strains of *Tubercle bacilli*, combinations of two or more drugs such as streptomycin, isoniazid, and para-aminosalicylic acid, are recommended. A combination of chloramphenicol with erythromycin is also indicated to delay emergence of resistance of *staphylococci* to the antibiotic.

Combination therapy with two or more active drugs has been found to be more effective than monotherapy for the treatment of resistant pathogens, especially *Enterobacteraceae* and *A. baumannii*. Other drug

TABLE 1.8 Drug combination for influenza virus infection and their in vitro and in vivo effects.

Drug combinations	Influenza strain (subtype)	Method of analysis	Drug interactions	Ref.
Two-drug combination				
Rimantadine + ribavirin	A/Chick/Germany/27 (FPV Weybridge) (H7N7)[a]	Virus yield inhibition (PFU)	Additive/synergistic	Galegov, Pushkarskaya, Obrosova-Serova, and Zhdanov (1977)
	A/USSR/77 (H1N1) A/Texas/77 (H3N2)	Virus yield inhibition (PFU)	Additive/synergistic	Hayden, Douglas, and Simons (1980)
	A/England/80 (H1N1)[b] A/Aichi/68 (H3N2) B/Lee/40	Virus yield inhibition (PFU)	Subadditive/ additive-synergistic	Madren, Shipman, and Hayden (1995)
	A/Virginia/87 (H1N1) A/Virginia/88 (H3N2)	Virus yield inhibition (ELISA)	Additive	Smee, Hurst, Wong, Bailey, and Morrey (2009)
Amantadine + ribavirin	A/USSR/90/77 (H1N1) A/Texas/1/77 (H3N2) A/New Jersey/76 (H1N1)	Virus yield inhibition (PFU)	Enhanced inhibitory effect	Hayden et al. (1980)
	A/Duck/MN/1525/81 (H5N1)	Virus yield inhibition ($TCID_{50}$)	Synergistic[c]	Nguyen et al. (2010)
	A/Duck/MN/1525/81 (H5N1)[d]	Virus yield inhibition ($TCID_{50}$)	No added benefit[c]	Nguyen et al. (2010)
	A/California/04/09 (H1N1)[e] A/California/05/09 (H1N1)[e] A/California/10/09 (H1N1)[e]	Virus yield inhibition (NR staining)	Additive[c]	Smee, Bailey, Morrison, and Sidwell (2002)

Zanamivir + ribavirin	A/Virginia/87 (H1N1) A/Virginia/88 (H3N2)	Virus replication inhibition (ELISA)	Additive	Govorkova, Fang, Tan, and Webster (2004)
Zanamivir + rimantadine	A/New Caledonia/20/99 (H1N1) A/Panama/2007/99 (H3N2)	Virus ($TCID_{50}$, MDCK) and cell-associated yield inhibition (ELISA)	Additive/synergistic[c]	Ilyushina, Bovin, Webster, and Govorkova (2006)
Oseltamivir carboxylate + amantadine	A/Nanchang/1/99 (H1N1) A/Panama/2007/99 (H3N2) A/Hong Kong/156/97 (H5N1)	Virus yield inhibition (PFU)	Enhanced inhibitory effect	Nguyen et al. (2009)
	A/Duck/MN/1525/81 (H5N1)	Virus yield inhibition ($TCID_{50}$)	Synergistic[c]	Nguyen et al. (2010)
	A/Duck/MN/1525/81 (H5N1)[d]	Virus yield inhibition ($TCID_{50}$)	No added benefit[c]	Nguyen et al. (2010)
Oseltamivir carboxylate + ribavirin	A/California/04/09 (H1N1)[e] A/California/05/09 (H1N1)[e] A/California/10/09 (H1N1)[e]	Virus yield inhibition (staining with NR)	Additive[c]	Smee et al. (2002)
Oseltamivir carboxylate + rimantadine	A/New Caledonia/20/99 (H1N1) A/Panama/2007/99 (H3N2)	Virus ($TCID_{50}$) and cell-associated yield inhibition (ELISA)	Additive/synergistic[c]	Ilyushina et al. (2006)
Peramivir + rimantadine	A/New Caledonia/20/99 (H1N1) A/Panama/2007/99 (H3N2)	Virus ($TCID_{50}$) and cell-associated yield inhibition (ELISA)	Additive/synergistic[c]	Ilyushina et al. (2006)
Peramivir + ribavirin	A/NWS/33 (H1N1)	Virus yield inhibition ($TCID_{50}$)	Synergistic[c]	Govorkova et al. (2004)
Oseltamivir carboxylate + favipiravir	A/Québec/144147/09 (H1N1)	Virus ($TCID_{50}$, MDCK) PCR	Enhanced inhibitory effect	Baz, Carbonneau, Rhéaume, Cavanagh, and Boivin (2018)

Continued

TABLE 1.8 Drug combination for influenza virus infection and their in vitro and in vivo effects—cont'd

Drug combinations	Influenza strain (subtype)	Method of analysis	Drug interactions	**Ref.**
Favipiravir + peramivir	A/Korea/2785/2009 (H1N1)	IC_{50}, ED_{50}, DBA/2 mice model	Synergistic	Park et al. (2014)
Favipiravir + oseltamivir	A/California/07/2009 (H1N1) A/Hong Kong/2369/2009 (H1N1)	Virus yield reduction (VYR) ($CCID_{50}$, MDCK)	Synergistic[c]/additive	Tarbet et al. (2014)
Favipiravir + peramivir	A/California/07/2009 (H1N1) A/Hong Kong/2369/2009 (H1N1)		Synergistic[c]/additive	
Favipiravir + zanamivir	A/California/07/2009 (H1N1) A/Hong Kong/2369/2009 (H1N1)		Synergistic[c]/additive	
Favipiravir + oseltamivir phosphate or laninamivir octanoate	A/California/04/2009 (H1N1)	BALB/c (BALB/c CrSlc) and nude (BALB/c-nu/nu) mice models	Enhanced efficacy	Kiso et al. (2018)
Nitazoxanide + oseltamivir or zanamivir	A/Puerto Rico/8/1934 (H1N1) A/WSN/1933 (H1N1) A/chicken/Italy/9097/1997 (H5N9)	Cell culture-based assays ($TCID_{50}$, MDCK)	Synergistic	Belardo, Cenciarelli, Frazia, Rossignol, and Santoroa (2015)
Carrageenan + zanamivir	A/Hansa Hamburg/01/09 (H1N1(09)pdm) A/Teal/Germany/Wv632/05 (H5N1) A/Turkey/Germany/R11/01 (H7N7) A/Aichi/2/68 (H3N2)	Cell culture-based assays, C57BL/6 mice models	Synergistic, higher efficacy	Morokutti-Kurz et al. (2015)

Oseltamivir + T-705	HPAI A/Turkey/15/2006 (H5N1)	Cell culture-based assays, IHC staining, BALB/c mice models	Highly protective efficacy	Marathe et al. (2016)
Oseltamivir + zanamivir	A/Mexico/4108/2009 (MX/09) (H1N1) A/Hong Kong/2369/2009 (H1N1) A/Brazil/1633/2008 (H1N1)	Hollow Fiber Infection Model (HFIM)	Higher efficacy, prevented spread of drug-resistant viruses	Camilly et al. (2018)
Three drug combination				
Oseltamivir carboxylate + amantadine + ribavirin	A/New Caledonia/20/99 (H1N1) A/Sydney/05/97 (H3N2) A/Duck/MN/1525/81 (H5N1)	Virus yield inhibition ($TCID_{50}$), NR staining, RNA copies	Highly synergistic[c]	Nguyen et al. (2010)
	A/California/04/09 (H1N1)[e] A/California/05/09 (H1N1)[e] A/California/10/09 (H1N1)[e]	Virus yield inhibition (NR staining)	Synergistic[c]	Smee et al. (2002)
	A/NewCaledonia/20/99(H1N1)[d] A/Wisconsin/67/05 (H3N2)[d] A/Duck/MN/1525/81 (H5N1)[d]	Virus yield inhibition (NR staining)	Additive[c]	Smee et al. (2002)
	A/Mississippi/3/01 (H1N1)[f] *A/Hawaii/21/07 (H1N1)*[f]	Virus yield inhibition (NR staining)	Synergistic[c]	Smee et al. (2002)

[a] *Studies were done in chick embryo fibroblasts cells.*
[b] *Studies were done in primary rhesus monkey kidney cells.*
[c] *Drug-drug interactions were analyzed by the three-dimensional model of Prichard and Shipman (1990) using the MacSynergyTM II software program.*
[d] *Amantadine-resistant influenza virus variant (shown in italic).*
[e] *Alternative subtype designation (H1N1 pdm 2009).*
[f] *Oseltamivir-resistant influenza virus variant (shown in italic).*

Note: Drug interactions were evaluated based on inhibition of extracellular (in some experiments cell-associated) virus yield or virus replication in Madin Darby canine kidney (MDCK) cells (unless otherwise indicated). Unless indicated, the viruses were sensitive to all drugs used in the study.
Abbreviations: *ELISA*, enzyme linked immunosorbent assay; *NR*, neutral red; *PFU*, plaque forming units; *Ref.*, references; *TCID*, tissue culture infectious dose.

combinations are currently under investigation or in the last stage of development, although they present some limitations such as lack of new mechanisms of action and consistent gaps in their spectrum against MBL-producing gram negative pathogens. Among the new compounds, ceftolozane-tazobactam, avibactam combinations, and plazomicin have shown wider spectrum of activity. Overall, careful and tailored use of currently available antibiotics either alone or in combinations along with adequate infection-prevention procedures and antimicrobial stewardship remains mandatory to minimize the infection burden caused by MDR bacteria.

Combination therapy has been found to be an effective strategy to tackle MDR- and XDR-TB. The use of existing anti-TB drugs with different mechanisms of action, given simultaneously, has significantly restrained or stopped the emergence of drug-resistant strains of *M. tuberculosis*. Using medications with different mechanisms of action targets the bacteria in different ways and makes treatment more effective. In fact, the idea of delaying drug resistance by combination of drugs with different targets was first pioneered in the treatment of tuberculosis and subsequently adopted in cancer and HIV infection treatment.

Combination of standard Pt-based anticancer chemotherapy drugs with histone deacetylases (HDACs), such as suberoylanilide hydroxamic acid (SAHA), and with gemcitabine monophosphate (GMP) have been used against nonsmall-cell lung cancer (NSCLC), cervical cancer, and pancreatic cancer. Lapatinib in combination with trastuzumab has been used against ErbB2-positive breast cancer. Several other combinations, including sabutoclax, a pan-Bcl-2 inhibitor, in combination with minocycline, an antibiotic that has previously shown anticancer activity, was found to be effective against pancreatic ductal adenocarcinoma compared to monotherapy. Thus, by means of a drug combination that targets different pathways, a synergistic effect could yield significant anticancer results.

The World Health Organization (WHO) has recommended a combination of antiretroviral drugs for people starting HIV treatment. People on ART (antiretroviral treatment) take a combination of HIV medicines (called an HIV treatment regimen) every day. A person's initial HIV regimen generally includes three HIV medicines from at least two different HIV drug classes used to treat HIV.

The broad host range of influenza virus and interspecies transmission are critical factors for its continuous circulation and evolution in nature. The intermediate hosts such as pigs, birds, and horses play a crucial role in maintaining the viruses and their transfer to humans. Therefore, in addition to developing new medicines or a universal vaccine, an effective management of such hosts to restrict the circulation and generation of new and more virulent variants of the viruses is urgently needed.

1.5 Future perspectives

The development of combination therapies or regimens is more complicated than the development of single drugs or monotherapy. However, the case is compelling given that monotherapy gives rise to resistance rapidly because of the one drug-one target approach and finding new antibiotics has almost failed for over a quarter of a century. A very obvious threat is that all antibiotics will at some point be compromised by increasing levels of resistance. Furthermore, as alternative antimicrobial agents such as antivirulence compounds and phage cocktails become increasingly attractive, it is very likely that these will be used in combination with antibiotics to curb the growing menace of drug resistance. Given the proven success of antibiotic-adjuvant combinations and congruous antibiotic and antiviral combinations in the clinic today, there is an excellent reason to speculate that the future of medicinal chemistry may be dominated by combination therapeutics.

The unnecessary antibiotic exposure in instances of unregulated combinations and the lack of rapid, reliable diagnostics to guide clinicians in initial therapy are issues that can be managed by more stringent regulations and innovations in molecular diagnostics.

Since combination drugs have been proven to increase efficacy and suppress resistance, compared to monotherapy, additional efforts to identify fixed-dose combinations should be explored. Such combinations will need to be powered by well-designed clinical trials, and the issue of who will sponsor such trials or pursue the development of effective drug formulations when most of the drugs in question are off-patent is an important policy question.

Another bigger challenge to the successful development of combination drugs as new medicines lies in the complex pharmacology of antibiotic action. Achieving the correct therapeutic levels or doses and the duration for a single antibiotic is already exceedingly difficult. Achieving these goals for two or more compounds in the combination that must be matched in terms of their pharmacokinetics and dynamics to maintain synergy considerably increases the complexity of drug development. Of course, before clinical trials, the toxicity of each component and the combination must be thoroughly investigated in case there are unexpected drug-drug interactions. For higher-order combinations in which more than two drugs are involved there is more complexity. One solution to this is the synthesis of single-agent hybrids that combine, in one molecule, the bioactive domains of each component. Such hybrids can suppress resistance and even gain new modes of action. Indeed, some hybrids such as MCB3837 and cadazoid, both oxazolidinone-quinolone hybrids, and cefilavancin, a glycopeptides-cephalosporin heterodimer, are currently in clinical trials. These hybrids may, however, exhibit difficulties in cell permeation, especially in the case of gram-negative bacteria in which porin exclusion limits the penetration of small molecules that are > 600 Da. Although not insurmountable, the synthesis of bigger compounds with more functionality is not a straightforward recipe for success. Despite these challenges, the time is right for renewed interest and efforts in developing new congruous and syncretic drug combinations to address the antibiotic resistance crisis.

The enormous number of drug combinations that would need to be tested to find the best match for each patient is daunting. To overcome this, researchers are developing methods to systematically evaluate large numbers of potential drug combinations. In one approach the effects of multiple combinations of FDA-approved drugs on different cancer cell lines are simultaneously tested, a method called high-throughput screening.

New drug-delivery technologies that could enable efficient and sustained delivery of combination drugs at synergistic ratios to the target sites would revolutionize combination therapy as a strategy to combat multidrug resistance. For example, as a drug delivery technology, the CombiPlex platform determines synergistic drug ratios in vitro and identifies an appropriate nanoscale carrier, such as liposomes or polymeric nanoparticles, to maintain that ratio in vivo and enhance its delivery to the site of action. CPX-351, a dual drug liposomal encapsulation of cytarabine and daunorubicin, is the first clinical proof-of-concept of the CombiPlex platform. It is likely that with such technologies capable of enhancing targeted delivery to specific tumor types, a greater variety of drug combinations may be possible for the treatment of a broad spectrum of malignancies in the decades to come.

It seems practically impossible to eradicate the influenza virus (IAV) from its zoonistic hosts using existing knowledge and approaches. There is an urgent need to gain molecular and genetic understanding of why some animals, such as sheep and rabbits, are resistant to IAV infection and some (e.g., ducks) are resistant to IAV disease and identify the genes that confer such resistance. Then, by inserting those genes or using gene-editing methods (e.g., RNA interference and CRISPR-Cas), the immediate host could potentially be made resistant to IAV infection. This will reduce and potentially eliminate IAV from the intermediate hosts and consequently its maintenance and transfer to humans. Initial proof-of-concept studies in this direction have already begun to restrict IAV transmission or replication in transgenic animals.

References

Ashley, E. D., & Johnson, M. D. (2011). Combination antifungal therapy. In C. Kauffman, P. Pappas, J. Sobel, & W. Dismukes (Eds.), *Essentials of clinical mycology*. New York, NY: Springer.

Bali, P., Pranpat, M., Swaby, R., Fiskus, W., Yamaguchi, H., Balasis, M., et al. (2005). Activity of suberoylanilide hydroxamic acid against human breast cancer cells with amplification of Her-2. *Clinical Cancer Research*, *11*, 6382–6389. https://doi.org/10.1158/1078-0432.CCR-05-0344.

Barchiesi, F., Falconi Di Francesco, L., & Scalise, G. (1997). In vitro activities of terbinafine in combination with fluconazole and itraconazole against isolates of Candida albicans with reduced susceptibility to azoles. *Antimicrobial Agents and Chemotherapy, 41*(8), 1812–1814.

Batirel, A., Balkan, I. I., Karabay, O., et al. (2014). Comparison of colistin-carbapenem, colistin-sulbactam, and colistin plus other antibacterial agents for the treatment of extremely drug-resistant *Acinetobacter baumannii* bloodstream infections. *European Journal of Clinical Microbiology & Infectious Diseases, 33,* 1311–1322. https://link.springer.com/article/10.1007/s10096-014-2070-6.

Baz, M., Carbonneau, J., Rhéaume, C., Cavanagh, M. H., & Boivin, G. (2018). Combination therapy with oseltamivir and favipiravir delays mortality but does not prevent oseltamivir resistance in immunodeficient mice infected with pandemic A(H1N1) influenza virus. *Viruses, 10,* 610. https://doi.org/10.3390/v10110610.

Belardo, G., Cenciarelli, O., Frazia, S. L., Rossignol, J. F., & Santoroa, M. G. (2015). Synergistic effect of nitazoxanide with neuraminidase inhibitors against influenza A viruses in vitro. *Antimicrobial Agents and Chemotherapy, 59,* 1061–1069. https://doi.org/10.1128/AAC.03947-14.

Bennett, J. E., Dismukes, W. E., et al. (1979). A comparison of amphotericin B alone and combined with flucytosine in the treatment of cryptococcal meningitis. *The New England Journal of Medicine, 301,* 126–131. https://doi.org/10.1056/NEJM197907193010303.

Blackwell, K. L., Burstein, H. J., Storniolo, A. M., Rugo, H., Sledge, G., & Koehler, M. (2010). Randomized study of lapatinib alone or in combination with trastuzumab in women with ErbB2-positive, trastuzumab-refractory metastatic breast cancer. *Journal of Clinical Oncology, 28,* 1131–1138. https://doi.org/10.1200/JCO.2008.21.4437.

Boguski, M. S., Mandl, K. D., & Sukhatme, V. P. (2009). Drug discovery. Repurposing with a difference. *Science, 324,* 1394–1395. https://doi.org/10.1126/science.1169920.

Brousseau, M. E., et al. (2004). Effects of an inhibitor of cholesteryl ester transfer protein on HDL cholesterol. *The New England Journal of Medicine,* (15), 1505–1515. https://doi.org/10.1056/NEJMoa031766.

Camilly, P. P., George, L. D., Jonathan, R. A., Matthew, S., Robert, K., & Ashley, N. B. (2018). Oseltamivir-zanamivir combination therapy suppresses drug-resistant H1N1 influenza A viruses in the hollow fiber infection model (HFIM) system. *European Journal of Pharmaceutical Sciences, 111,* 443–449. https://doi.org/10.1016/j.ejps.2017.10.027.

Campbell, J., Singh, A. K., Santa Maria, J. P., Kim, Y., Brown, S., Swoboda, J. G., et al. (2011). Synthetic lethal compound combinations reveal a fundamental connection between wall teichoic acid and peptidoglycan biosynthesis in *Staphylococcus aureus*. *ACS Chemical Biology, 6,* 106–116. https://doi.org/10.1021/cb100269f.

Centers for Disease Control and Prevention (CDC). (December 07, 2018). *Factsheet: What you should know about influenza (Flu) antiviral drugs.* Available in: https://www.cdc.gov/flu/pdf/freeresources/updated/antiviral-factsheet-updated.pdf.

Chen, Y.-L., Lehman, V. N., Averette, A. F., Perfect, J. R., & Heitman, J. (2013). Posaconazole exhibits in vitro and in vivo synergistic antifungal activity with caspofungin or FK506 against *Candida albicans*. *PLoS One, 8,* 57672. https://doi.org/10.1371/journal.pone.0057672.

Chiang, C. Y., Centis, R., & Migliori, G. B. (2010). Drug-resistant tuberculosis: Past, present, future. *Respirology, 15,* 413–432. https://doi.org/10.1111/j.1440-1843.2010.01738.x.

Chong, C. R., & Sullivan, D. J. (2007). New uses for old drugs. *Nature, 448,* 645–646. https://www.nature.com/articles/448645a.

Crandon, J. L., & Nicolau, D. P. (2013). Human simulated studies of aztreonam and aztreonam-avibactam to evaluate activity against challenging gram-negative organisms, including metallo-β-lactamase producers. *Antimicrobial Agents and Chemotherapy, 57,* 3299–3306. https://doi.org/10.1128/AAC.01989-12.

Cui, J., Ren, B., Tong, Y., Dai, H., & Zhang, L. (2015). Synergistic combinations of antifungals and anti-virulence agents to fight against *Candida albicans*. *Virulence, 6*, 362–371. https://doi.org/10.1080/21505594.2015.1039885.

Daikos, G. L., Petrikkos, P., Psichogiou, M., et al. (2009). Prospective observational study of the impact of VIM-1 metallo-β-lactamase on the outcome of patients with *Klebsiella pneumoniae* bloodstream infections. *Antimicrobial Agents and Chemotherapy, 53*, 1868–1873. https://doi.org/10.1128/AAC.00782-08.

Diyabalanage, H. V. K., Granda, M. L., & Hooker, J. M. (2013). Combination therapy: Histone deacetylase inhibitors and platinum-based chemotherapeutics for cancer. *Cancer Letters, 329*, 1–8. https://doi.org/10.1016/j.canlet.2012.09.018.

Durante-Mangoni, E., Signoriello, G., Andini, R., et al. (2013). Colistin and rifampicin compared with colistin alone for the treatment of serious infections due to extensively drug-resistant *Acinetobacter baumannii*: A multicenter, randomized clinical trial. *Clinical Infectious Diseases, 57*, 349–358. https://doi.org/10.1093/cid/cit253.

Farha, M. A., Leung, A., Sewell, E. W., D'Elia, M. A., Allison, S. E., Ejim, L., et al. (2013). Inhibition of WTA synthesis blocks the cooperative action of PBPs and sensitizes MRSA to β-lactams. *ACS Chemical Biology, 8*, 226–233. https://doi.org/10.1021/cb300413m.

FDA. (July 17, 2019). *FDA approves new treatment for complicated urinary tract and complicated intra-abdominal infections (news release)*. Silver Spring, MD: FDA. https://www.fda.gov/news-events/press-announcements/fda-approves-new-treatment-complicated-urinary-tract-and-complicated-intra-abdominal-infections.

Finland, M. (1974). Combinations of antimicrobial drugs: Trimethoprim-sulfamethoxazole. *The New England Journal of Medicine*, (12), 624–627. https://www.nejm.org/doi/pdf/10.1056/NEJM197409192911209.

Fitzgerald, J. B., Schoeberl, B., Nielsen, U. B., & Sorger, P. K. (2006). Systems biology and combination therapy in the quest for clinical efficacy. *Nature Chemical Biology, 2*, 458–466. https://www.nature.com/articles/nchembio817.

Frei, E., Karon, M., Levin, R. H., Freireich, E. J., Taylor, R. J., Hananian, J., et al. The effectiveness of combinations of antileukemic agents in inducing and maintaining remission in children with acute leukemia. *Blood, 26*, (1965). 642–656. https://www.ncbi.nlm.nih.gov/pubmed/5321112.

Galegov, G. A., Pushkarskaya, N. L., Obrosova-Serova, N. P., & Zhdanov, V. M. (1977). Combined action of ribavirin and rimantadine in experimental myxovirus infection. *Experientia, 33*, 905–906. https://doi.org/10.1007/BF01951273.

Garnacho-Montero, J., Sa-Borges, M., Sole-Violan, J., et al. (2007). Optimal management therapy for *Pseudomonas aeruginosa* ventilator-associated pneumonia: An observational, multicenter study comparing monotherapy with combination antibiotic therapy. *Critical Care Medicine, 35*, 1888–1895. https://doi.org/10.1097/01.CCM.0000275389.31974.22.

Govorkova, E. A., Fang, H. B., Tan, M., & Webster, R. G. (2004). Neuraminidase inhibitor-rimantadine combinations exert additive and synergistic anti-influenza virus effects in MDCK cells. *Antimicrobial Agents and Chemotherapy, 48*, 4855–4863. https://doi.org/10.1128/AAC.48.12.4855-4863.2004.

Guerra, C. R., Ishida, K., Nucci, M., & Rozenta, S. (2012). Terbinafine inhibits *Cryptococcus neoformans* growth and modulates fungal morphology. *Memórias do Instituto Oswaldo Cruz, 107*, 582–590. https://doi.org/10.1590/S0074-02762012000500003.

Hayden, F. G. (1986). Combinations of antiviral agents for treatment of influenza virus infections. *Journal of Antimicrobial Chemotherapy, 18*, 77–83. https://doi.org/10.1093/jac/18.Supplement_B.177.

Hayden, F. G., Douglas, R. G., & Simons, R. (1980). Enhancement of activity against influenza viruses by combinations of antiviral agents. *Antimicrobial Agents and Chemotherapy, 18*, 536–541. https://doi.org/10.1128/AAC.18.4.536.

Huang, P., Wang, D., Y, S., Huang, W., Zhou, Y., Cui, D., et al. (2014). Combination of small molecule prodrug and nanodrug delivery: Amphiphilic drug–drug conjugate for cancer therapy. *Journal of the American Chemical Society*, *136*, 11748–11756. https://doi.org/10.1021/ja505212y.

Hussain, M., Galvin, H. D., Haw, T. Y., Nutsford, A. N., & Husain, M. (2017). Drug resistance in influenza A virus: The epidemiology and management. *Infection and Drug Resistance*, *10*, 121–134. https://doi.org/10.2147/IDR.S105473.

Ilyushina, N. A., Bovin, N. V., Webster, R. G., & Govorkova, E. A. (2006). Combination chemotherapy, a potential strategy for reducing the emergence of drug-resistant influenza A variants. *Antiviral Research*, *70*, 121–131. https://doi.org/10.1016/j.antiviral.2006.01.012.

Jassal, M., & Bishai, W. R. (2009). Extensively drug-resistant tuberculosis. *The Lancet Infectious Diseases*, *9*, 19–30. https://doi.org/10.1016/S1473-3099(08)70260-3.

Jin, K. L., Park, J. Y., Noh, E. J., Hoe, K. L., Lee, J. H., Kim, J. H., et al. (2010). The effect of combined treatment with cisplatin and histone deacetylase inhibitors on HeLa cells. *Journal of Gynecologic Oncology*, *21*, 262–268. https://doi.org/10.3802/jgo.2010.21.4.262.

Jones, A. W. (2011). Early drug discovery and the rise of pharmaceutical chemistry. *Drug Testing and Analysis*, *3*, 337–344. https://doi.org/10.1002/dta.301.

Joosting, A. C., Head, B., Bynoe, M. L., & Tyrrell, D. A. (1968). Production of common colds in human volunteers by influenza C virus. *British Medical Journal*, *4*, 153–154. https://doi.org/10.1136/bmj.4.5624.153.

Katharina, G., Manfred, P. D., Kristian, P., Miglioli, P. A., & Allerberger, F. (2001). *In vitro* activity of fosfomycin in combination with various antistaphylococcal substances. *Journal of Antimicrobial Chemotherapy*, *48*, 209–217. https://doi.org/10.1093/jac/48.2.209.

Kauppila, J., Ronkko, E., Juvonen, R., Saukkoriipi, A., Saikku, P., Bloigu, A., et al. (2014). Influenza C virus infection in military recruits—Symptoms and clinical manifestation. *Journal of Medical Virology*, *86*, 879–885. https://doi.org/10.1002/jmv.23756.

Kiso, M., Lopes, T. J. S., Yamayoshi, S., Ito, M., Yamashita, M., Nakajima, N., et al. (2018). Combination therapy with neuraminidase and polymerase inhibitors in nude mice infected with influenza virus. *The Journal of Infectious Diseases*, *217*, 887–896. https://doi.org/10.1093/infdis/jix606.

Koul, A., Arnoult, E., Lounis, N., Guillemont, J., & Andries, K. (2011). The challenge of new drug discovery for tuberculosis. *Nature*, *469*, 483–490. https://www.nature.com/articles/nature09657.

Loy, D. E., Liu, W., Li, Y., Learn, G. H., Plenderleith, L. J., Sundararaman, S. A., et al. (2017). Out of Africa: Origins and evolution of the human malaria parasites *Plasmodium falciparum* and *Plasmodium vivax*. *International Journal for Parasitology*, *47*, 87–97. https://doi.org/10.1016/j.ijpara.2016.05.008.

Madren, L. K., Shipman, C., & Hayden, F. G. (1995). In vitro inhibitory effects of combinations of antiinfluenza agents. *Antiviral Chemistry & Chemotherapy*, *25*, 53–57. https://doi.org/10.1177/095632029500600205.

Marathe, B. M., Wong, S.-S., Vogel, P., Garcia-Alcalde, F., Webster, R. G., Webby, R. J., et al. (2016). Combinations of oseltamivir and T-705 extend the treatment window for highly pathogenic influenza A(H5N1) virus infection in mice. *Scientific Reports*, *6*, 26742. https://www.nature.com/articles/srep26742.

Mehta, R. S., Barlow, W. E., et al. (2012). Combination anastrozole and fulvestrant in metastatic breast cancer. *The New England Journal of Medicine*, *367*, 435–444. https://doi.org/10.1056/nejmoa1201622.

Morokutti-Kurz, M., König-Schuster, M., Koller, C., Graf, C., Graf, P., Kirchoff, N., et al. (2015). The intranasal application of zanamivir and carrageenan is synergistically active against influenza A virus in the murine model. *PLoS One*, *10*, e0128794 https://doi.org/10.1371/journal.pone.0128794.

Mosnier, A., Caini, S., Daviaud, I., Nauleau, E., Bui, T. T., Debost, E., et al. (2015). Clinical characteristics are similar across type A and B influenza virus infections. *PLoS One*, *10*, e0136186 https://doi.org/10.1371/journal.pone.0136186.

Nakazawa, H., Kikuchi, Y., Honda, T., Isago, T., & Nozaki, M. (2003). Enhancement of antimicrobial effects of various antibiotics against methicillin-resistant *Staphylococcus aureus* (MRSA) by combination with Fosfomycin. *Journal of Infection and Chemotherapy, 9,* 304–309. https://link.springer.com/article/10.1007/s10156-003-0266-2.

Nakornchai, S., & Konthiang, P. (2006). Activity of azithromycin or erythromycin in combination with antimalarial drugs against multidrug-resistant *Plasmodium falciparum* in vitro. *Acta Tropica, 100,* 185–191. https://doi.org/10.1016/j.actatropica.2006.10.008.

Nguyen, J. T., Hoopes, J. D., Le, M. H., Smee, D. F., Patick, A. K., Faix, D. J., et al. (2010). Triple combination of amantadine, ribavirin, and oseltamivir is highly active and synergistic against drug resistant influenza virus strains in vitro. *PLoS One, 5,* e9332 https://doi.org/10.1371/journal.pone.0009332.

Nguyen, J. T., Hoopes, J. D., Smee, D. F., Prichard, M. N., Driebe, E. M., Engelthaler, D. M., et al. (2009). Triple combination of oseltamivir, amantadine, and ribavirin displays synergistic activity against multiple influenza virus strains in vitro. *Antimicrobial Agents and Chemotherapy, 53,* 4115–4126. https://doi.org/10.1128/AAC.00476-09.

Nissen, S. E., et al. (2007). Effect of torcetrapib on the progression of coronary atherosclerosis. *The New England Journal of Medicine, 8,* 835. https://doi.org/10.1056/nejmoa070635.

Nzila, A. M., Nduati, E., Mberu, E. K., Sibley, C. H., Monks, S. A., Winstanley, P. A., et al. (2000). Molecular evidence of greater selective pressure for drug resistance exerted by the long-acting antifolate pyrimethamine/sulfadoxine compared with the shorter-acting chlorproguanil/dapsone on Kenyan *Plasmodium falciparum. The Journal of Infectious Diseases, 181,* 2023–2028. https://doi.org/10.1086/315520.

Olliaro, P. L., & Taylor, W. R. (2004). Developing artemisinin based drug combinations for the treatment of drug resistant falciparum malaria: A review. *Journal of Postgraduate Medicine.* http://www.jpgmonline.com/text.asp?2004/50/1/40/6652.

Ozols, R. F., Bundy, B. N., Greer, B. E., et al. (2003). Phase III trial of carboplatin and paclitaxel compared with cisplatin and paclitaxel in patients with optimally resected stage III ovarian cancer: A gynecologic oncology group study. *Clinical Oncology, 21,* 3194–3200. https://doi.org/10.1200/JCO.2003.02.153.

Pacorel, B., Leung, S. C., Stachulski, A. V., Davies, J., Vivas, L., Lander, H., et al. (2010). Modular synthesis and in vitro and in vivo antimalarial assessment of C-10 pyrrole Mannich Base derivatives of artemisinin. *Journal of Medicinal Chemistry, 53,* 633–640. https://doi.org/10.1021/jm901216v.

Park, S., Kim, J. I., Lee, I., Lee, S., Hwang, M. W., Bae, J. Y., et al. (2014). Combination effects of peramivir and favipiravir against oseltamivir-resistant 2009 pandemic influenza A(H1N1) infection in mice. *PLoS One, 9,* e101325 https://doi.org/10.1371/journal.pone.0101325.

Peters, W. (1990). The prevention of antimalarial drug resistance. *Pharmacology & Therapeutics, 47,* 499–508. https://doi.org/10.1016/0163-7258(90)90067-C.

Peters, W., & Robinson, B. L. (2016). Further studies on the retardation of drug resistance by the use of a triple combination of mefloquine, pyrimethamine and sulfadoxine in mice infected with *P. berghei* and '*P. berghei* NS'. *Annals of Tropical Medicine and Parasitology, 78,* 459–466. https://doi.org/10.1080/00034983.1984.11811850.

Poon, C., He, C., Liu, D., Lu, K., & Lin, W. (2015). Self-assembled nanoscale coordination polymers carrying oxaliplatin and gemcitabine for synergistic combination therapy of pancreatic cancer. *Journal of Controlled Release, 201,* 90–99. https://doi.org/10.1016/j.jconrel.2015.01.026.

Prichard, M. N., Shipman, C., Jr. (1990). A three-dimensional model to analyze drug-drug interactions. *Antiviral Research, 14,* 181–205.

Quinn, B. A., Dash, R., Sarkar, S., Azab, B., Bhoopathi, P., Das, S. K., et al. (2015). Pancreatic cancer combination therapy using a BH3 mimetic and a synthetic tetracycline. *Cancer Research, 75,* 2305–2315. https://doi.org/10.1158/0008-5472.CAN-14-3013.

Ramalingam, S. S., Maitland, M. L., Frankel, P., et al. (2010). Carboplatin and paclitaxel in combination with either vorinostat or placebo for first-line therapy of advanced non–small-cell lung cancer. *Clinical Oncology, 28*, 56–62. https://doi.org/10.1200/JCO.2009.24.9094.

Rationalizing combination therapies. (2017). *Nature Medicine, 23*, 1113. https://doi.org/10.1038/nm.4426.

Sader, H. S., Fritsche, T. R., Kaniga, K., Ge, Y., & Jones, R. N. (2005). Antimicrobial activity and spectrum of PPI-0903M (T-91825), a novel cephalosporin, tested against a worldwide collection of clinical strains. *Antimicrobial Agents and Chemotherapy*, 3501–3512. https://doi.org/10.1128/AAC.49.8.3501-3512.2005.

Samonis, G., Maraki, S., Karageorgopoulos, D. E., et al. (2012). Synergy of fosfomycin with carbapenems, colistin, netilmicin, and tigecycline against multidrug resistant *Klebsiella pneumoniae, Escherichia coli*, and *Pseudomonas aeruginosa* clinical isolates. *European Journal of Clinical Microbiology & Infectious Diseases, 31*, 695–701. https://link.springer.com/article/10.1007/s10096-011-1360-5.

Sheard, D. E., O'Brien-Simpson, M. N., Wade, J. D., & Separovic, F. (2019). Combating bacterial resistance by combination of antibiotics with antimicrobial peptides. *Pure and Applied Chemistry, 91*, 199–209. https://doi.org/10.1515/pac-2018-0707.

Sieradzki, K., & Tomasz, A. (1997). Suppression of beta-lactam antibiotic resistance in a methicillin-resistant *Staphylococcus aureus* through synergic action of early cell wall inhibitors and some other antibiotics. *Journal of Antimicrobial Chemotherapy, 39*, 47–51. https://doi.org/10.1093/jac/39.suppl_1.47.

Smee, D. F., Bailey, K. W., Morrison, A. C., & Sidwell, R. W. (2002). Combination treatment of influenza a virus infections in cell culture and in mice with the cyclopentane neuraminidase inhibitor RWJ270201 and ribavirin. *Chemotherapy, 48*, 88–93. https://doi.org/10.1159/000057668.

Smee, D. F., Hurst, B. L., Wong, M. H., Bailey, K. W., & Morrey, J. D. (2009). Effects of double combinations of amantadine, oseltamivir, and ribavirin on influenza A(H5N1) virus infections in cell culture and in mice. *Antimicrobial Agents and Chemotherapy, 53*, 2120–2128. https://doi.org/10.1128/AAC.01012-08.

Tamma, P. D., Cosgrove, S. E., & Maragakis, L. L. (2012). Combination therapy for treatment of infections with gram-negative bacteria. *Clinical Microbiology Reviews, 25*, 450–470. https://doi.org/10.1128/CMR.05041-11.

Tarbet, E. B., Vollmer, A. H., Hurst, B. L., Barnard, D. L., Donald, Y. F., & Smee, F. (2014). In vitro activity of favipiravir and neuraminidase inhibitor combinations against oseltamivir-sensitive and oseltamivir-resistant pandemic influenza A (H1N1) virus. *Archives of Virology, 159*, 1279–1291. https://doi.org/10.1007/s00705-013-1922-1.

Tobias, J. S., Wrigley, P. F. M., & Shaw, E. (1976). Combination antifungal therapy for cryptococcal meningitis. *Postgraduate Medical Journal, 52*, 305–308. https://doi.org/10.1136/pgmj.52.607.305.

Tzouvelekis, L. S., Markogiannakis, A., Piperaki, E., et al. (2014). Treating infections caused by carbapenemase-producing Enterobacteriaceae. *Clinical Microbiology and Infection, 20*, 862–872. https://doi.org/10.1111/1469-0691.12697.

U.S. Department of Health and Human Services. (n.d.). https://aidsinfo.nih.gov/understanding-hiv-aids/infographics/25/fda-approval-of-hiv-medicines.

Vestergaard, M., Paulander, W., Marvig, R. L., Clasen, J., Jochumsen, N., Molin, S., et al. (2016). Antibiotic combination therapy can select for broad-spectrum multidrug resistance in *Pseudomonas aeruginosa. International Journal of Antimicrobial Agents, 47*, 48–55. https://doi.org/10.1016/j.ijantimicag.2015.09.014.

Wald, N. J., & Law, M. R. (2003). A strategy to reduce cardiovascular disease by more than 80%. *BMJ, 326*, 1419. https://doi.org/10.1136/bmj.326.7404.1419.

White, N. J. (1998). Preventing antimalarial drug resistance through combinations. *Drug Resistance Updates, 1*, 3–9. https://doi.org/10.1016/S1368-7646(98)80208-2.

White, N. (1999). Antimalarial drug resistance and combination chemotherapy. *Philosophical Transactions of the Royal Society of London Series B, Biological Sciences, 354,* https://doi.org/10.1098/rstb.1999.0426.

White, N. J., & Pongtavornpinyo, W. (2003). The de novo selection of drug-resistant malaria parasites. *Proceedings of the Royal Society B, 270,* 545–554. https://doi.org/10.1098/rspb.2002.2241.

WHO. (2010). *Report: Global tuberculosis control.* Geneva: World Health Organization.

WHO. (2018). *Update of antibacterial agents in clinical development.* http://apps.who.int/medicinedocs/documents/s23564en/s23564en.pdf.

Worthington, R. J., & Melander, C. (2013). Combination approaches to combat multidrug-resistant bacteria. *Trends in Biotechnology, 31,* 177–184. https://doi.org/10.1016/j.tibtech.2012.12.006.

Xie, C., Duffy, A. G., Mabry-Hrones, D., Wood, B., Levy, E., Krishnasamy, V., et al. (2019). Tremelimumab in combination with microwave ablation in patients with refractory biliary tract cancer. *Hepatology, 69,* 2048–2060. https://doi.org/10.1002/hep.30482.

Yanbin, L., Qiang, W., Guisheng, Y., Wanxia, G., Yongquan, L., & Chen, X. (2014). DCDB 2.0: A major update of the drug combination database. *Database, 2014,* 1–6. https://doi.org/10.1093/database/bau124.

Yu, W., Shen, P., Bao, Z., Zhou, K., Zheng, B., Ji, J., et al. (2017). *In vitro* antibacterial activity of fosfomycin combined with other antimicrobials against KPC-producing *Klebsiella pneumoniae. International Journal of Antimicrobial Agents, 50,* 237–241. https://doi.org/10.1016/j.ijantimicag.2017.03.011.

Zarkotou, O., Pournaras, S., Tselioti, P., et al. (2011). Predictors of mortality in patients with bloodstream infections caused by KPC-producing *Klebsiella pneumoniae* and impact of appropriate antimicrobial treatment. *Clinical Microbiology and Infection, 17,* 1798–1803. https://doi.org/10.1111/j.1469-0691.2011.03514.x.

Zhirnov, O. P. (1987). High protection of animals lethally infected with influenza virus by aprotinin-rimantadine combination. *Journal of Medical Virology, 21,* 161–167. https://doi.org/10.1002/jmv.1890210208.

CHAPTER

2

Combination therapy against multidrug resistance

Musa Marimani

Department of Clinical Microbiology and Infectious Diseases, School of Pathology, Health Sciences, University of the Witwatersrand, Johannesburg, South Africa

2.1 Emergence of microbial drug resistance

Antibiotics are among the most widely used drugs to treat human diseases. Their application in the treatment of clinical conditions caused by a wide variety of pathogenic bacteria has led to a significant increase in the health and life expectancy of human patients. Apart from clinical institutions, antibiotics have also been employed in the agricultural and food industries (Petersen et al., 2002; Vidaver, 2002). This widespread use of antibiotics has enabled pathogenic microorganisms to adapt and develop various drug-resistant mechanisms (Abraham & Chain, 1940). Devastatingly, development of drug resistance has far-reaching implications in clinical institutions, as it compromises treatment plans and already-existing medical practices (WHO, 2000, 2010).

The main mechanism with which microorganisms acquire drug resistance is by developing genetic mutations (Martinez & Baquero, 2000) and through horizontal gene transfer (Andam, Fournier, & Gogarten, 2011). Acquisitions of different genetic mutations render the microbes resistant to therapy, while horizontal gene transfer allows transmission of resistant genes from a resistant strain to a susceptible one (Fig. 2.1). This dissemination of genetic material between pathogens promotes acquisition of drug-resistant genes by other members of the population. Thus the use of therapeutic agents across different industries further complicates control

https://doi.org/10.1016/B978-0-12-820576-1.00002-3

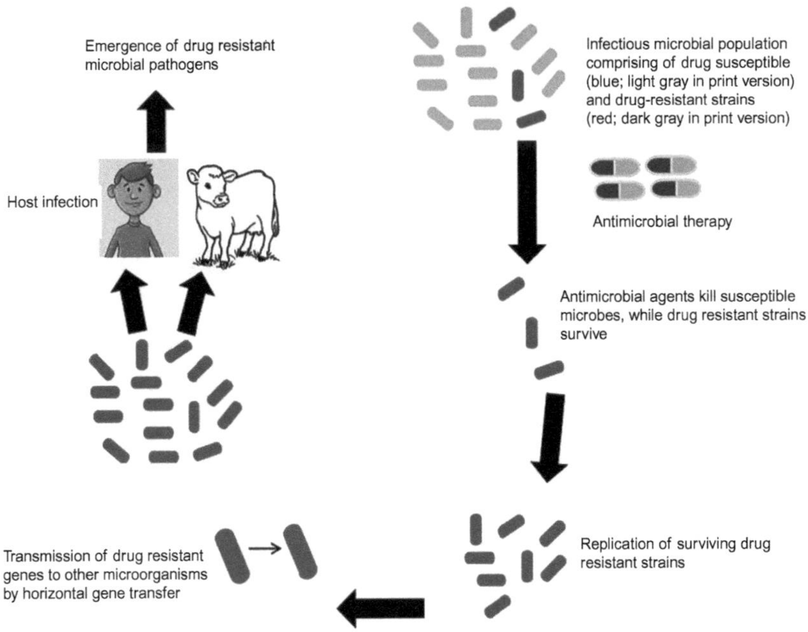

FIG. 2.1 Development and transmission of drug-resistant microorganisms. The microbial population is composed of susceptible and resistant strains. The susceptible strains are killed, while the resistant microbes survive and replicate following treatment with antimicrobial agents. Horizontal gene transfer facilitates dissemination of resistant genes from resistant strains to susceptible microbial species. Extensive host infection and transmission result in the emergence of various drug-resistant infections and diseases that are untreatable by conventional antimicrobial therapies.

measures. For example, transmission of drug-resistant microorganisms may occur from:

(a) an infected person to a healthy individual,
(b) an infected person to healthy livestock,
(c) infected livestock to a healthy animal, or
(d) infected livestock to a healthy person through zoonosis or consumption of infected animal products.

Furthermore, resistance genes may originate in both commensal (Sommer, Dantas, & Church, 2009) and noncommensal microbiota (Davies, 1997). Therefore complete analysis of the health, agricultural, and food industries as well as the ecosystem is crucial in identifying and understanding the transmission of drug-resistant genes between humans and microbial pathogens (Baquero, Alvarez-Ortega, & Martinez, 2009; Fajardo, Linares, & Martinez, 2009; Martinez, 2008). This chapter aims to review the various components that contribute to development of drug resistance, such as host and environmental factors, and to explore various mechanisms leading to microbial drug resistance. Furthermore, the application of combination therapy to combat the scourge of drug-resistant microbes is also discussed.

2.2 Host and environmental factors contributing to drug resistance

2.2.1 Host immune competence

Following the invasion of host cells and tissues, the microbial pathogen must either interact with or evade the host innate and adaptive immune responses. Upon establishing infection, the pathogen must aggressively replicate and disseminate to various host cells to accomplish a "disease state" within the host. The disease state is accompanied by manifestation of clinical symptoms consistent with the particular infecting microbial pathogen. Therefore host immunity is essential in recognizing and clearing the foreign microbial entity, thus preventing infection and subsequent disease.

A competent immune system may significantly decrease the microbial load within the host and subsequently reduce long treatment schedules (Geli et al., 2012; Iravani et al., 1995; Williams Jr. et al., 1995). Consequently, a strong immune system protects the host from infection with minimal drug usage or without the use of antimicrobial agents. Another important component is the understanding of interactions between the antimicrobial drugs and host immunity in combating infections. To this end, mathematical models of infection have been designed to explore the host immunity-drug dynamics in the progression of microbial diseases (Geli et al., 2012; Gjini & Brito, 2016; Handel, Margolis, & Levin, 2009). Data from these investigations have revealed a strong association between drug administration and host immune competence in accomplishing effective microbial clearance (Geli et al., 2012; Gjini & Brito, 2016; Handel et al., 2009). In a related study, a model was designed to study the effect of host immunity on the efficacy of an antimalaria drug (Gurarie & McKenzie, 2006). Similarly to previous studies, it was evidenced that host immune response was an important determinant in promoting or inhibiting the development of parasites harboring drug-resistant genes. Furthermore, the role of host immunity in combating infection was also evaluated in an infant with leukemia who was also infected with vancomycin-resistant *Enterococcus faecium* (VRE) (Honsa et al., 2017). The clinical profile of the patient did not change despite administration of front-line drugs to eradicate the VRE infection. DNA sequencing revealed the presence of a single mutation (L152F) within the *relA* gene, which led to overexpression of alarmone guanosine tetraphosphate. Relative to wildtype, the VRE strain containing the *relA* gene mutation formed smaller and less-defined biofilms

(Honsa et al., 2017). It was therefore concluded that the clinically relevant *relA* mutations emerged due to persistent VRE infection. This prolonged infection induced antibiotic tolerance and prevented efficient microbial eradication in an immunocompromised infant. Taken together, these studies indicate the importance of host immune competence in neutralizing and eradicating drug-resistant microorganisms. Additionally, they highlight the synergistic relationship between a strong immune response and drug administration in countering microbial pathogens (Honsa et al., 2017).

2.2.2 Incomplete adherence to treatment plan

Incomplete adherence to treatment schedule is one of the leading causes driving the development of antimicrobial resistance (Malfertheiner, 1993). Similarly, patients who use antimicrobial agents while consuming alcohol may potentially compromise their immune system (Fok et al., 2008; Lonnroth et al., 2008; Szabo, 1997), thus promoting emergence of drug-resistant strains. It has also been reported that some patients in developing countries use their prescribed drugs in conjunction with those obtained from traditional healers. Of primary concern is the fact that some of these herbal mixtures are of unknown composition, concentration, and efficacy (Lansang et al., 1990). These and other practices expose the surviving microbial population to subtherapeutic drug concentrations, advancing microbial drug resistance (Calva & Bojalil, 1996). Thus early cessation or improper use of prescribed drugs generally leads to resuscitation of infection and also compromises therapeutic efficacy of the treatment regimen (Nachega & Chaisson, 2003). For this reason, strict adherence to the drug prescription and treatment plan is a highly recommended strategy to promote microbial eradication and symptom alleviation and to prevent the emergence of pathogenic drug-resistant strains.

2.2.3 Overuse and improper prescription of therapeutics

Medical doctors, nurses, and other healthcare workers play pivotal roles in assessing and administering treatment to patients. However, they may compromise their patient's health if they:

(a) delay treatment,
(b) provide the wrong prescription drugs (Usluer, Ozgunes, & Leblebicioglu, 2005),
(c) offer unregulated clinical and medical practices, or.
(d) misdiagnose the patient's clinical condition.

In many developing countries, lack of adequate clinical information, resources, and infrastructure and the high patient-to-doctor ratio may be causes of these inappropriate clinical practices. For example, it was previously reported that in Lebanon more than half of clinical cases resulted in provision of wrong prescription doses (Saleh et al., 2015). In addition, more than 60% of medical doctors issued antibiotics that were not consistent with the correct treatment duration (Saleh et al., 2015). Health professionals in under-resourced communities may not have the capacity to perform proper diagnostic and antibiotic resistance assays, and thus they are more likely to administer broad-spectrum antibiotics to treat different bacterial infections. Extensive utilization of broad-spectrum antibiotics also exerts selection pressure that allows development of drug-resistant microbes. Furthermore, continuous use of these antibiotics also has an adverse effect, because they alter the composition of the existing patient microbiota (Calva & Bojalil, 1996). The inconsistency in the provision of antimicrobial therapy among medical doctors for the same medical condition is concerning. Ultimately, it is this type of discrepancy that exerts the selection pressure (Neu, 1992) that drives the emergence of drug-resistant microbial strains.

Therefore, provision of high-quality medical resources and infrastructure may facilitate adherence to proper clinical practices. Aggressive engagements among medical professionals from different geographical settings are urgently required to establish consistency in the provision of antimicrobial agents for the same medical condition. This may result in meaningful recommendations that lead to documentation and implementation of regulated and appropriate medical practices. Such medical practices may markedly decrease wrong prescriptions, inadequate treatment duration, and the use of broad-spectrum antibiotics. Importantly, training of junior medical staff by qualified and experienced medical personnel may substantially decrease the overuse and improper prescription of antimicrobial therapies. These regulated practices will go a long way towards preventing the selection pressure associated with drug tolerance and subsequent emergence of resistant pathogens.

2.2.4 Poor patient hygiene

Proper sanitation and hygiene practices are essential in combating the development and transmission of drug-resistant microorganisms. Environmental conditions such as contaminated water and crowding coupled with consumption of contaminated food products promotes circulation and transmission of these drug-resistant genotypes and phenotypes. Improving sanitation and personal hygiene practices decreases the emergence and spread of resistant microorganisms (Smith et al., 2004). Importantly, efficient infection control strategies in hospitals and the general public generally lead to a marked decline in the number of nosocomial and community-acquired resistant pathogens.

2.2.5 Inefficient infection control in clinical settings

Emergence of antimicrobial resistance is a global concern, particularly for nosocomial infections (Gaynes & Monnet, 1997; Monnet et al., 1998). Generally, microorganisms responsible for nosocomial infections are members of the normal flora that switch to "infectious mode" after multiplying in sterile sites. Nosocomial infections are mainly transmitted from infected patients to other patients that may potentially become infected with the drug-resistant pathogens (Bonten & Weinstein, 1996). In some hospitals, control measures implemented to counter the spread of resistant microbes include increased use of prophylactic agents (Boyce, 1996; Brun-Buisson et al., 1989), barrier precautions, and efficient hand washing (Gaynes et al., 1983). Another strategy involves the gradual use of one class of antimicrobial agents, which becomes upgraded to the second class once resistance to the first class has been demonstrated (Bonhoeffer, Lipsitch, & Levin, 1997; Kollef et al., 1997; McGowan Jr., 1986). However, it is generally difficult to evaluate and quantify the success of these interventions. Therefore mathematical models have been employed to provide quantitative predictions to evaluate the efficiency of these practices (Austin et al., 1999; Blower, Small, & Hopewell, 1996; Bonhoeffer et al., 1997). One such model described the structure and analysis of bacterial pathogens transmitted within a hospital setting (Lipsitch, Bergstrom, & Levin, 2000). This model predicted that the application of an antibiotic (without microbial resistance) is associated at the individual level with carriage of resistant bacteria, but not associated at the population level in causing prevalence (Lipsitch et al., 2000). The model also highlighted that nonspecific control measures that decrease the spread of all bacterial pathogens within a clinical setting will disproportionately lead to a decline in the colonization by drug-resistant bacteria (Lipsitch et al., 2000). Furthermore, changes caused by implementation of a successful control strategy will be observed after weeks or months in community-acquired infections. Importantly, drug resistance may decrease faster in a clinical setting even though it might not carry a fitness cost for the bacteria (Lipsitch et al., 2000). The fitness cost of the microorganism describes the ability of that particular pathogen to survive and replicate in a competitive environment.

In a separate study, an outbreak of hypervirulent *Klebsiella pneumoniae* strain was investigated in China (Gu et al., 2018). This outbreak occurred in the intensive care unit and was associated with the hospital ventilator. A total of 21 *K. pneumoniae* strains resistant to carbapenems were collected from patients within the intensive care unit (Gu et al., 2018). Thereafter, the antimicrobial susceptibility tests and genetic analysis of the resistant strains was assessed. Sequencing data indicated that all five resistant strains belonged to the ST11 group, which is the most dominant type of carbapenem-resistant *K. pneumoniae* in China. Genomic data indicated that these hypervirulent strains emerged after acquiring a virulent plasmid

DNA and were also identified in other regions of China (Gu et al., 2018). In light of these observations, adequate infection control measures in hospitals are critical for preventing emergence and transmission of drug-resistant pathogens within the clinical institution and subsequent spread to the general public.

2.2.6 Improper use of antimicrobial agents in agriculture

In the agricultural sector, antimicrobial agents are employed to prevent or eradicate diseases in animals and as growth promoters for animal breeding (Collignon et al., 2005; Petersen et al., 2002). These agents are also utilized as additives in commercialized vegetables and fruits and as antibiotic-containing manure for soil fertilization and other industrial procedures (Vidaver, 2002). The utilization of antimicrobial drugs particularly in food-producing animals has important clinical implications for animal and human health, as it may result in the emergence of drug-resistant strains. These resistant microbes may be transmitted from infected animals to humans via contact with animals, consumption of contaminated animal products, or through environmental transmission from contaminated water sources. Specifically, administration of antimicrobial agents in hospitals and agriculture produces selection pressure that may augment development and survival of drug-resistant strains (Witte, 1998). This selection pressure coupled with subsequent horizontal gene transfer ultimately leads to an increase in the number of resistant pathogens present in clinical institutions and the general population (Witte, 1998) (Fig. 2.1). It has been shown that there is a direct association between the increase in the number of drug-resistant bacteria and increased application of antimicrobial drugs across various institutions and industries (Aarestrup, 2012). Interestingly, it has been reported that the vast majority of antimicrobial drugs produced globally are mainly employed for food production (Mitema et al., 2001) as opposed to human consumption.

In many African countries, multidrug-resistant bacterial strains have been detected in fresh produce, meat products, and in humans who are in close proximity with livestock (Fortini et al., 2011; Kikuvi, Ombui, & Mitema, 2010). In Kenya, a high concentration of drug residues has been detected in commercialized meat products (Mitema et al., 2001) indicating that food-producing animals are the major source of drug-resistant microbes. These infected animals may directly or indirectly transmit resistant microorganisms to unsuspecting humans, particularly in underdeveloped and developing countries (Simango & Rukure, 1991). Indeed, the close proximity of humans and livestock in many developing countries promotes the spread of resistant microbes from infected animals to humans via animal handling (Simango & Rukure, 1991). A large proportion of people living in rural areas depend on unregulated methods for animal treatment due to the absence or high cost associated with regulated animal healthcare systems. These unregulated methods, coupled with poor personal hygiene and sanitation, result in suboptimal use of antimicrobial agents, thus providing a breeding ground for drug-resistant strains to flourish. Extensive administration of prophylactic drugs in aquaculture has been observed in some developing countries and is a major concern for animal and human health as well as the environment (Cabello, 2006). Consequently, adequate and responsible administration of antimicrobial agents across different sectors is crucial in preventing the emergence and transmission of drug-resistant pathogens. Furthermore, proper and regulated guidelines pertaining to personal hygiene, sanitation, and animal handling have to be documented and implemented, especially in developing countries, to decrease the high level of antimicrobial drug resistance reported in agriculture, hospitals, and other sectors.

2.3 Mechanisms of microbial drug resistance

The major mechanism employed by microbial pathogens to cause antimicrobial resistance is by introducing mutations at various sites in their genome. This is because many infectious microbes, particularly bacteria, have a high genetic plasticity (Munita & Arias, 2016) that enables them to survive under different host and environmental conditions. Interestingly, bacteria occupying the same ecological region with organisms that produce antimicrobial agents have developed mechanisms that allow them to flourish in the presence of powerful antibiotics that would otherwise kill them or inhibit their replication cycle. This acquired resistance and high genetic plasticity also allows bacteria to thrive even under high antibiotic concentrations in hospitals, agricultural industries, and environmental samples. Bacteria have evolved two methods that allow them to withstand, tolerate, and resist antibiotic onslaught:

(a) induction of mutations in genes serving as therapeutic drug targets for antibiotics, and
(b) acquiring foreign DNA (mainly plasmid DNA) containing resistant genes via horizontal gene transfer.

2.3.1 Genetic mutations

Genetic mutations in surviving bacterial cells from a susceptible strain advance drug tolerance and transmission of this trait to the next microbial generation. In the context of antimicrobial therapy, the therapeutic agent will eradicate the susceptible bacterial population while the resistant mutants will survive and predominate following replication. Generally, genetic mutations that confer drug resistance are at the expense of cell homeostasis (e.g., reduced bacterial fitness) and are only regulated when the bacteria are exposed to antibiotics. Mutations that confer drug resistance render the antimicrobial agent ineffective by (Munita & Arias, 2016):

(a) modifying the drug target and reducing therapeutic affinity,
(b) decreasing uptake of the antimicrobial agent,
(c) activating efflux pumps that facilitate removal of the therapeutic molecule, or
(d) altering the regulation of essential metabolic and immune signaling pathways (Munita & Arias, 2016).

Thus drug resistance conferred by genetic mutations is complex and necessary to ensure survival of bacterial pathogens under extreme conditions. For example, resistance to fluoroquinolone drugs may occur as a result of three independent biochemical pathways. Interestingly, these different pathways may exist simultaneously in the same bacteria and increase the severity of microbial drug resistance. Specifically, this accumulated resistant mechanism may:

(a) trigger overexpression of efflux pump genes and proteins leading to efficient removal of the antimicrobial molecule,
(b) induce mutations in genes coding for fluoroquinolone drug targets (i.e., topoisomerase and DNA gyrase), or
(c) decrease drug uptake by overexpression of *qnr* genes conferring fluoroquinolone resistance (Kaplan et al., 2013; Rezazadeh, Baghchesaraei, & Peymani, 2016; Yan et al., 2017a).

Notably, different bacterial pathogens have evolved different resistance mechanisms toward the same or a wide variety of antibiotics (Munita & Arias, 2016). Gram-positive bacteria confer resistance by modifying the penicillin binding site, while gram-negative bacterial pathogens accomplish drug resistance by overexpression of β-lactamase enzymes (Munita & Arias, 2016). The discrepancy in resistance mechanisms conferred by gram-positive and gram-negative bacteria has been attributed to variations in the bacterial cell envelope between the two bacterial groups (Munita & Arias, 2016).

2.3.2 Horizontal gene transfer

Horizontal gene transfer entails the transfer of foreign DNA from one organism to another. In terms of antimicrobial resistance, it involves transmission of foreign DNA from a drug-resistant microbial strain to a susceptible strain. This results in the acquisition and expression of resistant genes in the susceptible strain leading to development of drug resistance (Fig. 2.1). Continual bacterial replication and horizontal gene transfer ensures preservation of the drug-resistant genotype (Fig. 2.1). Bacterial pathogens with this drug-resistant trait may gradually acquire additional mutations so that they induce drug resistance to various antimicrobial agents (Aye et al., 2016; Gamngoen et al., 2018; Gurjav et al., 2016; Kim et al., 2018; Li et al., 2017; Uddin et al., 2018). There are three main mechanisms employed by bacteria to acquire foreign DNA:

(a) transformation, or acquisition and incorporation of exogenous DNA from the environment;
(b) conjugation, or transmission of foreign DNA via cell-to-cell contact; and
(c) transduction, or introduction of exogenous DNA into bacterial cells by a virus.

Of these three mechanisms, conjugation is the most widely reported method responsible for increased drug resistance in clinical settings. This observation may be attributed to the ease of transmission via cell-to-cell contact, which may easily occur between patients, particularly in crowded and under-resourced healthcare centers. Conjugation can facilitate horizontal gene transfer by utilizing mobile genetic elements to transfer genetic information or by direct transfer of genetic information between chromosomes (Manson, Hancock, & Gilmore, 2010). Plasmids and transposons are the leading mobile genetic elements responsible for spreading drug resistance among microbial pathogens implicated in persistent and chronic human infections (Munita & Arias, 2016). Integrons consist of genetic determinants of components and can trigger site-specific recombination that captures mobile gene cassettes (Hall & Collis, 1995). These elements are capable of efficient dissemination of drug resistance directly to bacterial chromosomes. Accumulation of microbial drug resistance induced by various horizontal gene transfer mechanisms significantly contributes to the increase in the scourge of drug-resistant diseases. This has devastating clinical implications for patients harboring drug-resistant strains and also compromises the competency of the public health system.

2.4 The use of combination therapy to counter multidrug resistance

The high adaptation, tolerance, and mutation rate of many microbial pathogens has facilitated their survival and replication in the presence of powerful antimicrobial agents (Lu et al., 2018a). Indeed, these properties have allowed many infectious pathogens of clinical importance to adapt and tolerate high concentrations of antimicrobial agents, induce genetic mutations, acquire resistant genes, and ultimately flourish and spread to other human and nonhuman hosts. This is of critical clinical importance, because infection by these drug-resistant pathogens leads to the development of diseases that do not respond to conventional therapies or may require long treatment duration (Bjorke et al., 2018; Hu et al., 2018). This has devastating global public health implications, as it results in increased morbidity and mortality rates. Long-term monotherapy (Hosono et al., 2018; Zhang et al., 2018a), coupled with inappropriate clinical practices and improper use of therapeutic molecules, has advanced the evolution of drug-resistant pathogens (Llor & Bjerrum, 2014; Zapata-Cachafeiro et al., 2014). Of high clinical and social concern is that this high level of antimicrobial resistance exists across different sectors including healthcare (Cornejo-Juarez et al., 2015; Liu et al., 2018), agriculture (Keen et al., 2018; Rangasamy et al., 2017), and other industries. Clinically, the presence of drug-resistant pathogens necessitates the use of more effective and expensive antimicrobial agents to alleviate disease symptoms.

What is even more concerning is the emergence of microbial pathogens that are resistant to broad-spectrum antibiotics (Li et al., 2013; Tran-Dien et al., 2018; Vestergaard et al., 2016). Broad-spectrum antibiotics include antibacterial agents such as ampicillin, amoxicillin, streptomycin, chloramphenicol, and tetracycline. This class of antibiotics is capable of treating diseases caused by a wide range of bacterial pathogens across different geographical regions. Therefore resistance to these therapeutic agents represents a critical global public health concern. Moreover, development and dissemination of resistant strains against carbapenems is a major public health concern with dire clinical implications (Fuste et al., 2013; Yan et al., 2017b; Ye et al., 2018). These antibiotic molecules include drugs such as doribax, doripenem, meropenem, ertapenem, and primaxin and are employed to treat chronic and multidrug-resistant bacterial infections (Hawkey et al., 2018; Shaker & Shaaban, 2017). As a result, carbapenems are drugs of last resort in the treatment of many human infections caused by pathogenic bacteria and represent the most expensive and powerful class of antimicrobial agents. Thus resistance to this group of therapeutic compounds has a devastating impact on the healthcare system across the globe. Therefore innovative strategies are urgently required to treat and counter the development and dissemination of drug-resistant pathogens. Ideally, drug development research must catch up with or surpass the rate at which the scourge of drug-resistant microbes is emerging. Consequently, this necessitates the production of new, effective, and safe therapeutic molecules capable of combating the occurrence and evolution of drug-resistant microbes across different institutions and industries. Globally, this will have a positive impact as it would inevitably eradicate drug-resistant diseases and alleviate the clinical and social burden associated with emergence and treatment of resistant microbial strains.

Development of novel antimicrobial therapy requires efficacy and safety studies to be conducted in cultured mammalian cells and in animal and human clinical trials. This exercise may be time-consuming and expensive and does not urgently address the devastating clinical crisis caused by drug-resistant pathogens. Therefore combination therapy involving administration of multiple licensed therapeutic agents has been employed as an alternative strategy to treat microbial diseases that do not respond to conventional drugs. This method of employing different antimicrobial agents to treat diseases caused by drug-resistant pathogens is advantageous, as different drugs are directed against different therapeutic targets simultaneously. This approach generally leads to a substantial decrease in microbial proliferation and gives the host immunity enough time to clear infection. Consequently, efficient microbial eradication removes the pathogen from circulation and allows symptom alleviation, which promotes host recovery. Conveniently, the simultaneous onslaught of various microbial drug targets afforded by combination therapy as opposed to monotherapy significantly decreases the development and spread of drug-resistant strains. Therefore combination therapy frequently leads to efficient microbial neutralization and clearance, provided proper clinical practices pertaining to drug usage are strictly followed. Moreover, this effective strategy has been widely implemented to treat life-threatening human diseases caused by various drug-resistant microbial pathogens (de Knegt et al., 2017; Touil, Boucherit-Otmani, & Boucherit, 2018; Yang et al., 2017).

2.4.1 Combination therapy against *Enterobacteriaceae* species

In a study conducted by Thwaites et al. (2018), plazomicin, a recently approved antimicrobial compound, was used in combination with tazobactam/piperacillin or ceftazidime against multidrug-resistant *Enterobacteriaceae* species (Thwaites et al., 2018). These bacterial pathogens were resistant to β-lactam and aminoglycoside antibiotics and included *Citrobacter freundii*, *Escherichia coli*, and various *Klebsiella* and *Enterobacter* species (Thwaites et al., 2018). Time-kill assays and checkerboard studies indicated that simultaneous application of antimicrobial agents led to a significant decline in bacterial growth and highlighted the synergistic and therapeutic advantage of applying combination therapy to combat multidrug-resistant pathogens (Thwaites et al., 2018). In a related study, the advantage of administering plazomicin in conjunction with other antimicrobial compounds was evaluated against a drug-resistant *K. pneumoniae* strain (Rodriguez-Avial et al., 2015). This strain was rendered drug resistant by virtue of producing carbapenemase enzymes, which catalyze the hydrolysis of antibiotics. Checkerboard assays demonstrated that administration of plazomicin with fosfomycin or meropenem significantly inhibited bacterial replication. Impressively, coadministration of plazomicin and colistin led to a more pronounced decrease in the proliferation of the drug-resistant *K. pneumoniae* pathogen (Rodriguez-Avial et al., 2015). Data from this investigation highlighted the antibacterial efficacy and versatility of employing plazomicin in combination with other therapeutic compounds to counter replication of drug-resistant pathogens (Rodriguez-Avial et al., 2015).

Additionally, the impact of administering plazomicin, dalbavancin, and ceftobiprole was assessed against methicillin-resistant *Staphylococcus aureus* strains (Lopez Diaz et al., 2017). These pathogens conferred drug resistance by secreting aminoglycoside-modifying enzymes that decrease the efficacy of aminoglycoside antibiotics. The combination of plazomicin and meropenem produced the highest synergistic effect, resulting in a substantial decline in bacterial growth (Lopez Diaz et al., 2017). Collectively, these studies indicated the efficacy and utility of the novel antimicrobial agent plazomicin in inhibiting the replication of various drug-resistant bacterial pathogens in combination with other antibacterial drugs (Lopez Diaz et al., 2017). Thus this strengthens the use of combination therapy to counter the activity and dissemination of clinically relevant microbial pathogens that do not respond to treatment by conventional antimicrobial therapies.

2.4.2 Combination therapy against *Mycobacterium tuberculosis*

Combination therapy has also been employed to combat various forms of drug-resistant tuberculosis (TB) infections (de Knegt et al., 2017). The activities of the antitubercular drugs moxifloxacin and linezolid, the antiefflux pump molecules timcodar and verapamil, as well as antibacterial compounds that destabilize the *Mycobacterium tuberculosis* (*Mtb*) cell wall colistin and SQ109 were tested against multidrug-resistant (MDR) bacterial strains. Time-kill assays revealed that administration of moxifloxacin at 0.125 mg/L resulted in 99% *Mtb* growth inhibition at day 6 following treatment, while linezolid demonstrated moderate anti-TB activity during the same time frame. In contrast to this, a high dosage (256 mg/L) of SQ109 was required to completely eradicate bacterial cells at day 1 after exposure, while 1 mg/L of the drug inhibited *Mtb* replication by 99% at day 6 following administration (de Knegt et al., 2017). The other antibacterial agents colistin, verapamil, and timcodar only exerted anti-TB activity when administered at the highest drug concentration. Disappointingly, their anti-TB efficacy was less than that achieved by moxifloxacin or SQ109 monotherapy during the same period. Impressively, simultaneous administration of moxifloxacin, linezolid, and colistin achieved the highest anti-*Mtb* activity, thus indicating the importance of combination therapy in countering pathogenic MDR TB strains (de Knegt et al., 2017).

Another study evaluated the antibacterial activity of three drug combinations against susceptible and MDR TB clinical isolates: combination 1, levofloxacin/linezolid/ethambutol; combination 2, levofloxacin/amikacin/ethambutol; and combination 3, levofloxacin/linezolid/amikacin

(Lopez-Gavin et al., 2016). Data emanating from this investigation illustrated that all the drug combinations significantly improved anti-TB activity in both susceptible and MDR strains relative to administration of each individual anti-TB drug and highlighted the beneficial synergistic effect afforded by combination therapy (Lopez-Gavin et al., 2016). Similarly, the same research group had previously conducted an investigation to evaluate the utility of employing three different combinations of anti-TB drugs (Lopez-Gavin et al., 2016). The tested drug combinations included: combination 1, clofazimine/pretomanid/levofloxacin; combination 2, clofazimine/pretomanid/moxifloxacin; and combination 3, clofazimine/pretomanid/UB-8902 (Lopez-Gavin et al., 2015).

These antibacterial compounds were tested against susceptible and MDR *Mtb* clinical isolates. Encouragingly, all the different drug combinations exerted the same anti-TB activity against both susceptible and MDR *Mtb* strains, thus demonstrating the beneficial effects of combination therapy in inhibiting bacterial growth irrespective of the drug sensitivity or resistant profile of the *Mtb* pathogens (Lopez-Gavin et al., 2015). Moreover, the antibacterial activity of amikacin and doxycycline were assessed following exposure to susceptible, isoniazid-resistant, streptomycin/isoniazid-resistant, rifampicin-resistant, MDR and extensively drug-resistant (XDR) *Mtb* clinical isolates (Gonzalo et al., 2015). Promising findings from this study indicated that coadministration of amikacin and doxycycline effected a marked decrease in bacterial replication in 18 of the 29 tested strains. In the context of TB infection, combination therapy has been generally demonstrated to significantly retard bacterial replication in susceptible, drug-resistant, MDR, and XDR clinical isolates, as compared to administration of a single anti-TB agent (Gonzalo et al., 2015). Therefore the multitargeting strategy provided by combination therapy is generally superior in neutralizing a broader range of *Mtb* clinical isolates as compared to monotherapy (Gonzalo et al., 2015).

2.4.3 Combination therapy against HIV

Human immunodeficiency virus (HIV) is one of the leading causes of morbidity and mortality. Accordingly, development of HIV drug-resistant strains constitutes a major public health crisis, particularly in high endemic regions such as sub-Saharan Africa. This warranted the assessment of combination therapy as an approach to neutralizing emerging drug-resistant HIV strains. Indeed, a number of compelling studies have been performed to test this strategy in countering the activity of clinically relevant HIV drug-resistant strains. The use of HIV-1 nucleocapsid inhibitor *S*-acyl-2-mercaptobenzamide thioester-10 (SAMT10) together with the viral fusion inhibitor sifuvirtide has been interrogated (Yang et al., 2017). These anti-HIV molecules were tested against HIV-1 in cultured mammalian cells and ex vivo using the mucosal explant model. A substantial decline in markers of viral replication was reported in vitro and ex vivo following coadministration of sifuvirtide and SAMT10 in both pseudotyped and laboratory HIV-1 strains (Yang et al., 2017). The anti-HIV activity exerted by combination therapy surpassed that effected by administration of a single antiviral agent (Yang et al., 2017) and validated the use of multiple therapeutic molecules to suppress viral replication and emergence of mutant HIV-1 strains.

Currently, combinatorial use of 3BNC117 and 10-1074 broadly neutralizing antibodies has been reported in two related phase 1b clinical trials against HIV-1 (Bar-On et al., 2018; Mendoza et al., 2018). In both trials, the participants were treated with 30 mg/kg (Bar-On et al., 2018; Mendoza et al., 2018) of each broadly neutralizing antibody at 0, 3, and 6 weeks (Mendoza et al., 2018). Data collected from these clinical trials indicated that the viral load was significantly suppressed for about 21 weeks (Mendoza et al., 2018) and 3 months (Bar-On et al., 2018) in participants that received both antibody formulations. Importantly, no development of drug-resistant variants was observed in either trial (Bar-On et al., 2018; Mendoza et al., 2018), confirming the importance of combination therapy in countering and preventing the emergence of drug-resistant microbial variants.

Furthermore, it has been established that HIV-2 is inherently resistant to most anti-HIV1 therapies (Campbell-Yesufu & Gandhi, 2011; Cavaco-Silva et al., 2014; Menendez-Arias & Alvarez, 2014; Ren et al., 2002). To this end, Hu et al. (2017) tested the ability of a single pill composed of abacavir and lamivudine in suppressing HIV-2 infection in humanized mice (Hu et al., 2017). Relative to control mice, complete viral inhibition and protection was observed in animals treated with a combination of abacavir and lamivudine. This consolidated the anti-HIV efficacy, utility, and versatility of combination therapy in neutralizing both HIV-1 and HIV-2 drug-resistant HIV strains (Hu et al., 2017).

2.4.4 Combination therapy against *Candida albicans*

Candida species are the leading cause of opportunistic nosocomial fungal infections. Depending on the host immune competence, these infections may lead to the development of superficial and chronic systemic infections (Lu et al., 2018b). *Candida albicans* is the primary fungal pathogen responsible for the majority of clinical cases. Fluconazole, a triazole antifungal compound, is one of the most effecting drugs employed for the treatment of fungal infections. However, emergence of drug-resistant fungal pathogens has rendered this first-generation drug as well as other antifungal compounds ineffective in the treatment of opportunistic fungal infections. To augment the efficacy of antifungal drugs, combination therapy has been utilized in the treatment of fungal diseases that do not respond to conventional antimicrobial drugs. Indeed, the combination of fluconazole and dexamethasone has been employed to inhibit the replication of drug-resistant *C. albicans* (Sun et al., 2017). In this study, application of both antifungal compounds induced an additive effect, which enhanced their antifungal activity against drug-resistant strains in cultured mammalian cells and animals (Sun et al., 2017).

Gentamycin is an aminoglycoside molecule that displays weak antifungal efficacy against *Fusarium* species, the primary fungal plant pathogen in the agricultural industry (Lu et al., 2018b). This drug was used together with fluconazole to enhance their activity in inhibiting the growth of drug-resistant *C. albicans* strains. Simultaneous application of both antimicrobial compounds significantly improved their antifungal activity and led to a marked decline in the fungal load (Lu et al., 2018b). Furthermore, this observation was also replicated in an animal model of *Galleria mellonella* following coadministration of both antimicrobial drugs. Data from this investigation indicated that gentamycin and fluconazole performed synergistically to counter the drug-resistant *C. albicans* pathogens by deactivating the efflux pump activity responsible for decreasing drug uptake and efficacy (Lu et al., 2018b).

In an independent study, the use of fluconazole and 2-isopropyl-5-methylphenol (thymol) was evaluated against resistant clinical isolates of *Candida* species (Sharifzadeh et al., 2018). Combinatorial administration of both antifungal compounds displayed a synergistic effect against all *C. albicans* isolates (Sharifzadeh et al., 2018). Additionally, this combination also displayed improved efficacy against *C. krusei* and *C. glabrata* strains. Positive results reported in this study indicated that combinatorial use of thymol and fluconazole may potentially be employed as an effective strategy to combat various drug-resistant *Candida* pathogens (Sharifzadeh et al., 2018). Another investigation examined the antifungal activity induced following treatment of resistant sessile and planktonic *C. albicans* cells with a combination of different antifungal agents (Touil et al., 2018). The drug combinations evaluated in this study were amphotericin B/caspofungin and amphotericin B/voriconazole. Both drug combinations led to a significant decrease in the minimum inhibitory concentration (MIC) values required to treat sessile and planktonic cells (Touil et al., 2018). Importantly, a marked decline in the fungal load was reported, which illustrated that a combination of amphotericin B with either caspofungin or voriconazole may be employed as a therapeutic method of treating both planktonic and sessile *C. albicans* species (Touil et al., 2018).

2.4.5 Combination therapy against *Plasmodium falciparum*

Plasmodium falciparum is the causative agent of malaria. This protozoan parasite is transmitted to humans following a bite by a female *Anopheles* mosquito. Antimalaria drugs are widely employed to prevent *P. falciparum* infection and treat disease symptoms. Unfortunately, the development of drug-resistant *P. falciparum* phenotypes against antimalaria agents such as chloroquine (Kemirembe, Cabrera, & Cui, 2017) and artemisinin (Tasai et al., 2017) has hampered the significant milestones achieved in malaria research. For this reason, combination therapy has been evaluated as a therapeutic method to treat and regulate the spread of malaria. This includes the utilization of tafenoquine and chloroquine along with six artemisinin drugs as combination therapy against *P. falciparum* (Kemirembe et al., 2017). Findings from this investigation showed that tafenoquine and most artemisinin drugs displayed improved efficacy toward the malaria parasite (Kemirembe et al., 2017). These preliminary data suggest that these drug combinations may potentially be used to inhibit asexual development stages of *P. falciparum* (Kemirembe et al., 2017).

The advantage of applying a combination of tafenoquine and a low dosage of artesunate has also been assessed (Tasai et al., 2017). This drug combination was tested in an animal model against replication, development, and transmission of *Plasmodium gallinaceum* to the mosquito vector *Aedes aegypti*. A 5-day daily treatment of tafenoquine decreased mortality rate, while administration of a high dose (30 mg/kg) of the same therapeutic agent halted parasite transmission (Tasai et al., 2017). Conversely, complete eradication of *P. gallinaceum* blood stages and efficient gametocyte clearance was observed in chickens treated with a combination of tafenoquine and artesunate. These outcomes indicated that tafenoquine is capable of decreasing malaria spread in treated animals, and may be employed in conjunction with a low dose of artesunate to induce enhanced antimalaria activity (Tasai et al., 2017). Recently, a research study conducted by Sharma et al. (2018) aimed to explore the antimalaria effects triggered by combinatorial administration of bischalcones and artemisinin against susceptible and chloroquine-resistant *P. falciparum* parasites (Sharma et al., 2018). The bischalcone compounds 9, 11, and 13 displayed antimicrobial activity. However, compound 13 induced a significant number of clinically desirable antimalaria properties. Encouragingly, improved anti-*plasmodium* activity was observed in cultured human erythrocytes treated with a combination of bischalcone compound 13 and artemisinin (Sharma et al., 2018). This promising data illustrated that coadministration of these antimalaria compounds may be utilized as an effective mechanism to enhance treatment of both susceptible and drug-resistant *P. falciparum* pathogens (Sharma et al., 2018).

2.5 Conclusion

In conclusion, the evolution of drug-resistant microbial pathogens has caused enormous medical challenges in many public health institutions across the world. The high mutagenesis rate of many human pathogens of clinical importance such as HIV and *Mtb* has rendered currently licensed therapeutics ineffective in the treatment of diseases caused by drug-resistant strains. Additionally, various forms of horizontal gene transfer mechanisms have enhanced the development and transmission of resistant microbes. Poor clinical practices and wrong prescription of antimicrobial agents by health professionals as well as improper use of therapeutic compounds by patients have also been implicated in the increase in the number of drug-resistant infections.

Furthermore, the misuse of therapeutic agents in the agricultural and other sectors has also contributed massively to the emergence of microbes harboring resistant genes. Continual application of a single therapeutic compound (i.e., monotherapy) has allowed infectious agents to exert selection pressure and ultimately led to development of drug-resistant microorganisms. Therefore, implementation of proper medical practices by health professionals and strict adherence to the treatment schedule by patients may significantly reduce the occurrence of drug-resistant cases. Moreover, effective medical interventions capable of eliminating drug-resistant infections have to be urgently implemented. In the long run, development of novel and potent therapeutic compounds may mitigate drug-resistant infections. However, this practice may require a long period of time before a positive clinical outcome is realized.

Consequently, the combination of already-licensed therapeutic agents has been utilized as a drastic measure of treating and preventing the dissemination of a wide array of drug-resistant microbial species. To date, combination therapy has been successfully utilized to counter the replication and spread of clinically relevant infectious microbes that do not respond to conventional treatment regimens (Gonzalo et al., 2015; Hu et al., 2017; Kemirembe et al., 2017; Lopez-Gavin et al., 2015; Lu et al., 2018b; Mendoza et al., 2018; Sharma et al., 2018; Sun et al., 2017; Thwaites et al., 2018). The positive clinical outcome following administration of this procedure has been mainly attributed to its ability to target and suppress different therapeutic drug targets simultaneously. This multitargeting strategy has a negative impact on the numerous critical microbial signaling pathways required to establish host infection. This, in turn, significantly reduces the amount of selection pressure exerted on the drugs by infectious microbes. Ultimately, this has profound clinical implications as it prevents the development and transmission of drug-resistant mutant strains. The application of combination therapy generally decreases the drug MIC values required to inhibit microbial growth and also displays synergistic effects that improve antimicrobial therapy.

Apart from combination therapy, alternative therapeutic methods, such as clustered regularly interspaced short palindromic repeat (CRISPR)/CRISPR-associated protein 9 (cas9) and RNA interference (RNAi), can be used to counter the evolution of drug-resistant strains by suppressing the expression of drug-resistant genes. The CRISPR-cas9 gene editing procedure may be used to cleave the target microbial DNA (Arranz-Solis et al., 2018; Ryu et al., 2018; Wang et al., 2018). Cleavage of target DNA induces double-stranded breaks and repeated DNA binding and cleavage events are activated following DNA repair mechanisms. Homology directed repair (HDR) and nonhomologous end joining (NHEJ) are responsible for target DNA repair. The HDR mechanism results in precise repair of the cleaved target DNA sequence, while the NHEJ process is prone to errors and results in insertions and deletions being introduced in the cleaved DNA. As a result, this technique may be utilized to weaken microbial replication after repeated cleavage cycles. This induces mutations in target DNA and the subsequent decline in microbial replication may allow the host to mount an immune response, leading to eradication of drug-resistant microbial pathogens.

Conversely, the RNAi procedure exists naturally and regulates gene expression in various eukaryotic organisms. The double-stranded molecules, such as primary-microRNAs (pri-miRNAs), precursor miRNAs (pre-miRNAs), and microRNAs (miRNAs), participate at different stages of the RNAi mechanism. Complete base-pairing between the guide strand of these molecules and target mRNA leads to target mRNA degradation, while incomplete base-pairing results in translational suppression. Interestingly, introduction of synthetic molecules that mimic pri-miRNAs, pre-miRNAs or miRNAs into a cell leads to activation of the RNAi pathway (Hoffmann et al., 2018; Wang, Gu, & Knipple, 2018; Zhang et al., 2018b). As a result, this procedure may be conveniently employed to target and inhibit the expression of drug-resistant genes, thus decreasing the transmission of drug-resistant pathogens.

Bacteriophages may be used to regulate the development and dissemination of drug-resistant microorganisms (Chang et al., 2011; Tinoco et al., 2017). These agents may be employed to infect and eradicate pathogenic bacteria from the environment and infected host tissues. However, additional safety precautions have to be followed to prevent host infection by bacteriophages designed to clear drug-resistant bacterial species. Quorum sensing (QS) is another feasible strategy that could be utilized to circumvent development of drug-resistant microbes. Many pathogenic microorganisms rely on cell communication that unleashes a cascade of signaling pathways that lead to establishment of infection within host cells and tissues. Accordingly, application of anti-QS molecules inhibits cell-to-cell communication, thus preventing host infection and subsequent disease manifestation (Haque et al., 2018; Welsh et al., 2015). Finally, the development of immune-stimulatory molecules is another innovative strategy to combat drug resistance. The application of such compounds may trigger the host immune signaling pathways to mount a strong immune response, resulting in efficient clearance of infectious pathogens.

Acknowledgments

We gratefully acknowledge financial assistance provided by the National Research Foundation of South Africa (NRF) Research Development Grant for Y-Rated Researchers (RDYR180418322304; Grant No. 116339), the Department of Science and Technology (DST)/NRF Innovation Postdoctoral Fellowship (Grant No. 112051), the University of the Witwatersrand Faculty of Health Sciences Research Committee (FRC, Grant No. 001 254 8464101 5121105 000000 0000000000 5254), and the National Health Laboratory Service Research grant (NHLSRT-1/3/17-1/9/19).

Author contributions

Musa Marimani compiled the book chapter with the assistance of Aijaz Ahmad and Adriano Duse.

Competing interests

We have no competing interests to declare.

References

Aarestrup, F. (2012). Sustainable farming: Get pigs off antibiotics. *Nature, 486*(7404), 465–466.

Abraham, E. P., & Chain, E. (1940). An enzyme from bacteria able to destroy penicillin. *Nature, 146*, 837.

Andam, C. P., Fournier, G. P., & Gogarten, J. P. (2011). Multilevel populations and the evolution of antibiotic resistance through horizontal gene transfer. *FEMS Microbiology Reviews, 35*(5), 756–767.

Arranz-Solis, D., et al. (2018). Toxoplasma CRISPR/Cas9 constructs are functional for gene disruption in *Neospora caninum*. *International Journal for Parasitology, 48*(8), 597–600.

Austin, D. J., et al. (1999). Vancomycin-resistant enterococci in intensive-care hospital settings: Transmission dynamics, persistence, and the impact of infection control programs. *Proceedings of the National Academy of Sciences of the United States of America, 96*(12), 6908–6913.

Aye, K. S., et al. (2016). Genotypic characterization of multi-drug-resistant *Mycobacterium tuberculosis* isolates in Myanmar. *Journal of Infection and Chemotherapy: Official Journal of the Japan Society of Chemotherapy, 22*(3), 174–179.

Baquero, F., Alvarez-Ortega, C., & Martinez, J. L. (2009). Ecology and evolution of antibiotic resistance. *Environmental Microbiology Reports, 1*(6), 469–476.

Bar-On, Y., et al. (2018). Safety and antiviral activity of combination HIV-1 broadly neutralizing antibodies in viremic individuals. *Nature Medicine, 24*(11), 1701–1707.

Bjorke, A. B., et al. (2018). Evaluation of long-term antiepileptic drug use in patients with temporal lobe epilepsy: Assessment of risk factors for drug resistance and polypharmacy. *Seizure, 61*, 63–70.

Blower, S. M., Small, P. M., & Hopewell, P. C. (1996). Control strategies for tuberculosis epidemics: New models for old problems. *Science, 273*(5274), 497–500.

Bonhoeffer, S., Lipsitch, M., & Levin, B. R. (1997). Evaluating treatment protocols to prevent antibiotic resistance. *Proceedings of the National Academy of Sciences of the United States of America, 94*(22), 12106–12111.

Bonten, M. J., & Weinstein, R. A. (1996). The role of colonization in the pathogenesis of nosocomial infections. *Infection Control and Hospital Epidemiology, 17*(3), 193–200.

Boyce, J. M. (1996). Treatment and control of colonization in the prevention of nosocomial infections. *Infection Control and Hospital Epidemiology, 17*(4), 256–261.

Brun-Buisson, C., et al. (1989). Intestinal decontamination for control of nosocomial multiresistant gram-negative bacilli. Study of an outbreak in an intensive care unit. *Annals of Internal Medicine, 110*(11), 873–881.

Cabello, F. C. (2006). Heavy use of prophylactic antibiotics in aquaculture: A growing problem for human and animal health and for the environment. *Environmental Microbiology, 8*(7), 1137–1144.

Calva, J., & Bojalil, R. (1996). Antibiotic use in a periurban community in Mexico: A household and drugstore survey. *Social Science & Medicine, 42*(8), 1121–1128.

Campbell-Yesufu, O. T., & Gandhi, R. T. (2011). Update on human immunodeficiency virus (HIV)-2 infection. *Clinical Infectious Diseases: An Official Publication of the Infectious Diseases Society of America, 52*(6), 780–787.

Cavaco-Silva, J., et al. (2014). HIV-2 integrase polymorphisms and longitudinal genotypic analysis of HIV-2 infected patients failing a raltegravir-containing regimen. *PLoS One, 9*(3), e92747.

Chang, K. C., et al. (2011). Genomic analysis of bacteriophage varphiAB1, a varphiKMV-like virus infecting multidrug-resistant *Acinetobacter baumannii*. *Genomics, 97*(4), 249–255.

Collignon, P., et al. (2005). The routine use of antibiotics to promote animal growth does little to benefit protein undernutrition in the developing world. *Clinical Infectious Diseases: An Official Publication of the Infectious Diseases Society of America, 41*(7), 1007–1013.

Cornejo-Juarez, P., et al. (2015). The impact of hospital-acquired infections with multidrug-resistant bacteria in an oncology intensive care unit. *International Journal of Infectious Diseases: IJID: Official Publication of the International Society for Infectious Diseases, 31*, 31–34.

Davies, J. E. (1997). Origins, acquisition and dissemination of antibiotic resistance determinants. *Ciba Foundation Symposium, 207*, 15–27 [discussion 27–35].

de Knegt, G. J., et al. (2017). Activity of moxifloxacin and linezolid against *Mycobacterium tuberculosis* in combination with potentiator drugs verapamil, timcodar, colistin and SQ109. *International Journal of Antimicrobial Agents, 49*(3), 302–307.

Fajardo, A., Linares, J. F., & Martinez, J. L. (2009). Towards an ecological approach to antibiotics and antibiotic resistance genes. *Clinical Microbiology and Infection: The Official Publication of the European Society of Clinical Microbiology and Infectious Diseases, 15*(Suppl. 1), 14–16.

Fok, A., et al. (2008). Risk factors for clustering of tuberculosis cases: A systematic review of population-based molecular epidemiology studies. *The International Journal of Tuberculosis and Lung Disease : The Official Journal of the International Union against Tuberculosis and Lung Disease, 12*(5), 480–492.

Fortini, D., et al. (2011). Plasmid-mediated quinolone resistance and beta-lactamases in *Escherichia coli* from healthy animals from Nigeria. *The Journal of Antimicrobial Chemotherapy, 66*(6), 1269–1272.

Fuste, E., et al. (2013). Carbapenem-resistance mechanisms of multidrug-resistant *Pseudomonas aeruginosa*. *Journal of Medical Microbiology, 62*(Pt 9), 1317–1325.

Gamngoen, R., et al. (2018). A comparison of Rv0559c and Rv0560c expression in drug-resistant *Mycobacterium tuberculosis* in response to first-line antituberculosis drugs. *Tuberculosis, 108*, 64–69.

Gaynes, R., & Monnet, D. (1997). The contribution of antibiotic use on the frequency of antibiotic resistance in hospitals. *Ciba Foundation Symposium, 207*, 47–56 [discussion 56–60].

Gaynes, R. P., et al. (1983). Control of aminoglycoside resistance by barrier precautions. *Infection Control: IC, 4*(4), 221–224.

Geli, P., et al. (2012). "One-size-fits-all"? Optimizing treatment duration for bacterial infections. *PLoS One, 7*(1), e29838.

Gjini, E., & Brito, P. H. (2016). Integrating antimicrobial therapy with host immunity to fight drug-resistant infections: Classical vs. adaptive treatment. *PLoS Computational Biology, 12*(4), e1004857.

Gonzalo, X., et al. (2015). Combination of amikacin and doxycycline against multidrug-resistant and extensively drug-resistant tuberculosis. *International Journal of Antimicrobial Agents, 45*(4), 406–412.

Gu, D., et al. (2018). A fatal outbreak of ST11 carbapenem-resistant hypervirulent *Klebsiella pneumoniae* in a Chinese hospital: a molecular epidemiological study. *The Lancet Infectious Diseases, 18*(1), 37–46.

Gurarie, D., & McKenzie, F. E. (2006). Dynamics of immune response and drug resistance in malaria infection. *Malaria Journal, 5,* 86.

Gurjav, U., et al. (2016). Transmission of multi-drug resistant tuberculosis in Mongolia is driven by Beijing strains of *Mycobacterium tuberculosis* resistant to all first-line drugs. *Tuberculosis, 101,* 49–53.

Hall, R. M., & Collis, C. M. (1995). Mobile gene cassettes and integrons: Capture and spread of genes by site-specific recombination. *Molecular Microbiology, 15*(4), 593–600.

Handel, A., Margolis, E., & Levin, B. R. (2009). Exploring the role of the immune response in preventing antibiotic resistance. *Journal of Theoretical Biology, 256*(4), 655–662.

Haque, S., et al. (2018). Developments in strategies for Quorum Sensing virulence factor inhibition to combat bacterial drug resistance. *Microbial Pathogenesis, 121,* 293–302.

Hawkey, P. M., et al. (2018). Treatment of infections caused by multidrug-resistant Gram-negative bacteria: Report of the British Society for Antimicrobial Chemotherapy/Healthcare Infection Society/British Infection Association Joint Working Party. *The Journal of Antimicrobial Chemotherapy, 73*(Suppl. 3), iii2–iii78.

Hoffmann, D. B., et al. (2018). Effects of RANKL knockdown by virus-like particle-mediated RNAi in a rat model of osteoporosis. *Molecular Therapy. Nucleic acids, 12,* 443–452.

Honsa, E. S., et al. (2017). RelA mutant *Enterococcus faecium* with multiantibiotic tolerance arising in an immunocompromised host. *MBio, 8*(1).

Hosono, Y., et al. (2018). The association between erythromycin monotherapy for *Mycobacterium avium* complex lung disease and cross-resistance to clarithromycin: A retrospective case-series study. *Journal of Infection and Chemotherapy: Official Journal of the Japan Society of Chemotherapy, 24*(5), 353–357.

Hu, S., et al. (2017). A humanized mouse model for HIV-2 infection and efficacy testing of a single-pill triple-drug combination anti-retroviral therapy. *Virology, 501,* 115–118.

Hu, B., et al. (2018). Role of TGF-beta1/Smad3-mediated fibrosis in drug resistance mechanism of prolactinoma. *Brain Research, 1698,* 204–212.

Iravani, A., et al. (1995). Short-course ciprofloxacin treatment of acute uncomplicated urinary tract infection in women. The minimum effective dose. The Urinary Tract Infection Study Group [corrected]. *Archives of Internal Medicine, 155*(5), 485–494.

Kaplan, E., et al. (2013). Characterization of fluoroquinolone resistance and qnr diversity in Enterobacteriaceae from municipal biosolids. *Frontiers in Microbiology, 4,* 144.

Keen, P. L., et al. (2018). Seasonal dynamics of tetracycline resistance gene transport in the Sumas River agricultural watershed of British Columbia, Canada. *The Science of the Total Environment, 628–629,* 490–498.

Kemirembe, K., Cabrera, M., & Cui, L. (2017). Interactions between tafenoquine and artemisinin-combination therapy partner drug in asexual and sexual stage *Plasmodium falciparum. International Journal for Parasitology. Drugs and Drug Resistance, 7*(2), 131–137.

Kikuvi, G. M., Ombui, J. N., & Mitema, E. S. (2010). Serotypes and antimicrobial resistance profiles of Salmonella isolates from pigs at slaughter in Kenya. *Journal of Infection in Developing Countries, 4*(4), 243–248.

Kim, S., et al. (2018). In vitro activity of collinin isolated from the leaves of *Zanthoxylum schinifolium* against multidrug- and extensively drug-resistant *Mycobacterium tuberculosis. Phytomedicine: International Journal of Phytotherapy and Phytopharmacology, 46,* 104–110.

Kollef, M. H., et al. (1997). Scheduled change of antibiotic classes: A strategy to decrease the incidence of ventilator-associated pneumonia. *American Journal of Respiratory and Critical Care Medicine*, *156*(4 Pt 1), 1040–1048.

Lansang, M. A., et al. (1990). Purchase of antibiotics without prescription in Manila, the Philippines. Inappropriate choices and doses. *Journal of Clinical Epidemiology*, *43*(1), 61–67.

Li, W., et al. (2013). Mechanism of tetracycline resistance by ribosomal protection protein Tet(O). *Nature Communications*, *4*, 1477.

Li, D., et al. (2017). Screening mutations in drug-resistant *Mycobacterium tuberculosis* strains in Yunnan, China. *Journal of Infection and Public Health*, *10*(5), 630–636.

Lipsitch, M., Bergstrom, C. T., & Levin, B. R. (2000). The epidemiology of antibiotic resistance in hospitals: Paradoxes and prescriptions. *Proceedings of the National Academy of Sciences of the United States of America*, *97*(4), 1938–1943.

Liu, Y., et al. (2018). Analysis on clinical characteristics and drug resistance of *Candida parapsilosis* bloodstream infections in West China Hospital, China, from 2012 to 2015. *Journal de Mycologie Medicale*, *28*(1), 222–226.

Llor, C., & Bjerrum, L. (2014). Antimicrobial resistance: Risk associated with antibiotic overuse and initiatives to reduce the problem. *Therapeutic Advances in Drug Safety*, *5*(6), 229–241.

Lonnroth, K., et al. (2008). Alcohol use as a risk factor for tuberculosis—A systematic review. *BMC Public Health*, *8*, 289.

Lopez Diaz, M. C., et al. (2017). In-vitro activity of several antimicrobial agents against methicillin-resistant *Staphylococcus aureus* (MRSA) isolates expressing aminoglycoside-modifying enzymes: Potency of plazomicin alone and in combination with other agents. *International Journal of Antimicrobial Agents*, *50*(2), 191–196.

Lopez-Gavin, A., et al. (2015). In vitro activity against *Mycobacterium tuberculosis* of levofloxacin, moxifloxacin and UB-8902 in combination with clofazimine and pretomanid. *International Journal of Antimicrobial Agents*, *46*(5), 582–585.

Lopez-Gavin, A., et al. (2016). In vitro time-kill curves study of three antituberculous combinations against *Mycobacterium tuberculosis* clinical isolates. *International Journal of Antimicrobial Agents*, *47*(1), 97–100.

Lu, J., et al. (2018a). Non-antibiotic antimicrobial triclosan induces multiple antibiotic resistance through genetic mutation. *Environment International*, *118*, 257–265.

Lu, M., et al. (2018b). Gentamicin synergises with azoles against drug-resistant *Candida albicans*. *International Journal of Antimicrobial Agents*, *51*(1), 107–114.

Malfertheiner, P. (1993). Compliance, adverse events and antibiotic resistance in *Helicobacter pylori* treatment. *Scandinavian Journal of Gastroenterology. Supplement*, *196*, 34–37.

Manson, J. M., Hancock, L. E., & Gilmore, M. S. (2010). Mechanism of chromosomal transfer of *Enterococcus faecalis* pathogenicity island, capsule, antimicrobial resistance, and other traits. *Proceedings of the National Academy of Sciences of the United States of America*, *107*(27), 12269–12274.

Martinez, J. L. (2008). Antibiotics and antibiotic resistance genes in natural environments. *Science*, *321*(5887), 365–367.

Martinez, J. L., & Baquero, F. (2000). Mutation frequencies and antibiotic resistance. *Antimicrobial Agents and Chemotherapy*, *44*(7), 1771–1777.

McGowan, J. E., Jr. (1986). Minimizing antimicrobial resistance in hospital bacteria: Can switching or cycling drugs help? *Infection Control: IC*, *7*(12), 573–576.

Mendoza, P., et al. (2018). Combination therapy with anti-HIV-1 antibodies maintains viral suppression. *Nature*, *561*(7724), 479–484.

Menendez-Arias, L., & Alvarez, M. (2014). Antiretroviral therapy and drug resistance in human immunodeficiency virus type 2 infection. *Antiviral Research*, *102*, 70–86.

Mitema, E. S., et al. (2001). An assessment of antimicrobial consumption in food producing animals in Kenya. *Journal of Veterinary Pharmacology and Therapeutics*, *24*(6), 385–390.

Monnet, D. L., et al. (1998). Antimicrobial use and resistance in eight US hospitals: Complexities of analysis and modeling. Intensive Care Antimicrobial Resistance Epidemiology Project and National Nosocomial Infections Surveillance System Hospitals. *Infection Control and Hospital Epidemiology*, *19*(6), 388–394.

Munita, J. M., & Arias, C. A. (2016). Mechanisms of antibiotic resistance. *Microbiology Spectrum*, *4*(2).

Nachega, J. B., & Chaisson, R. E. (2003). Tuberculosis drug resistance: A global threat. *Clinical Infectious Diseases: An Official Publication of the Infectious Diseases Society of America*, *36*(Suppl. 1), S24–S30.

Neu, H. C. (1992). The crisis in antibiotic resistance. *Science*, *257*(5073), 1064–1073.

Petersen, A., et al. (2002). Impact of integrated fish farming on antimicrobial resistance in a pond environment. *Applied and Environmental Microbiology*, *68*(12), 6036–6042.

Rangasamy, K., et al. (2017). Emergence of multi drug resistance among soil bacteria exposing to insecticides. *Microbial Pathogenesis*, *105*, 153–165.

Ren, J., et al. (2002). Structure of HIV-2 reverse transcriptase at 2.35-A resolution and the mechanism of resistance to non-nucleoside inhibitors. *Proceedings of the National Academy of Sciences of the United States of America*, *99*(22), 14410–14415.

Rezazadeh, M., Baghchesaraei, H., & Peymani, A. (2016). Plasmid-mediated quinolone-resistance (qnr) genes in clinical isolates of *Escherichia coli* collected from several hospitals of Qazvin and Zanjan provinces, Iran. *Osong Public Health and Research Perspectives*, *7*(5), 307–312.

Rodriguez-Avial, I., et al. (2015). In vitro activity of the next-generation aminoglycoside plazomicin alone and in combination with colistin, meropenem, fosfomycin or tigecycline against carbapenemase-producing Enterobacteriaceae strains. *International Journal of Antimicrobial Agents*, *46*(6), 616–621.

Ryu, N., et al. (2018). Effective PEI-mediated delivery of CRISPR-Cas9 complex for targeted gene therapy. *Nanomedicine: Nanotechnology, Biology, and Medicine*, *14*(7), 2095–2102.

Saleh, N., et al. (2015). Evaluation of antibiotic prescription in the Lebanese community: A pilot study. *Infection Ecology & Epidemiology*, *5*, 27094.

Shaker, M. A., & Shaaban, M. I. (2017). Formulation of carbapenems loaded gold nanoparticles to combat multi-antibiotic bacterial resistance: In vitro antibacterial study. *International Journal of Pharmaceutics*, *525*(1), 71–84.

Sharifzadeh, A., et al. (2018). Potential effect of 2-isopropyl-5-methylphenol (thymol) alone and in combination with fluconazole against clinical isolates of *Candida albicans, C. glabrata* and *C. krusei*. *Journal de Mycologie Medicale*, *28*(2), 294–299.

Sharma, U. K., et al. (2018). Facile synthesis of vanillin-based novel bischalcones identifies one that induces apoptosis and displays synergy with Artemisinin in killing chloroquine resistant *Plasmodium falciparum*. *European Journal of Medicinal Chemistry*, *155*, 623–638.

Simango, C., & Rukure, G. (1991). Potential sources of *Campylobacter* species in the homes of farmworkers in Zimbabwe. *The Journal of Tropical Medicine and Hygiene*, *94*(6), 388–392.

Smith, M. A., et al. (2004). Health promotion and disease prevention through sanitation education in South African Zulu and Xhosa women. *Journal of Transcultural Nursing: Official Journal of the Transcultural Nursing Society*, *15*(1), 62–68.

Sommer, M. O. A., Dantas, G., & Church, G. M. (2009). Functional characterization of the antibiotic resistance reservoir in the human microflora. *Science*, *325*(5944), 1128–1131.

Sun, W., et al. (2017). Strong synergism of dexamethasone in combination with fluconazole against resistant *Candida albicans* mediated by inhibiting drug efflux and reducing virulence. *International Journal of Antimicrobial Agents*, *50*(3), 399–405.

Szabo, G. (1997). Alcohol's contribution to compromised immunity. *Alcohol Health and Research World*, *21*(1), 30–41.

Tasai, S., et al. (2017). Artesunate-tafenoquine combination therapy promotes clearance and abrogates transmission of the avian malaria parasite *Plasmodium gallinaceum*. *Veterinary Parasitology*, *233*, 97–106.

Thwaites, M., et al. (2018). Activity of plazomicin in combination with other antibiotics against multidrug-resistant Enterobacteriaceae. *Diagnostic Microbiology and Infectious Disease*, *92*(4), 338–345.

Tinoco, J. M., et al. (2017). Antibacterial effect of genetically-engineered bacteriophage varphiEf11/varphiFL1C(Delta36)P(nisA) on dentin infected with antibiotic-resistant *Enterococcus faecalis*. *Archives of Oral Biology*, *82*, 166–170.

Touil, H. F. Z., Boucherit-Otmani, Z., & Boucherit, K. (2018). In vitro activity of antifungal combinations against planktonic and sessile cells of *Candida albicans* isolated from medical devices in an intensive care department. *Journal de Mycologie Medicale*, *28*(3), 414–418.

Tran-Dien, A., et al. (2018). Early transmissible ampicillin resistance in zoonotic *Salmonella enterica* serotype Typhimurium in the late 1950s: A retrospective, whole-genome sequencing study. *The Lancet. Infectious diseases*, *18*(2), 207–214.

Uddin, R., et al. (2018). Identification and characterization of potential druggable targets among hypothetical proteins of extensively drug resistant *Mycobacterium tuberculosis* (XDR KZN 605) through subtractive genomics approach. *European Journal of Pharmaceutical Sciences: Official Journal of the European Federation for Pharmaceutical Sciences*, *114*, 13–23.

Usluer, G., Ozgunes, I., & Leblebicioglu, H. (2005). A multicenter point-prevalence study: Antimicrobial prescription frequencies in hospitalized patients in Turkey. *Annals of Clinical Microbiology and Antimicrobials*, *4*, 16.

Vestergaard, M., et al. (2016). Antibiotic combination therapy can select for broad-spectrum multidrug resistance in *Pseudomonas aeruginosa*. *International Journal of Antimicrobial Agents*, *47*(1), 48–55.

Vidaver, A. K. (2002). Uses of antimicrobials in plant agriculture. *Clinical Infectious Diseases: An Official Publication of the Infectious Diseases Society of America*, *34*(Suppl. 3), S107–S110.

Wang, J., Gu, L., & Knipple, D. C. (2018). Evaluation of some potential target genes and methods for RNAi-mediated pest control of the corn earworm *Helicoverpa zea*. *Pesticide Biochemistry and Physiology*, *149*, 67–72.

Wang, S., et al. (2018). Genome engineering of *Clostridium difficile* using the CRISPR-Cas9 system. *Clinical Microbiology and Infection : The Official Publication of the European Society of Clinical Microbiology and Infectious Diseases*, *24*(10), 1095–1099.

Welsh, M. A., et al. (2015). Small molecule disruption of quorum sensing cross-regulation in *Pseudomonas aeruginosa* causes major and unexpected alterations to virulence phenotypes. *Journal of the American Chemical Society*, *137*(4), 1510–1519.

Williams, J. W., Jr., et al. (1995). Randomized controlled trial of 3 vs 10 days of trimethoprim/sulfamethoxazole for acute maxillary sinusitis. *JAMA*, *273*(13), 1015–1021.

Witte, W. (1998). Medical consequences of antibiotic use in agriculture. *Science*, *279*(5353), 996–997.

World Health Organization. (2000). Communicable diseases cluster. In *Overcoming antimicrobial resistance*. Geneva: World Health Organization. http://www.who.int/iris/handle/10665/66672. [Accessed 25 October 2018].

World Health Organization. (2010). *Antimicrobial resistance*. Geneva: World Health Organization. http://www.who.int/mediacentre/factsheets/fs194/en/. [Accessed 25 October 2018].

Yan, L., et al. (2017a). Bacterial plasmid-mediated quinolone resistance genes in aquatic environments in China. *Scientific Reports*, *7*, 40610.

Yan, J., et al. (2017b). Multidrug resistance mechanisms of carbapenem resistant *Klebsiella pneumoniae* strains isolated in Chongqing, China. *Annals of Laboratory Medicine*, *37*(5), 398–407.

Yang, Y., et al. (2017). A novel preventive strategy against HIV-1 infection: Combinatorial use of inhibitors targeting the nucleocapsid and fusion proteins. *Emerging Microbes & Infections*, *6*(6), e40.

Ye, Y., et al. (2018). Mechanism for carbapenem resistance of clinical Enterobacteriaceae isolates. *Experimental and Therapeutic Medicine*, *15*(1), 1143–1149.

Zapata-Cachafeiro, M., et al. (2014). Determinants of antibiotic dispensing without a medical prescription: A cross-sectional study in the north of Spain. *The Journal of Antimicrobial Chemotherapy*, *69*(11), 3156–3160.
Zhang, Y. Q., et al. (2018a). Ceftriaxone combination therapy versus respiratory fluoroquinolone monotherapy for community-acquired pneumonia: A meta-analysis. *The American Journal of Emergency Medicine*, *36*(10), 1759–1765.
Zhang, B., et al. (2018b). Integration of RNAi and RNA-seq uncovers the immune responses of *Epinephelus coioides* to L321_RS19110 gene of *Pseudomonas plecoglossicida*. *Fish & Shellfish Immunology*, *81*, 121–129.

CHAPTER

3

Multidrug resistance and the prospects of combination therapy

A. Balakrishna[a], *G. Sravya*[b], *T.V. Surendra*[a], *C. Suresh Reddy*[c], *Grigory V. Zyryanov*[b,d], *N. Bakthavatchala Reddy*[b]

[a]Department of Chemistry, Rajeev Gandhi Memorial College of Engineering and Technology (Autonomous), Nandyal, Andhra Pradesh, India, [b]Ural Federal University, Chemical Engineering Institute, Yekaterinburg, Russian Federation, [c]Department of Organic Chemistry, Sri Venkateswara University, Tirupati, Andhra Pradesh, India, [d]I. Ya. Postovskiy Institute of Organic Synthesis, Ural Division of the Russian Academy of Sciences, Yekaterinburg, Russian Federation

Abbreviations

ADCC	antibody-dependent cell-mediated cytotoxicity
AIDS	acquired immunodeficiency syndrome
AST	antimicrobial susceptibility testing
BMI	body mass index
BOC	boceprevir
CART	classification and regression tree
cART	combination antiretroviral therapy
CD4	cluster of differentiation 4
Ch-MLNPs	chitosan-multilayered nanoparticles
CSF	cerebrospinal fluid
DM	diabetes mellitus
DOX	doxorubicin
EASL	European Association for the Study of the Liver
ER	estrogen receptor
HBV	hepatitis B
HCV	hepatitis C virus
HDACi	histone deacetylase inhibitor
HIV	human immunodeficiency virus
ITT	intention to treat
MDR	multidrug resistance

https://doi.org/10.1016/B978-0-12-820576-1.00003-5

MM	multiple myeloma
MMAE	monomethylauristatin E
MMT	multiple-medications therapy
MTB	*Mycobacterium tuberculosis*
MTDLs	multitarget-directed ligands
NK	natural killer
PAI	photoacoustic imaging
PDCs	polymer-drug conjugates
PDT	photodynamic therapy
PR	progesterone receptor
PTT	partial thromboplastin time
PTX	paclitaxel
RMD	romidepsin
RNA	ribonucleic acid
SIV	simian immunodeficiency virus
SSA	sub-Saharan Africa
TB	tuberculosis
TDZ	thidiazuron
TLV	telaprevir
TNBC	triple-negative breast cancer
WHO	World Health Organization
XDR	extensively drug-resistant
XG	xyloglucan

3.1 Introduction

Of the millions of deaths worldwide due to different disease conditions, cancer and infectious diseases caused more than 13 million deaths in 2018. Multidrug resistance is undoubtedly posing a great threat to global health, and if nothing is done to halt the rise of multidrug resistance, global deaths could rise at an alarming rate. Among the various strategies, combination therapy, which involves the combination of two or more drugs used simultaneously to treat a disease condition, has shown great promise. This strategy has been successfully used to treat various diseases such as tuberculosis, leprosy, cancer, malaria, and HIV/AIDS, where multidrug resistance has already wreaked havoc. There are more than 10,000 ongoing clinical trials currently registered in the United States alone exploring combination therapies for cancer, infectious diseases, and metabolic, cardiovascular, autoimmune, and neurological disorders. Recently, interest has increased in combination therapy possessing synergistic mechanisms of action. One major advantage of combination therapies is that they reduce development of drug resistance since a pathogen or tumor is less likely to have resistance to multiple drugs simultaneously (Elizabeth, Buchbinder, et al., 2016). The rationale for combination therapy is to use drugs that work by different mechanisms, thereby decreasing the likelihood that resistance would develop. When drugs with different effects are combined, each drug can be used at its optimal dose, without intolerable side effects (Hanahan et al., 2011). It has been proven that the benefits of combination therapies are long-term survival for a broader population of patients. In this chapter, we present detailed information and statistics on various drug combinations for overcoming MDR with combination therapy.

3.2 Combination therapy: Cancer studies

Interdependent pathological changes lead to multigenic and multifactorial diseases such as cancer in humans. Many synthetic and traditional cytotoxic drugs have been developed to control cancer, although they have many limitations such as lack of efficacy, severe toxicity, and drug resistance. Hence, interest in developing more efficacious and safer targeted therapies is growing.

According to the literature, while curing cancer, high dosages of chemotherapeutic drugs such as doxorubicin (DOX) and paclitaxel (PTX) lead to high toxicity. Considering this, Huang et al. employed a kind of dual-drug conjugate, which makes good use of two drugs to fight cancer with less off-target toxicity. Here two drugs were used and delivered simultaneously. In this new nanodrug delivery system, DOX works as a hydrophobic core and xyloglucan (XG) works as a hydrophilic shell; they form stable nanoparticles under aqueous solution. PTX is encapsulated by the XG-DOX conjugate (PTX nano-DDS). Their identification has demonstrated that the conjugated drug possesses outstanding advantages over ordinary systems, including precise control of the molar ratio of the drugs and high hepatic targeting. This type of system is reasonable to be developed as a new type of targeted dual-drug delivery system in combination therapy (Huang et al., 2016).

Although a great deal of research has been aimed at developing systemic treatments for human breast carcinomas, their lack of responsiveness to hormonal therapies due to the absence of molecular targets, such as an estrogen receptor (ER), progesterone receptor (PR), or C-erbB-2 (HER2), makes it challenging. For triple-negative breast cancer (TNBC) cells with negative phenotypes, due to the lack of a recognized molecular target and poor prognosis of late-stage patients with TNBC, frequent chemo-resistance makes it extremely challenging to advance the treatment regimen. Kampert et al. validated that phenotypically stratified carbon nanoparticles are very effective in delivering a novel combination triple-drug formulation for synergistic regression of TNBC in vitro and in vivo (Kampert et al., 2017). Multiple myeloma (MM) is a mature B cell neoplasm that results in multiorgan failure. The median age of onset, diverse clinical manifestations, heterogeneous survival rate, clonal evolution, and intrinsic and acquired drug resistance all impact on the therapeutic management of the disease. Specifically, the emergence of multidrug resistance (MDR) during the course of treatment contributes significantly to treatment failure. The introduction of immune modulatory agents and proteasome inhibitors has led to an increase in overall patient survival for the majority of patients; however, relapse remains inevitable with evidence that these agents, like the conventional chemotherapeutics, are also subject to the development of MDR. Thalidomide and its derivatives are currently approved for use across all phases of MM therapy (Fig. 3.1). These drugs possess immunomodulatory, antiangiogenic, anti-inflammatory, and antiproliferative capacity (Kampert et al., 2017).

In cancer therapy, the efficiency of delivering anticancer drugs to tumor sites via nanoparticles (NPs) is of great benefit. Much effort has been devoted to optimizing NP delivery systems and, so far, many systems have aimed to effectively deliver and release a single drug to treat various tumors. Lou et al. (2018) have developed a multilayered nanosystem to serve as a multifunctional platform for the treatment of drug-resistant breast cancers by employing a poly(lactic-*co*-glycolic acid) core, a liposome second layer, and a chitosan third layer. The chitosan-multilayered nanoparticles (Ch-MLNPs) can codeliver three chemotherapeutic agents: doxorubicin (DOX), paclitaxel (PTX), and silybin. The three drugs are released from the multilayered NPs in a controlled and sequential manner upon internalization and localization in the cellular endosomes (Lou et al., 2018). The presence of a chitosan layer allows the nanosystem to target a well-characterized MDR breast cancer biomarker, the CD44s receptor. An in vitro cytotoxicity study showed that the nanosystem loaded with triple drugs, DOX-PTX-silybin, resulted in better antitumor efficacy than the single-drug or dual-drug nanoformulations. Likely attributed to the MDR inhibition effect of silybin, the codelivered DOX and PTX exhibited

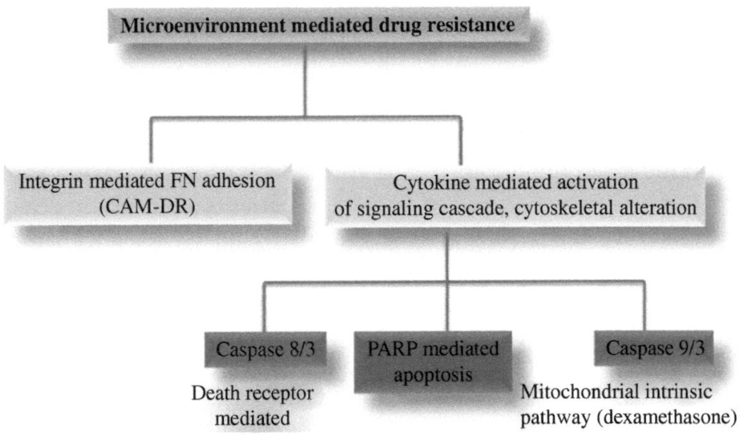

FIG. 3.1 Microenvironment-mediated drug-resistance pathways in MM.

a better synergistic effect on MDR breast cancer cells than on non-MDR breast cancer cells (Lou et al., 2018).

Pancreatic cancer is a tumor with a robust immune escape mechanism and chemotherapy resistance, as if it had armor or a shield. It also displays great power under the shield and armor. Comprehensive treatment combining chemotherapy, immunotherapy, and other new treatments is important for improving the prognosis of pancreatic cancer patients. An expert committee has analyzed the situation and suggested that what is necessary in the future for cancer vaccine therapy to improve the prognosis of pancreatic cancer is (i) its use in combination with immune-checkpoint inhibitors; (ii) a new combination strategy, such as a drug that acts on the tumor microenvironment (TME); (iii) a precise cancer vaccine using neoantigen; and (iv) a combination with strong multiple drugs such as cyclophosphamide (Berd et al., 1986), capecitabine (Duffy et al., 2013), FOLFIRINOX or nab-paclitaxel plus gemcitabine (Siolas et al., 2016). Many studies have revealed that the resistance to artemisinin-based combination therapy might emerge in Kenya and sub-Saharan Africa (SSA) in the same pattern as with chloroquine and sulfadoxine-pyrimethamine (Siolas et al., 2016).

Peninah et al. (2018) studied optimized niosome nanoparticles comprising the chemotherapeutic agent doxorubicin and chemosensitizer curcumin in terms of surfactant content. A novel biocompatible structure (LipoNiosome, a combination of niosome and liposome) containing Tween 60:cholesterol:DPPC (at 55:30:15:3) with 3% DSPE-mPEG was designed and developed to serve as a model for selective codelivery of hydrophilic and hydrophobic drugs to cancerous cells. The proposed formulation provided potential benefits, including pH-sensitive sustained release, smooth globular surface morphology, high entrapment efficiency, and small diameter (Peninah et al., 2018). Exposure of cancer cells to LipoNiosome-doxorubicin-curcumin has shown excellent performance of specific cellular internalization and synergistic toxic effect (> 40% as compared to free drugs and > 23% when compared to single doxorubicin delivery) against Saos-2, MG-63, and KG-1 cell lines (Naderinezhad et al., 2017).

Wang et al. (2009) developed size- and shape-tunable multifunctional $Cu_{2-X}Te$ nanocubes that present a strong absorbance in the NIR region, ideal for photoacoustic (PAI) and X-ray contrast imaging. The nanocubes serve as highly effective chemo-photothermal-photodynamic cancer therapeutic combinatorial treatment agents that can overcome hypermethylated cancer cell resistance to chemotherapeutic drugs (Poulose et al., 2016). Synergistic combination therapy with modules like chemo, PTT, and PDT is usually attained by hybrid/core-shell/manually assembled nanostructures that carry drug cargos (Wang et al., 2009). Not many reports are available when it comes to use of a single NC with the intrinsic abilities to serve imaging cum multiple therapeutic options such as PDT and PTT. Carbon nanohybrid structures that consist of graphene and carbon units were reported to serve as PDT and PTT agents (Hu et al., 2015).

Also, multiple-medications therapy (MMT) or combination of drugs is an important alternative to treat disease factors when a single medicine is not sufficient to effectively treat a disease. The approach of compilations of different drugs might have disadvantages for patients (Tony, Mok, et al., 2009). Also, in recent years multitarget-directed ligands (MTDLs) have been developed to overcome undesirable clinical effects and for the treatment of multifaceted neurodegenerative disease. The strategy of MTDLs is to incorporate pharmacophores of drugs into another compound via stable chemical bonds and generate one multifunctional drug that works as a single unit. Another strategy is to incorporate the pharmacophores via chemically degradable bonds to form molecules that can be converted to two or more active ingredients after administration (Arteaga et al., 2014).

Songwen et al. reported the design and synthesis of a series of novel compounds by incorporating the functionality of phosphoramide mustard into the quinazoline scaffold of EGFR/HER2 inhibitors (Lin et al., 2017). The synthesized compounds were evaluated as multitarget-directed ligands against tumor cells. Also, the MTDLs were analyzed in vitro with enzymatic assay. The results obtained have shown that the tumor cell lines with a high HER2 level are more sensitive to the compounds than tumor cells with a low HER2 level. The IC_{50} on the tumor cell inhibition was calculated as 7.4 nM and 82 nM against EGFR and HER2, respectively. Also, the screening of antitumor activity on the H522 xenograft model cell line showed excellent inhibitory activity, with a TGI of 68% at a dose of 100 mg/kg daily for 28 days without causing any significant body weight loss. Further, the synthesized MTDLs were confirmed as potential lead compounds for the treatment of lung cancer.

It is suggested the synthesis of stimuli-responsive polymer-drug conjugates (PDCs) based on the dendritic polyglycerol sulfate and monomethylauristatin E for inhibition of tumor cells. The PDCs were synthesized with the highly potent antimitotic agent monomethylauristatin E conjugated to dendritic polyglycerol and dendritic polyglycerol sulfate via a reductively cleavable, self-immolative disulfide linker. The cell viability test performed on human cancer cell lines A549 (lung carcinoma) and HeLa (cervix carcinoma) revealed that the synthesized conjugated polymer drugs effectively worked on the reduction of cell proliferation. Also, the results obtained stated that sulfated conjugated polymer drugs were more effective than nonsulfated polymer conjugates. The real-time cell analysis was done using flow cytometry and confocal laser scanning microscopy, which showed the retarded drug released from the polymers, with a much later cytotoxic response after treatment with the nonsulfated conjugates due to less cellular uptake (Nadine et al., 2019).

The applicability of the synthesized PDCs against a variety of cancer types was also studied. The four different human cancer cell lines belonging to the most common cancer types were selected for the study. The cell viability was examined by a CCK-8 proliferation and cytotoxicity assay following 48h incubation of A549 (lung carcinoma), HeLa (cervix carcinoma), MCF-7 (breast adenocarcinoma), and Caco-2 (colon adenocarcinoma) cells with the PDCs as well as free MMAE. Nontreated cells served as the control, set to 100% cell viability. Conjugated sulfated drugs showed good cytotoxicity against A549 and HeLa cells in the nanomolar range, which is more effective than the nonsulfated ones (Nadine et al., 2019).

3.3 Combination therapy: HIV/AIDS studies

Achievement of early viral conquest is significant in patients with chronic HCV infection treated with telaprevir (TLV) or boceprevir (BOC) to avoid selection of drug resistance and accomplish a cure. No head-to-head studies paralleling TLV and BOC have been executed up to now (Benito et al., 2013). For their study of liver (EASL) treatment recommendations, the European Association for Hepatitis C no longer differentiates between HIV/hepatitis C virus (HCV)-coinfected and HCV-monoinfected patients. However, recent data from Spain are questioning these recommendations on the basis of the findings of higher revert rates and lower cure rates in HIV/HCV-infected subjects. The aim of this study was to equate HCV cure rates in monoinfected and coinfected patients from Germany (Bischoff et al., 2018). An HIV-1 infected infant started on combination antiretroviral therapy (cART) at 30h of life was recently reported to have no noticeable plasma viremia after terminating cART. This study examined the impact of early cART initiation on measures of HIV-1 reservoir size in HIV-1-infected children with persistent virologic defeat (Bitnun et al., 2014). HIV/AIDS remains a worldwide health problem and viral suppression has been an elusive objective. HIV + patients are presently treated with combination antiretroviral therapy (cART), which is not remedial. For many patients, cART is inaccessible, intolerable, or unaffordable. Therefore, a new class of therapeutics for HIV is required to overcome these limitations. Cell and gene therapy for HIV has been anticipated as a way to offer a functional cure for HIV in the form of a virus/infection-resistant immune system (Chung et al., 2013).

Replication-competent HIV persists in infected people despite suppressive combination antiretroviral therapy (cART) and represents a major obstacle to HIV functional cure or eradication. A model of cART-mediated viral suppression in SIVmac239-infected Indian rhesus macaques was used to evaluate the impact of the histone deacetylase inhibitor (HDACi) romidepsin (RMD) on viremia in vivo. Eight macaques were virologically suppressed to clinically relevant levels (< 30 viral RNA copies/mL plasma) using a three-class, five-drug cART. The regimen called for multiple intravenous infusions of either RMD ($n=5$) or saline ($n=3$) starting 31–54 weeks after cART initiation. In vivo RMD treatment resulted in significant transient increases in acetylated histone levels in CD4 + T cells. RMD-treated animals demonstrated plasma viral load measurements for each 2-week treatment cycle that were significantly higher than in saline control treated animals during periods of treatment, suggestive of RMD-induced viral reactivation. However, plasma virus rebound was indistinguishable between RMD-treated and control animals for a subset of animals released from cART. These findings suggest that HDACi drugs such as RMD can reactivate residual virus in the presence of suppressive antiviral therapy and may be a valuable component of a comprehensive HIV functional cure/eradication strategy (Del Prete et al., 2015).

Chronic hepatitis B continues as a significant public health problem despite the availability of safe and effective antivirals and a highly effective protective vaccine. Current therapy, however, rarely leads to cure and lifelong therapy is often required, contributing to poor uptake and ongoing morbidity. New insights into the hepatitis B viral life cycle and the host immune response have expanded the potential targets for drug therapies with interesting antiviral candidates and novel immunotherapeutic approaches in early-stage development. Yet, HBV persistence is multifactorial due to an intrahepatic reservoir and ongoing HBV-mediated immune dysregulation, making "cure" unlikely to be realized through even the most efficacious monotherapy. Building on the success seen in the treatment of hepatitis C (HCV) and human immunodeficiency virus (HIV), combination therapy may be an essential strategy to improve efficacy and decrease viral breakthrough. Combinations acting on immune and viral targets are particularly attractive. However, creating synergy while balancing efficacy and safety remains a clear challenge. Various approaches to combination therapy are reviewed, highlighting strengths and challenges of each potential strategy. Overall, combination therapies are attractive as the next step toward a cure and are a key strategy for achieving treatment with finite durations and durable endpoints (Emery et al., 2017).

Ten percent of all respondents agreed or strongly agreed with the statement that "AIDS is now very nearly cured" and 13% felt that the threat of AIDS is less serious than in the past. HIV-positive men were more likely to perceive AIDS as a less-serious threat or as very nearly cured. Overall, 8% of men in the sample indicated that they practice safer sex less often since new AIDS treatments came along; 18% of HIV-positive men on combination therapy regimens said they practice safer sex less frequently since treatments have advanced. Regardless of serostatus, nearly 20% of men indicated they would stop practicing safer sex if an AIDS cure was announced (Kelly et al., 1998). Combination antiretroviral therapy (cART) has not allowed the cure of HIV. The main obstacle to HIV eradication is the existence of quiescent reservoirs. Several other limitations of cART have been described, such as strict lifelong treatment and high costs, restricting it to Western countries, as well as the development of multidrug resistance. Given these limitations and the impetus to find a cure, the development of new treatments is necessary (Douce et al., 2016).

Definitive treatment of HIV infection remains a critical but elusive goal, with persistence of residual virus even in the face of prolonged administration of suppressive combination antiretroviral treatment (cART) providing a source for recrudescent infection if treatment is stopped. Characterization of the residual virus and devising strategies to target it for eradication are key goals in HIV treatment research. Indian rhesus macaques (In-RM) infected with SIVmac have been widely used in such research. However, it has proven challenging to achieve and sustain clinically relevant levels of suppression (< 30 vRNA copies/mL plasma) with cART in such models. Ease of viral suppression by cART is related to pretreatment levels of viral replication, and levels of replication of SIVmac239/251 are lower in Chinese rhesus macaques (Ch-RM) than in In-RM. Studies have been undertaken evaluating cART administration to SIVmac-infected Ch-RM as a potential model for studies of residual virus and eradication strategies. Four SIVmac239-infected Ch-RM received cART including reverse transcriptase inhibitors PMPA/FTC and integrase inhibitor L-870812 daily for 8 weeks. Plasma viral loads were promptly reduced to < 30copies/mL upon initiation of cART. Cell-associated SIV DNA levels in lymphocytes from the gut were also significantly reduced. Jejunal and colonic CCR5+CD4+ mucosal memory T cells increased significantly; restoration of these cells was associated with reductions in immune activation. In conclusion, cART effectively suppressed viral replication to < 30vRNAcopies/mL in SIVmac239-infected Ch-RM, reducing immune activation and restoring mucosal immune cell populations. SIVmac239-infected Ch-RM may be a useful model for studying responses to cART and persistent tissue reservoirs and evaluating candidate eradication strategies to cure HIV infection (Ling et al., 2013).

Globally, 240,000 infants are newly infected with HIV-1 each year and 3.2 million children are living with the infection. Combination antiretroviral therapy (cART) has reduced HIV-1-related disease and mortality in children but is not curative, owing to the early generation of a latent reservoir of long-lived memory CD4 + T cells bearing replication-competent HIV-1 provirus integrated into cellular DNA. Effective combination antiretroviral therapy studies investigated Vacc-4x/rhuGM-CSF therapeutic vaccination prior to HIV latency reversal using romidepsin. The main finding was a statistically significant reduction from baseline in viral reservoir measurements. HIV-specific functional T-cell responses following Vacc-4x/rhuGM-CSF immunotherapy in relation to virological outcomes on the HIV reservoir were evaluated (Luzuriaga et al., 2016).

Elimination of the latent HIV reservoir remains the biggest hurdle to achieving HIV cure. In order to specifically eliminate HIV-infected cells, they must be distinguishable from uninfected cells. CD2 was recently identified as a potential marker enriched in the HIV-1 reservoir on CD4 + T cells, the largest, longest-lived, and best-characterized constituent of the HIV reservoir (Hojen et al., 2017). To reduce the HIV reservoir in CD2hi CD4 + memory T cells, a humanized a-CD2 fusion protein, the FDA-approved alefacept, was employed. It was noticed that alefacept can specifically target and reduce CD2hi HIV-infected cells in vitro. This study is being broadened by clinical trials to explore a variety of natural killer (NK) cells as mediators of antibody-dependent cell-mediated cytotoxicity (ADCC) including primary NK cells, expanded NK cells, and the CD16 transduced NK-92 cell lines (Tomalka et al., 2018). Based on the investigations, CD16.NK-92 has a natural priority to kill CD2hi CD45RA – memory T cells in both healthy and HIV + patient samples as well as to reduce HIV DNA from HIV + samples from donors well controlled on antiretroviral therapy (Priyanka et al., 2015).

3.4 Combination therapy: Tuberculosis

TB was confirmed microbiologically (90%) or histologically (10%) in 69 patients (71% male; median age 43 years; 54% born in Africa). The median time between TB treatment initiation and antiretroviral therapy was 8 weeks (range 1–22 weeks). At baseline, median HIV-1 RNA was 5.4 log10 copies/mL and median CD4 cell count 74 cells/mm^3. In the ITT analysis, combined success at week 48 was achieved in 57/69 patients (83%, 95% CI 74–92). Twelve patients did not achieve virological success, and TB was not cured in one of them. Among the 47 patients who fully adhered to the strategy, the success rate was 96% (95% CI 90–100) and was not affected by low rifampicin and isoniazid serum concentrations. Forty-nine serious adverse events were reported in 31 patients (45%), and 11 led to antiretroviral drug interruption. All adverse events resolved. The immune reconstitution inflammatory syndrome occurred in 23 patients (33%, 95% CI 22–44), and was associated with a low baseline BMI ($P = .03$) and a low hemoglobin level ($P = .02$) (Lortholary et al., 2016).

Drug-resistance testing, or antimicrobial susceptibility testing (AST), is mandatory for *Mycobacterium tuberculosis* (MTB) in cases of failure to standard therapy. Research studies showed different methods and techniques of phenotypic and genotypic approaches. Although multi- and extensively drug resistant (MDR-XDR) tuberculosis is present worldwide, AST-MTB is still mainly performed according to the resources available rather than the drug-resistance rates. Phenotypic methods, i.e., culture-based AST, are commonly used in high-income countries to confirm susceptibility of new cases of tuberculosis. They are also performed to detect resistance in tuberculosis cases with risk factors, in combination with genotypic tests. In low-income countries, genotypic methods screening hot-spot mutations known to confer resistance were found to be easier to perform because they avoid the culture and biosafety constraint (Schon et al., 2017). Given that genotypic tests can rapidly detect the prominent mechanisms of resistance, such as the *rpoB* mutation for rifampicin resistance, we are facing new challenges with the observation of false-resistance (mutations not conferring resistance) and false-susceptibility (mutations different from the common mechanism) results. Phenotypic and genotypic approaches are thus complementary for obtaining a high sensitivity and specificity for detecting drug resistances and susceptibilities to accurately predict MDR/XDR cure and to gather relevant data for resistance surveillance. Although AST-MTB was established in the 1960s, there is no consensus reference method for MIC determination against which the numerous AST-MTB techniques can be compared. This information is necessary for assessing the in vitro activity and setting breakpoints for future antituberculosis agents (Schon et al., 2017). Treatment of multidrug-resistant tuberculosis (MDR-TB) is challenging, because it mostly relies on drugs with lower efficacy and greater toxicity than those used for drug-susceptible TB (Gualano et al., 2019).

Thioridazine (TDZ) in combination with antibiotics to which extensively drug-resistant *Mycobacterium tuberculosis* (XDR-TB) is initially resistant yields a cure. This is due to the fact that TDZ enhances the killing of intracellular *M. tuberculosis* by nonkilling macrophages, inhibits the genetic expression of efflux pumps of *M. tuberculosis* that extrude antibiotics prior to reaching their intended targets, and inhibits the activity of existing efflux pumps that contribute to the multidrug-resistant phenotype of *M. tuberculosis*. The combination of these effects of TDZ probably contributes to the successful recent cures of XDR-TB cases, in which the phenothiazine TDZ is used in combination with antibiotics to which the patient with XDR-TB was initially unresponsive (Amaral et al., 2012a).

Over a period of 14 years, Amaral et al. (2012a, 2012b) have studied thioridazine, an old neuroleptic, that has been shown to have in vitro activity against intracellular *M. tuberculosis*; regardless of its antibiotic resistance status, thioridazine cures infected mice of antibiotic-susceptible and multidrug-resistant tuberculosis (TB) infections and, when used in combination with antibiotics used for TB therapy, renders the organism significantly more susceptible. This chapter describes the mechanisms by which thioridazine alone and in combination with antibiotics cures an extensively drug-resistant infection and why it is expected to cure totally drug-resistant TB infections as well. The concepts presented are entirely new; because they focus on the activation of killing by nonkilling macrophages, where *M. tuberculosis* normally resides during infection, and are coupled to the inhibition of efflux pumps that contribute to the antibiotic-resistant status, effective therapy of any antibiotic-resistant TB infection is possible (Amaral et al., 2012b). Delamanid (Deltyba), a nitroimidazo-oxazole derivative, is a new antituberculosis (TB) drug that exhibits potent in vitro and in vivo antitubercular activity against drug-susceptible and -resistant strains of *M. tuberculosis*. It is approved in several countries, including Japan and those of the European Union, for use as part of an appropriate combination regimen in adults with multidrug-resistant tuberculosis (MDR-TB), when an effective treatment regimen cannot otherwise be composed due to resistance or tolerability. In a robust phase II trial in adult patients with MDR-TB, oral delamanid 100 mg twice daily for 2 months plus an optimized background regimen improved sputum culture conversion rates to a significantly greater extent than placebo (Amaral et al., 2012b). In a 6-month extension study, long-term (B8 months) treatment with delamanid was associated with a higher incidence of favorable outcomes (i.e., cured or completed all treatment) than short-term (B2 months) treatment, with an accompanying reduction in unfavorable outcomes as defined by the WHO (i.e., prespecified proportion of TB-positive sputum cultures, death, or treatment discontinuation for C2 months without medical approval). Delamanid was not associated with clinically relevant drug-drug interactions, including with antiretroviral drugs and those commonly used in treating TB. Delamanid was generally well tolerated in patients with MDR-TB, with gastrointestinal adverse events and insomnia reported most commonly. Although the incidence of QT interval prolongation was higher with delamanid-based therapy, it was not associated with clinical symptoms such as syncope and arrhythmia. In conclusion, delamanid is a useful addition to the treatment options currently available for patients with MDR-TB (Blair et al., 2015).

In vitro studies have shown good activity of linezolid against *M. tuberculosis*, including multidrug-resistant strains. However, clinical experience with linezolid in tuberculosis is scarce (Fortun et al., 2010). Cavitary disease and bilateral lesions are among the risk factors for poor outcome of pulmonary tuberculosis (TB) (Marfina et al., 2018). Monte Carlo experiments (MCEs) were used to identify the dose that would achieve the target exposure in 10,000 adult patients with meningeal or pulmonary MDR-TB. The optimal doses identified were validated using probit analyses of clinical data from two prospective clinical trials of patients with pulmonary and meningeal tuberculosis. Classification and regression-tree (CART) analyses were used to identify the gatifloxacin minimal inhibitory concentration (MIC) below which patients failed or relapsed on combination therapy (Deshpande et al., 2018).

Few studies have examined outcomes for children treated for multidrug-resistant tuberculosis (MDR-TB), including those receiving concomitant treatment for MDR-TB and HIV coinfection. In Lesotho, where the adult HIV seroprevalence is estimated to be 24%, we sought to measure outcomes and adverse events in a cohort of children treated for MDR-TB using a community-based treatment delivery model (Satti et al., 2012).

Extensively drug-resistant (XDR) tuberculosis (TB) and multidrug resistant (MDR)-TB with additional resistance to injectable agents or fluoroquinolones are challenging to treat due to lack of available, effective drugs. Linezolid is one of the few drugs that have shown promise in treating these conditions. Long-term linezolid use is associated with toxicities such as peripheral and optic neuropathies. Diabetes mellitus (DM), especially when uncontrolled, can also result in peripheral neuropathy. The global burden of DM is increasing, and DM has been associated with a three-fold increased risk of developing TB disease (Swaminathan et al., 2017). TB and DM can be a challenging combination to treat. DM can inhibit the host immune response to tuberculosis infection; and TB and some anti-TB drugs can worsen glycemic control. A child experiencing neuropathy that is a possible complication of both DM and linezolid used to treat TB has not been reported previously, but peripheral neuropathy is now evidently reported in a 15-year-old boy with type 1 DM, diagnosed with MDR-TB and additional resistance to injectable TB medications (Swaminathan et al., 2017).

3.5 Conclusions

As technology introduces new combination products, a number of new combined treatments, therapies, and regimens have been tested side by side and have proven to help patient populations. Sometimes combination drug therapy is used not to cure but to reduce symptoms and prolong life. As technology advances and experimentation with existing approved entities broadens, providing safe and longer-lasting options for patients will become a primary goal for all future product development. Combination antibiotic therapy is widely practiced on the Indian subcontinent. Combination antibiotic therapy is used in critically ill patients due to widespread emergence of multidrug-resistant organisms (MDR). Conditions treated with combination therapy include tuberculosis, leprosy, cancer, malaria, and HIV/AIDS. One major benefit of combination therapies is that they reduce development of drug resistance, since a pathogen or tumor is less likely to have resistance to multiple drugs simultaneously.

References

Amaral, L., et al. (2012a). Why thioridazine in combination with antibiotics cures extensively drug-resistant *Mycobacterium tuberculosis* infections. *International Journal of Antimicrobial Agents*, *39*(5), 376–380.

Amaral, L., et al. (2012b). Why and how thioridazine in combination with antibiotics to which the infective strain is resistant will cure totally drug-resistant tuberculosis. *Expert Review of Anti-Infective Therapy, 10*(8), 869–873.

Arteaga, C. L., et al. (2014). ERBB receptors: From oncogene discovery to basic science to mechanism-based cancer therapeutics. *Cancer Cell, 25*, 282–303.

Benito, J. M., et al. (2013). Triple combination therapy for hepatitis C with telaprevir exhibits greater early antiviral activity than with boceprevir. *Antiviral Therapy, 18*, 709–715.

Berd, D., et al. (1986). Induction of cell-mediated immunity to autologous melanoma cells and regression of metastases after treatment with a melanoma cell vaccine preceded by cyclophosphamide. *Cancer Research, 46*, 2572–2577.

Bischoff, J., et al. (2018). Rates of sustained virological response 12 weeks after the scheduled end of direct-acting antiviral (DAA)-based hepatitis C virus (HCV) therapy from the National German HCV registry: Does HIV coinfection impair the response to DAA combination therapy? *HIV Medicine, 19*(4), 299–307.

Bitnun, A., et al. (2014). 23rd annual Canadian conference on HIV/AIDS research. *Clinical Infectious Diseases*.

Blair, H. A., et al. (2015). Delamanid: A review of its use in patients with multidrug-resistant tuberculosis. *Drugs, 75*(1), 91–100.

Chung, J., et al. (2013). Combinatorial RNA-based gene therapy for the treatment of HIV/AIDS. *Expert Opinion on Biological Therapy, 13*(3), 437–445.

Del Prete, G. Q., et al. (2015). Elevated plasma viral loads in romidepsin-treated simian immunodeficiency virus-infected rhesus macaques on suppressive combination antiretroviral therapy. *Antimicrobial Agents and Chemotherapy, 60*(3), 1560–1572.

Deshpande, D., et al. (2018). Gatifloxacin pharmacokinetics/pharmacodynamics-based optimal dosing for pulmonary and meningeal multidrug-resistant tuberculosis. *Clinical Infectious Diseases, 67*, S274–S283.

Douce, V. L., et al. (2016). Improving combination antiretroviral therapy by targeting HIV-1 gene transcription. *Expert Opinion on Therapeutic Targets, 20*(11), 1311–1324.

Duffy, A., et al. (2013). Immunological off-target effects of standard treatments in gastrointestinal cancers. *Annals of Oncology, 25*, 24–32.

Elizabeth, I., Buchbinder, M. D., et al. (2016). CTLA-4 and PD-1 pathways similarities, differences, and implications of their inhibition. *American Journal of Clinical Oncology, 39*, 98–106.

Emery, J. S., et al. (2017). Treatment of hepatitis B virus with combination therapy now and in the future. *Best Practice & Research Clinical Gastroenterology, 31*(3), 347–355.

Fortun, J., et al. (2010). Linezolid in the treatment of multidrug-resistant tuberculosis. *Clinical Infectious Diseases, 50*(1), 49–55.

Gualano, G., et al. (2019). Putting in harm to cure: Drug related adverse events do not affect outcome of patients receiving treatment for multidrug-resistant tuberculosis. Experience from a tertiary hospital in Italy. *PLoS One, 14*(2), e0212948.

Hanahan, D., et al. (2011). Hallmarks of cancer: The next generation. *Cell, 144*(5), 646–674.

Hojen, T. G., et al. (2017). Sequential Vacc-4x and romidepsin during combination antiretroviral therapy (cART): Immune responses to Vacc-4x regions on p24 and changes in HIV reservoirs. *Journal of Infection, 75*(6), 555–571.

Hu, Z., et al. (2015). Functionalized graphene/C60 nanohybrid for targeting photothermally enhanced photodynamic therapy. *RSC Advances, 5*, 654–664.

Huang, Z., et al. (2016). PTX encapsulated by an XG-DOX conjugate for combination therapy against multi-drug resistance. *RSC Advances, 6*(109), 107606–107612.

Kampert, T., et al. (2017). Phenotypically screened carbon nanoparticles for enhanced combinatorial therapy in triple negative breast cancer. *Cellular and Molecular Bioengineering, 10*(5), 371–386.

Kelly, J. A., et al. (1998). Protease inhibitor combination therapies and perceptions of gay men regarding AIDS severity and the need to maintain safer sex. *AIDS, 12*, F91–F95.

Lin, S., et al. (2017). Design, synthesis and biological evaluation of quinazoline-phosphoramidate mustard conjugates as anticancer drugs. *European Journal of Medicinal Chemistry, 127*, 442–458.

Ling, B., et al. (2013). Effect of combination antiretroviral therapy on Chinese rhesus macaques of simian immunodeficiency virus infection. *AIDS Research and Human Retroviruses, 29*(11), 1465–1474.

Lortholary, O., et al. (2016). Tenofovir DF/emtricitabine and efavirenz combination therapy for HIV infection in patients treated for tuberculosis: The ANRS 129 BKVIR trial. *Journal of Antimicrobial Chemotherapy, 71*(3), 783–793.

Lou, S., et al. (2018). Multifunctional nanosystem for targeted and controlled delivery of multiple chemotherapeutic agents for the treatment of drug-resistant breast cancer. *ACS Omega, 3*(8), 9210–9219.

Luzuriaga, K., et al. (2016). Early combination antiretroviral therapy limits HIV-1 persistence in children. *Annual Review of Medicine, 67*, 201–213.

Marfina, G. Y., et al. (2018). Bilateral cavitary multidrug- or extensively drug-resistant tuberculosis: Role of surgery. *European Journal of Cardio-Thoracic Surgery, 53*(3), 618–624.

Naderinezhad, S., et al. (2017). Co-delivery of hydrophilic and hydrophobic anti cancer drugs using bio-compatible pH-sensitive lipid-based nano-carriers for multidrug- resistance cancers. *RSC Advances, 7*(48), 30008–30019.

Nadine, R., et al. (2019). Reductively cleavable polymer-drug conjugates based on dendritic polyglycerol sulfate and monomethyl auristatin E as anticancer drugs. *Journal of Controlled Release, 300*, 13–21.

Peninah, M., et al. (2018). Selective sweeps and genetic lineages of *Plasmodium falciparum* multi-drug resistance (pfmdr1) gene in Kenya. *Malaria Journal, 17*, 398.

Poulose, A. C., et al. (2016). Multifunctional $Cu_{2-x}Te$ nanocubes mediated combination therapy for multi-drug resistant MDA MB 453. *Science Reports, 6*, 35961.

Priyanka, U., et al. (2015). Cell-associated HIV-1 DNA and RNA decay dynamics during early combination antiretroviral therapy in HIV-1-infected infants. *Clinical Infectious Diseases, 61*(12), 1862–1870.

Satti, H., et al. (2012). Outcomes of comprehensive care for children empirically treated for multidrug-resistant tuberculosis in a setting of high HIV prevalence. *PLoS One, 7*(5), 37114.

Schon, T., et al. (2017). *Mycobacterium tuberculosis* drug resistance testing: Challenges, recent developments and perspectives. *Clinical Microbiology and Infection, 23*(3), 154–160.

Siolas, D., et al. (2016). Antitumor activity and immune response in CD40 immunotherapy with gemcitabine and nab-paclitaxel in an orthotopic pancreatic cancer mouse model. *Journal of Clinical Oncology, 34*, 271.

Swaminathan, A., et al. (2017). Peripheral neuropathy in a diabetic child treated with linezolid for multidrugresistant tuberculosis: A case report and review of the literature. *BMC Infectious Diseases, 17*, 417–422.

Tomalka, A. G., et al. (2018). In vitro evidence that combination therapy with CD16-bearing NK-92 cells and FDA-approved alefacept can selectively target the latent HIV reservoir in CD4+ CD2hi memory T cells. *Frontiers in Immunology, 5*(9), 2552.

Tony, S., Mok, M. D., et al. (2009). Gefitinib or carboplatin-paclitaxel in pulmonary adenocarcinoma. *The New England Journal of Medicine, 361*, 947–957.

Wang, C., et al. (2009). Gold nanorod/Fe_3O_4 nanoparticle "nano-pearl-necklaces" for simultaneous targeting, dual-mode imaging, and photothermal ablation of cancer cells. *Angewandte Chemie, 121*, 2797–2801.

CHAPTER

4

Combination therapy against human infections caused by *Candida* species

Indresh Kumar Maurya[a], *Ruchi Badoni Semwal*[b], *Deepak Kumar Semwal*[c]

[a]Department of Microbial Biotechnology, Panjab University, Chandigarh, India, [b]Department of Chemistry, Pt. Lalit Mohan Sharma Government Postgraduate College, Rishikesh, Uttarakhand, India, [c]Department of Phytochemistry, Faculty of Biomedical Sciences, Uttarakhand Ayurved University, Dehradun, India

4.1 Introduction

During the past few decades, fungal resistance to antifungal drugs has increased worldwide, due to the prolonged and widespread use of these drugs, which makes fungal treatment more challenging (Cannon et al., 2009; Perlin, Richardson, & Alastruey-Izquierdo, 2017). Fungal pathogens cause approximately 1.5 million human deaths globally every year (Brown et al., 2012). The major fungal pathogens, including *Candida albicans* and non-*albicans* species (i.e., *C. glabrata, C. tropicalis,* and *C. parapsilosis*), are the fourth leading cause of bloodstream infections and are responsible for 50%–70% of all infections in many parts of the world (Kullberg & Arendrup, 2015). Further, the steady increase in antifungal drug resistance is a major cause of the high rates of morbidity and mortality for systemic fungal infections in immunocompromised patients (Geddes-McAlister & Shapiro, 2019).

Despite steep growth in fungal infections, only four major classes of antifungal drugs (i.e., polyenes, azoles, echinocandins, and allylamines) are available clinically in oral, topical, and intravenous forms and they

Combination Therapy Against Multidrug Resistance
https://doi.org/10.1016/B978-0-12-820576-1.00004-7

have been developed over the last five to six decades (Hoban, Zhanel, & Karlowsky, 1999; Perfect, 2017). This limited repertoire of existing antifungal drugs also has several drawbacks: undesirable side effects on host tissue, lack of potency, inadequate drug safety profile, inadequate pharmacokinetic properties, and low number of targets (Spellberg et al., 2011). Recently, these challenges have driven efforts to determine the efficacy of combination therapy in the management and treatment of *Candida* spp. infections (Odds, 2003). Combination antifungal therapy has several potential benefits: broad-spectrum efficacy, higher potency at low concentrations than any of the antifungal drugs used alone, low levels of toxicity, good tolerability, and reduction of the emergence of drug-resistant *Candida* spp. (Lewis, Klepser, & Pfaller, 2000). However, the development process has been slow, limited to amphotericin B (Amp B) with 5-flucytosine (5FC) or 5FC with fluconazole (FLC). The use of azoles (e.g., voriconazole, or VOR) with the echinocandins (e.g., caspofungin, or CAS) has also shown promise in antifungal combination therapy (Louie et al., 1999). The combination of different antifungal agents can be synergistic, indifferent, additive, or antagonistic. The in vitro drug interactions are usually based on the valuations of the "no interaction" theory, which assumes that the different drugs in combination do not interact with each other. When the observed efficacy of the antifungal drug combination is greater than that determined from the "no interaction" theory, a synergistic effect is said to have occurred. Similarly, antagonistic activity of the antifungal drugs occurs when the observed effect is less than that predicted (Odds, 2003).

C. albicans is the most common pathogenic species infecting humans. It is the main cause of genital yeast infections and, with its rising medical and socioeconomic importance, it represents a serious public health challenge. *Candida* can cause serious mucosal and life-threatening infections in immune-compromised hosts, and this pathogen is known worldwide as accounting for the most common fungal infection in both HIV and non-HIV patients. This chapter focuses on the current uses and limitations of combination therapy against *Candida* infections.

4.2 Pathogenic *Candida* spp.

Candida spp. are major fungal pathogens affecting humans and are universally present in nature. Commonly, they exist as commensal microorganisms on several parts of the human body, such as skin, mouth, gastrointestinal tracts (GITs), and female genital tracts (Kullberg & Arendrup, 2015). Interestingly, these pathogen species are not limited to their human/animal reservoirs but can also be frequently found in the hospital environment, including its canteen foods, exhausts, floors, etc. (Pappas et al., 2009). They are found as yeasts with the unicellular mode of development and can form either true hyphae or pseudohyphae (Pfaller & Diekema, 2007). Furthermore, they are normally commensal on diseased skin and mucosal membranes of various parts, including the GIT. However, in persons with low immunity, they become an opportunistic pathogen, causing recurrent chronic oral and vaginal candidiasis as well as life-threatening systemic disease (Richardson & Lass-Flörl, 2008; Sobel, 2007). Due to the reasons listed, *Candida* spp. are globally considered as the most common cause of opportunistic mycoses (Miceli, Díaz, & Lee, 2011).

As previously mentioned, the increasing number of patients with mucosal and systemic candidiasis represents a serious and growing problem, with *Candida* spp. also attacking tissues that are resistant to invasion (Richardson & Lass-Flörl, 2008). They are considered to be true opportunistic pathogens in the sense that they exploit advanced mechanisms to enter the blood circulation and inner tissues. The severity of this invasion of *Candida* spp. mainly depends on the host's immune status (Chi et al., 2011). These pathogens cause a wide range of illnesses, including mucocutaneous disease and hepatosplenic candidiasis. *Candida* spp. have been reported to be the sixth most commonly isolated pathogen and the fourth most prevalent bloodstream pathogen isolated (Gudlaugsson et al., 2003). The treatment of progressive invasive candidiasis is a challenge for the practitioner, due to the lack of reliable diagnostic methods (Barnes, Stocking, Bowden, Poynton, & White, 2013).

Approximately 200 diverse yeast species have been reported in the genus *Candida*, whose common feature is the lack of any sexual cycle (Butler, 2010). It is noteworthy that a small number of *Candida* spp. are pathogenic and cause various diseases in humans. Among them, seven species in the genus of *Candida* are known to be opportunistic human pathogens; *C. albicans* accounts for the majority of the systemic infections and accounts for approximately 50%–60% of candidiasis in humans. Selected medically significant *Candida* spp. are listed in Table 4.1.

Furthermore, *C. glabrata* has recently shown increasing incidence, because of being less susceptible to azoles and Amp B. Various other species of *Candida* such as *C. lusitaniae*, *C. krusei*, and *C. guilliermondii* are also

TABLE 4.1 Selected medically significant *Candida* spp.

***Candida* spp.**	**% abundance**	***Candida* spp.**	**% abundance**
Candida albicans	~ 50–70	*Candida krusei*	~ 1–3
Candida glabrata	~ 15–30	*Candida lusitaniae*	< 5
Candida parapsilosis	~ 10–20	*Candida dubliniensis*	< 5
Candida tropicalis	~ 6–12	*Candida guilliermondii*	< 5
Candida kefyr	< 5		

important due to their resistance to FLC. Similarly, *C. krusei* is resistant to azoles and less susceptible to other drugs like Amp B, and hence has become a pathogen of clinical significance (Pfaller et al., 2004).

4.3 Pathogenicity mechanisms

Many factors identified over the past few decades contribute to the pathogenic potential of *Candida*. Various adhesins, enzymes, toxins, and pseudohypha were found to be key factors that influence the potency of *C. albicans* to communicate disease. The expression of adhesins can be found on the cell surface, assisting in adhesion to the skin and mucosa. Extracellular hydrolytic enzyme production, including aspartyl proteinases and serine proteinases, aids in tissue invasion. The glycoprotein extracts of the *Candida* cell wall are pyrogenic, similar to bacterial endotoxins. The pseudohypha is released from the hyphal tip and also helps in the invasion.

C. albicans has a unique property of phenotypic switching. It can transform itself between three polymorphic forms in the tissue: yeast (blastospores), pseudohyphae, and true hyphae. This allows it to adapt to changing conditions in the host and thereby assists the fungus in evading the host defense system. These pathogenic yeasts are considered lethal because of their unique morphology, capability of biofilm formation, and hydrolytic enzyme secretion. Moreover, various other properties like the ability to adjust to differences in pH, metabolic flexibility, strong nutrient gaining system, and a vigorous stress response mechanism also make them more lethal. Particularly, *C. albicans* virulence is due to adhesion, hyphal formation, phenotypic switching, and its ability to colonize and infect almost every tissue (Naglik, Albrecht, Bader, & Hube, 2004).

4.4 Clinical manifestations

A spectrum of infections, ranging from skin to mucosal infection to invasive and allergic infections, are caused by *Candida* spp. The characterization of different candidiasis types based on their infection site is given below.

4.4.1 Mucosal candidiasis

1. Oropharyngeal candidiasis
2. Vulvovaginitis candidiasis
3. Balanitis and balanoposthitis
4. Esophageal candidiasis
5. Angular and denture stomatitis

4.4.2 Cutaneous candidiasis

1. Intertrigo
2. Paronychia
3. Diaper candidiasis
4. Perianal candidiasis
5. Generalized disseminated candidiasis

4.4.3 Invasive candidiasis

1. Urinary tract infection
2. Pulmonary candidiasis
3. Septicemia
4. Arthritis and osteomyelitis
5. Meningitis
6. Ocular candidiasis
7. Hepatosplenic candidiasis
8. Disseminated candidiasis
9. Nosocomial candidiasis

4.4.4 Allergic candidiasis

1. Candidiasis hypersensitivity syndrome
2. Gastritis
3. Irritable bowel syndrome
4. Eczema

4.5 Antifungal drugs

The era of antifungal chemotherapy successfully began with the introduction in 1958 of Amp B, produced by the soil actinomycete *Streptomyces* (Roemer & Krysan, 2014). Although Amp B was the main treatment for severe fungal infections for more than five decades, severe adverse effects prompted the continued search for another potent antifungal equally effective and with fewer side effects, which could be used both orally as well as intravenously. Another effective antifungal drug, fluconazole (FLC), was introduced in 1990. FLC was effective for the treatment of oropharyngeal candidiasis in HIV patients and quickly became one of the most popular antifungal agents for mucosal as well as systemic fungal infections (Denning & Hope, 2010). However, the lack of activity against opportunistic molds (i.e., *Aspergillus*, *Mucorales*, and *Fusarium* spp.) and intrinsic resistance among some *Candida* spp. (e.g., *C. glabrata* and *C. krusei*) created an essential need for wide-spectrum

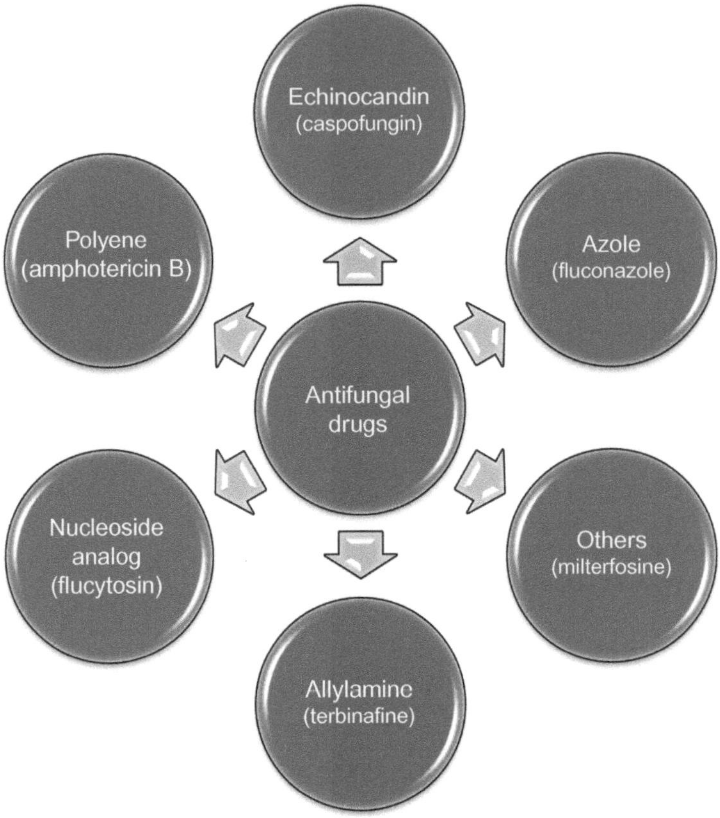

FIG. 4.1 Different classes of antifungal drugs.

alternatives. Itraconazole (ITR) was a partial solution to the limitations of FLC, because it had better activity against endemic fungi and *Aspergillus* spp. (Lewis, Klepser, & Pfaller, 1998). As per Campoy and Adrio (2017), different classes of antifungal drugs with their common examples are shown in Fig. 4.1.

The 20th century was a successful period for antifungal drugs, particularly in the area of new antifungal drug discoveries of the echinocandin class (Fig. 4.2) (Lewis, 2011). Echinocandin drugs are semisynthetic lipopeptides that have been found effective in damaging the fungal cell wall by preventing β-1, 3-D-glucan synthesis in the cell wall of susceptible fungi. Such agents are found to be effective with a very low level of toxicity in mammalian cells. The evaluation of efficacy of echinocandins against various common opportunistic yeasts like *Cryptococcus* spp. and less common molds like *Fusarium*, *Scedosporium*, and *Mucorales* is still lacking and needs validation.

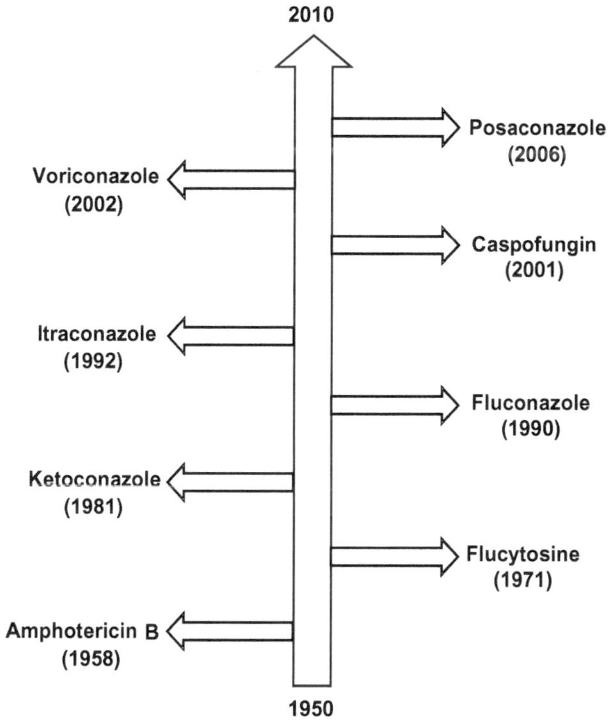

FIG. 4.2 Timeline of common antifungal drug development.

It is noteworthy that the safety profiles and tolerability of systemic antifungal drugs have remarkably improved in recent years. Hence, it has become easier for immunocompromised patients to receive long-term treatment. Now clinicians require a focus on dose-limiting toxicity, such as nephrotoxicity and hepatotoxicity.

Oral itraconazole can cause nausea and gastrointestinal (GI) disturbances associated with the cyclodextrin excipient, making it difficult to tolerate for prolonged treatment courses. Itraconazole has also been described as causing a unique triad of hypertension, hypokalemia, and edema that may be related to the negative inotropic effect of the drug or adrenal suppression (Table 4.2). Therefore, prolonged administration of ITR is not recommended in patients with a history of heart failure. Although a rash has been reported with all antifungal classes in 5%–15% of patients, VOR treatments in ambulatory patients have been associated with a unique retinoid-like phototoxic reaction that presents with cheilitis, erythema, and occasional blistering. However, recent reports have linked this phototoxic reaction to the subsequent development of squamous cell carcinoma and melanoma (Riahi & Cohen, 2011).

TABLE 4.2 List of antifungal drugs, mode of action, and their side effects in human.

Class	Mode of action	Side effects
Polyene (amphotericin B, nystatin, natamycin, etc.)	Cell membrane permeability (directly binds to ergosterol, leading to losing of membrane integrity)	Nephrotoxicity, hepatic toxicity, etc.
Azoles (fluconazole, itraconazole, posaconazole, voriconazole, ketoconazole, clotrimazole, miconazole, etc.)	Cell membrane permeability (ergosterol biosynthesis pathway by targeting 14α—lanosterol demethylase)	Anaphylaxis, nausea, gastrointestinal disturbance, phototoxicity, cardiomyopathy, etc.
Echinocandin (caspofungin, micafungin, anidulafungin, etc.)	Inhibit cell wall components (inhibitors of glucan synthase)	Hepatic toxicity, etc.
Nucleoside analogs (flucytosine, griseofulvin, etc.)	Inhibit nucleic acid synthesis, disrupt microtubule functions	Gastrointestinal disturbance, anemia associated with decreased erythropoietin production, etc.
Allylamines (naftifine, terbinafine, tolnaftate, etc.)	Inhibit the *ERG1* gene of ergosterol biosynthesis	Rash, photosensitivity, etc.
Others (aurones, ciclopirox, haloprogin, miltefosine, orotomide, etc.)	Inhibitors of cell membrane respiration processes, morphogenesis, ergosterol, etc.	Hepatic toxicity, rash, etc.

4.6 Drug-resistance mechanism in *Candida* spp.

From a clinical aspect, drug resistance is manifested by an infection continuously increasing in spite of proper treatment with a particular drug (Hwang & Gums, 2016). The drug resistance may be either primary in nature, when a fungus is already resistant before starting treatment, or secondary, when an initially sensitive fungus becomes resistant after receiving the drug (Cannon et al., 2009). Although the molecular basis of antifungal resistance in fungi is not very clear, evidence accumulated so far suggests that multidrug resistance (MDR) is a multifactorial phenomenon comprising a combination of several mechanisms (White, Marr, & Bowden, 1998). Many mechanisms may be involved during chemotherapy by which a drug is resisted by a pathogen. Among them, the main mechanism is alterations in the ergosterol biosynthetic pathway, which leads to reduced affinity to FLC (Perlin et al., 2017). In addition, another important mechanism is decreased intracellular drug accumulation in *Candida* cells of drug extraction pumps (ABC and MFS superfamilies of proteins). It is well-known that azole-resistant isolates of *C. albicans* produced transcriptional activation of genes encoding membrane proteins (ABC or MFS) (Cannon et al., 2009). The resistant *Candida* cells, normally exhibiting increased expression of efflux pump encoding genes, also express a simultaneous increase in the efflux of a drug (Prasad & Goffeau, 2012). Rapid efflux of an incoming drug prevents cells from accumulating a lethal concentration of the antifungal drug, enabling them to survive. Although drug inactivation is a common mechanism in bacteria, it has not yet been recorded in *Candida* cells (Delarze & Sanglard, 2015).

4.7 Currently used combination therapy

The in vitro studies on antifungal agents have shown that combinations can enlarge the coverage, increase the fungicidal effectiveness, and decrease the growth of fungal resistance. Combined antifungal agents can have synergistic activity with less toxicity and more effectiveness than single antifungals. However, on the other hand, the known studies have shown that different concentrations of each antifungal drug in combination can be associated with results that range from antagonistic values to synergistic values. The several host factors also affect the efficacy of an antifungal drug in in vivo settings and cannot always be affected in in vitro studies. Decreased toxicities or activity-related issues can also be shown when the combination therapies are studied in humans (Johnson, MacDougall, Ostrosky-Zeichner, Perfect, & Rex, 2004). Multiple concentrations of antifungal drug combinations in a clinical trial are very difficult due to the expensive clinical trials and the limited number of research candidates studied. Presently, the time duration and cost is the main limiting factor for clinical trials available on combination antifungals therapy (Rex et al., 2001). Currently, many fungicidal mechanisms for synergistic activity between antifungal drugs have been proposed. One combination is the use of terbinafine and azoles, which work on the same biochemical pathway by inhibiting ergosterol biosynthesis in the cell membrane and therefore inhibiting the fungal cell membrane function (Table 4.3).

Further, *C. glabrata* infection with low susceptibility to fluconazole (FLC) might also benefit from terbinafine added to fluconazole, voriconazole, or itraconazole (Perea, Gonzalez, Fothergill, Sutton, & Rinal, 2002). In this process, Amp B and the azoles damage the cell membrane and allow increased uptake of other agents like flucytosine (Yamamoto et al., 1997). Another mechanism of synergy includes inhibition of different fungal cell membranes and cell wall targets, giving synergism between echinocandins (i.e., caspofungine) and polyene (Amp B) (Petraitis et al., 2003). The antagonistic activity among antifungal drugs is one of the most important factors that determine the selection of any drug combination that might be useful in invasive fungal infection treatment. However, in some cases, an incorrect combination of antifungal drugs can decrease the fungicidal potency and often increase the toxicity level. The antagonistic mechanisms are also

TABLE 4.3 Synergistic and antagonistic activity of selected combinations of antifungal drugs.

Combination of antifungals	Possible mechanisms of actions	Effects
Amp B + flucytosine	Cell wall damage and increased uptake of flucytosine	Synergism/partial antagonism
Amp B + azole	Fungal cell wall modifications	Antagonism
Azole + flucytosine	Cell wall damage and increased uptake of flucytosine	Synergism
Terbinafine + azole	Inhibition of ergosterol biosynthesis	Synergism
Flucytosine + azole	Cell wall damage and uptake of flucytosine	Synergism

diverse in nature due to action of both antifungal agents on the same site, decreasing the availability of one or the other or modifying a target site in the fungal cell and, due to exposure to the other antifungal, making the fungal cell less susceptible to another antifungal agent (Scheven, Scheven, Hahn, & Senf, 1995; Sugar & Liu, 1998). These two mechanisms appear in the interaction between the azoles (e.g., FLC) and polyene (e.g., Amp B), where azoles inhibit the synthesis of the ergosterol pathway, usually causing the Amp B to bind and inactivate the cell membrane ergosterol. Similarly, polyene (Amp B) is also believed to have partial antagonistic activity with flucytosine by altering the *Candida* cell membrane permeability (Shadomy, Wagner, Espinel, & Davis, 1975), but this type of antagonistic activity of amphotericin and flucytosine is still not well known in the in vitro studies. Moreover, the echinocandin class of drugs with azole is another common combination for the treatment of invasive *Candida* infection. However, no clinical trials on this combination have been available. On the other hand, posaconazole combined with caspofungin or micafungin has been well-studied using both in vitro and in vivo models (Cui, Ren, Tong, Dai, & Zhang, 2015). Chen, Lehman, Averette, Perfect, and Heitman (2013) found that posaconazole exhibits the synergistic effect with caspofungin against echinocandin-resistant *Candida* spp. This finding was further confirmed by another laboratory study on *C. albicans*, *C. glabrata*, and *C. parapsillosis*, which showed that a *Candida* isolate with known resistance to either azole or echinocandin can be susceptible to the combination of two drugs, and, more notably, when posaconazole is combined with an echinocandin class of drugs perhaps due to synergistic effect.

4.8 Toxicity of antifungal drugs

Antifungal drugs have been found to have many adverse effects in human. Amp B with triazoles has been reported to produce hepatotoxicity,

TABLE 4.4 Common toxicities of antifungal agents.

Human-related toxicity	Antifungals
Hepatic toxicity	All azoles, Amp B, 5-FC, echinocandins
Renal toxicity	Amp B, voriconazole
Central nervous system	Voriconazole
Eye	Voriconazole
Cutaneous	Voriconazole
Gastrointestinal	Itraconazole, posaconazole and 5-FC
Cardiac toxicity	Azoles
Infusion reaction	Amp B, echinocandins
Bone marrow	Amp B, 5-FC

organ dysfunction, and cutaneous reactions in prolonged administration. Selected common toxicities exhibited by existing antifungal agents are shown in Table 4.4.

Oral itraconazole may cause nausea and GI-related problems with the cyclodextrin excipient, which makes it intolerable for long-term treatment. This drug also causes hypertension, hypokalemia, and edema, which may be related to a negative inotropic effect of the drug or adrenal suppression. Hence, long-term use of ITR is not recommended in patients with a history of heart failure. Although all these drugs show a rash problem in approximately 5%–15% of patients, the use of VOR in ambulatory patients has been associated with retinoid-like phototoxicity which found with cheilitis, erythema and infrequent blistering. This phototoxicity reaction has been linked to the subsequent development of seqamous cells carcinoma and melanoma (Riahi & Cohen, 2011).

4.9 Conclusions

Limited in vivo and clinical studies are available to support the application of combined antifungal drugs for some severe and most difficult invasive fungal infections. The combination of azoles (i.e., FLC) with polyenes (i.e., Amp B) needs more clinical trials before routine application in anti-*Candida* treatment. For many antifungal drug-resistant and invasive infections, the scientific community should test several combinations of antifungal drug-drug interactions before adding a second antifungal to the treatment procedure. However, mismatched combinations of antifungal drugs can lead to antagonistic activity and decrease the efficacy and increase adverse effects in a patient that is already immunosuppressed and in critical condition.

References

Barnes, R. A., Stocking, K., Bowden, S., Poynton, M. H., & White, P. L. (2013). Prevention and diagnosis of invasive fungal disease in high- risk patient within an integrative care pathway. *The Journal of Infection, 67*, 206–2014.

Brown, G. D., Denning, D. W., Gow, N. A. R., Levitz, S. M., Netea, M. G., & White, T. C. (2012). Hidden killers: Human fungal infections. *Science Translational Medicine, 4*, 165–178.

Butler, G. (2010). Fungal sex and pathogenesis. *Clinical Microbiology Reviews, 1*, 140–159.

Campoy, S., & Adrio, J. L. (2017). Antifungals. *Biochemical Pharmacology, 133*, 86–96.

Cannon, R. D., Lamping, E., Holmes, A. R., Niimi, K., Baret, P. V., & Keniya, M. V. (2009). Efflux-mediated antifungal drug resistance. *Clinical Microbiology Reviews, 22*, 291–321.

Chen, Y. L., Lehman, V. N., Averette, A. F., Perfect, J. R., & Heitman, J. (2013). Posaconazole exhibits in vitro and in vivo synergistic antifungal activity with caspofungin or FK506 against *Candida albicans*. *PLoS One, 8*(3), e57672.

Chi, H. W., Yang, Y. S., Shang, S. T., Chen, K. H., Yeh, K. M., Chang, F. Y., et al. (2011). *Candida albicans* versus non-albicans bloodstream infections: The comparison of risk factors and outcome. *Journal of Microbiology, Immunology, and Infection, 44*, 369–375.

Cui, J., Ren, B., Tong, Y., Dai, H., & Zhang, L. (2015). Synergistic combinations of antifungals and anti-virulence agents to fight against *Candida albicans*. *Virulence, 6*, 362–371.

Delarze, E., & Sanglard, D. (2015). Defining the frontiers between antifungal resistance, tolerance and the concept of persistence. *Drug Resistance Updates, 23*, 12–19.

Denning, D. W., & Hope, W. W. (2010). Therapy for fungal diseases: Opportunities and priorities. *Trends in Microbiology, 18*, 195–204.

Geddes-McAlister, J., & Shapiro, R. S. (2019). New pathogens, new tricks: Emerging, drug-resistant fungal pathogens and future prospects for antifungal therapeutics. *Annals of the New York Academy of Sciences, 1435*, 57–78.

Gudlaugsson, O., Gillespie, S., Lee, K., Vande, B. J., Hu, J., Messer, S., et al. (2003). Attributable mortality of nosocomial candidemia, revisited. *Clinical Infectious Diseases, 37*, 1172–1177.

Hoban, D. J., Zhanel, G. G., & Karlowsky, J. A. (1999). In vitro susceptibilities of *Candida* and *Cryptococcus neoformans* isolates from blood cultures of neutropenic patients. *Antimicrobial Agents and Chemotherapy, 43*, 1463–1464.

Hwang, A. Y., & Gums, J. G. (2016). The emergence and evolution of antimicrobial resistance: Impact on a global scale. *Bioorganic & Medicinal Chemistry, 24*, 6440–6445.

Johnson, M. D., MacDougall, C., Ostrosky-Zeichner, L., Perfect, J. R., & Rex, J. H. (2004). Combination antifungal therapy. *Antimicrobial Agents and Chemotherapy, 48*, 693–715.

Kullberg, B. J., & Arendrup, M. C. (2015). Invasive candidiasis. *The New England Journal of Medicine, 373*, 1445–1456.

Lewis, R. E. (2011). Current concepts in antifungal pharmacology. *Mayo Clinic Proceedings, 86*, 805–817.

Lewis, R. E., Klepser, M. E., & Pfaller, M. A. (1998). Update on clinical antifungal susceptibility testing for *Candida* species. *Pharmacotherapy, 18*, 509–515.

Lewis, R. E., Klepser, M. E., & Pfaller, M. A. (2000). In vitro pharmacodynamic characteristics of flucytosine determined by time-kill methods. *Diagnostic Microbiology and Infectious Disease, 36*, 101–105.

Louie, A., Liu, W. D., Miller, A., Sucke, A. C., Liu, Q. F., Drusano, G. L., et al. (1999). Efficacies of high-dose fluconazole plus amphotericin B and high-dose fluconazole plus 5-fluorocytosine versus amphotericin B, fluconazole, and 5-fluorocytosine monotherapies in treatment of experimental endocarditis, endophthalmitis, and pyelonephritis due to *Candida albicans*. *Antimicrobial Agents and Chemotherapy, 43*, 2831–2840.

Miceli, M. H., Díaz, J. A., & Lee, S. A. (2011). Emerging opportunistic yeast infections. *The Lancet Infectious Diseases, 11*, 142–151.

Naglik, J., Albrecht, A., Bader, O., & Hube, B. (2004). *Candida albicans* proteinases and host/pathogen interactions. *Cellular Microbiology, 6*(10), 915–919.

Odds, F. C. (2003). Synergy, antagonism, and what the chequerboard puts between them. *The Journal of Antimicrobial Chemotherapy*, *52*(1), 1.

Pappas, P. G., Kauffman, C. A., Andes, D., Benjamin, D. K., Calandra, T. F., & Edwards, J. E. (2009). Clinical practice guidelines for the management of candidiasis: 2009 update by the Infectious Diseases Society of America. *Clinical Infectious Diseases*, *48*, 503–535.

Perea, S., Gonzalez, G., Fothergill, A. W., Sutton, D. A., & Rinal, D. M. G. (2002). In vitro activities of terbinafine in combination with fluconazole, itraconazole, voriconazole, and posaconazole against clinical isolates of *Candida glabrata* with decreased susceptibility to azoles. *Journal of Clinical Microbiology*, *40*, 1831–1833.

Perfect, J. R. (2017). The antifungal pipeline: A reality check. *Nature Reviews Drug Discovery*, *16*, 603–616.

Perlin, D. S., Richardson, R. R., & Alastruey-Izquierdo, A. (2017). The global problem of antifungal resistance: Prevalence, mechanisms, and management. *Lancet Infectious Diseases*, *17*, 383–392.

Petraitis, V., Petraitiene, R., Sarafandi, A. A., Kelaher, A. M., Lyman, C. A., Casler, H. E., et al. (2003). Combination therapy in treatment of experimental pulmonary aspergillosis: Synergistic interaction between an antifungal triazole and an echinocandin. *The Journal of Infectious Diseases*, *187*, 1834–1843.

Pfaller, M. A., & Diekema, D. J. (2007). Epidemiology of invasive candidiasis: A persistent public health problem. *Clinical Microbiology Reviews*, *20*, 133–163.

Pfaller, M. A., Messer, S. A., Boyken, L., Hollis, R. J., Rice, C., Tendolkar, S., et al. (2004). In vitro activities of voriconazole, posaconazole and fluconazole against 4169 clinical isolates of *Candida* spp. and *Cryptococcus neoformans* collected during 2001 and 2002 in the ARTEMIS global antifungal surveillance program. *Diagnostic Microbiology and Infectious Disease*, *48*, 201–205.

Prasad, R., & Goffeau, A. (2012). Yeast ATP-binding cassette transporter conferring multidrug resistance. *Annual Review of Microbiology*, *66*, 39–63.

Rex, J. H., Walsh, T. J., Nettleman, M., Anaissie, E. J., Bennett, J. E., Bow, E. J., et al. (2001). Need for alternative trial designs and evaluation strategies for therapeutic studies of invasive mycoses. *Clinical Infectious Diseases*, *33*, 95–106.

Riahi, R. R., & Cohen, P. R. (2011). Voriconazole associated phototoxicity. *Dermatology Online Journal*, *17*, 15.

Richardson, M., & Lass-Flörl, C. (2008). Changing epidemiology of systemic fungal infection. *Clinical Microbiology and Infection*, *14*, 5–24.

Roemer, T., & Krysan, D. J. (2014). Antifungal drug development: Challenges, unmet clinical need and new approaches. *Cold Spring Harbor Perspectives in Medicine*, *4*, 1–14.

Scheven, M., Scheven, C., Hahn, K., & Senf, A. (1995). Post-antibiotic effect and post-expositional polyene antagonism of azole antifungal agents in *Candida albicans*: Dependence on substance lipophilia. *Mycoses*, *38*, 435–442.

Shadomy, S., Wagner, G., Espinel, I. E., & Davis, B. A. (1975). In vitro studies with combinations of 5-fluorocytosine and amphotericin B. *Antimicrobial Agents and Chemotherapy*, *8*, 117–121.

Sobel, J. D. (2007). Vulvovaginal candidosis. *Lancet*, *369*, 1961–1971.

Spellberg, B., Blaser, M., Guidos, R. J., Boucher, H. W., Bradley, J. S., Eisenstein, B. I., et al. (2011). Combating antimicrobial resistance: Policy recommendations to save lives. *Clinical Infectious Diseases*, *52*, 397–428.

Sugar, A. M., & Liu, X. P. (1998). Interactions of itraconazole with am photericin B in the treatment of murine invasive candidiasis. *The Journal of Infectious Diseases*, *177*, 1660–1663.

White, T. C., Marr, K. A., & Bowden, R. A. (1998). Clinical, cellular, and molecular factors that contribute to antifungal drug resistance. *Clinical Microbiology Reviews*, *11*, 382–402.

Yamamoto, Y., Maesaki, S., Kakeya, H., Yanagihara, K., Ohno, H., Ogawa, K., et al. (1997). Combination therapy with fluconazole and flucytosine for pulmonary cryptococcosis. *Chemotherapy*, *43*, 436–441.

Further reading

Dutcher, J. D. (1968). The discovery and development of amphotericin B. *Diseases of the Chest, 54*, 296–298.

Rodrigues, C. F., Silva, S., & Henriques, M. (2014). *Candida glabrata*: A review of its features and resistance. *European Journal of Clinical Microbiology & Infectious Diseases, 33*, 673–688.

Rukayadi, Y., Lee, K., Lee, M. S., Yong, D., & Hwang, J. K. (2009). Synergistic anticandidal activity of xanthorrhizol in combination with ketoconazole or amphotericin B. *FEMS Yeast Research, 9*, 1302–1311.

Sanglard, D. (2016). Emerging threats in antifungal-resistant fungal pathogens. *Frontiers in Medicine, 3*, 1–10.

Sastry, A. S., & Bhatt, S. (2018). *Essentials of medical microbiology* (2nd ed.). New Delhi: Jaypee Brothers Medical Publishers.

CHAPTER

5

Metallodrug-driven combination chemotherapy in cancer treatment

Afzal Hussain[a], Mohamed Fahad AlAjmi[a], Prince F. Iqbal[b], Waseem A. Wani[c]

[a]Department of Pharmacognosy, College of Pharmacy, King Saud University, Riyadh, Saudi Arabia, [b]Department of Chemistry, Government Degree College Boys, Pulwama, Jammu and Kashmir, India, [c]Department of Chemistry, Government Degree College, Tral, Jammu and Kashmir, India

Abbreviations

AGC	advanced gastric cancer
DNA	deoxyribonucleic acid
EGFR	epidermal growth factor receptor
FDA	Food and Drug Administration
KTZ	ketoconazole
5FU	5-fluorouracil

5.1 Introduction

Cancer is the most common human genetic disease, which progresses via multistep carcinogenesis involving a number of physiological systems of the human body. Initially, cancer is localized, but later it spreads (metastasizes) to distant and different sites of the human body. Both the factors—that is, the involvement of several physiological systems and the metastasis—make cancer highly difficult to treat and manage (Reichert & Wenger, 2008; Wani, Prashar, Shreaz, & Gómez-Ruiz, 2016; Zou, 2005). Merely hearing the word cancer can send shivers down the spine of many people, as its control and treatment have challenged human beings for a

https://doi.org/10.1016/B978-0-12-820576-1.00005-9

long time (Iqbal, Malik, & Wani, 2018). On a global level, cancer stands as the second most common fatal disease after cardiovascular diseases (Ali, Wani, & Saleem, 2011). Cancer has been recognized as a major public health issue in the United States, where one out of every four deaths is due to cancer. The estimates of Siegel et al., published by the American Cancer Society, predicted 1,762,450 new cancer cases and 606,880 cancer-related deaths in the United States by the end of 2018 (Siegel, Miller, & Jemal, 2018). In Europe, cancer is the second most noted cause of death after cardiovascular diseases. Europe has witnessed huge losses of life due to increasing reports of cancer-related deaths (about 20% of deaths). Malvezzi et al. estimated 1,410,000 cancer-related deaths in Europe for the year 2018 (Malvezzi et al., 2019). Cancer is the second most common fatal disease in India (Ali et al., 2011) and it is worryingly increasing on the Indian subcontinent, which may be attributed to poor living standards and inadequate medical facilities. The most common cancer types reported in India are cancer of the lungs, breasts, colon, rectum, stomach, and liver (Murthy & Mathew, 2004; Nandakumar, 1990–1996; Rao & Ganesh, 2008; Wynder, Covey, & Mabuchi, 1974).

In a nutshell, cancer continues to be a major challenge to the entirety of humanity. This has made the treatment of cancer a top priority for oncologists, medical chemists, and biochemists the world over (Ali, Wani, Haque, & Saleem, 2013). Several methods, including radiation therapy, chemotherapy, surgery, immunotherapy and monoclonal antibody therapy, are presently used for the treatment of cancer. Chemotherapy, the use of small molecule drugs (synthetic or natural), is one of the preferred methods for the treatment of cancer. Chemotherapy takes into account the metastatic nature of the majority of cancers. In addition, the anticancer drugs used during chemotherapy reach all the cells and tissues of the human body, where other treatment procedures cannot be used. Additionally, the availability of a large reservoir of anticancer drugs with diverse mechanisms of action and drug combination procedures makes chemotherapy a standout performer in cancer treatment procedures (Saleem, Wani, Haque, Milhotra, & Ali, 2013).

The medicinal properties of metals have been reported for 5000 years (Orvig & Abrams, 1999). There are reports of the utilization of several compounds containing metal ions/atoms for various medicinal purposes, such as the treatment of bacterial and fungal infections, viral attacks, and cancer (Alderden, Hall, & Hambley, 2006). Metal-containing molecules (metal complexes) exhibit biological activities due to their interesting binding properties with important biomolecules, such as proteins, hormones, enzymes, nucleic acids, etc., which play an important role in cancer growth and proliferation (Ali, Wani, Saleem, & Haque, 2013). Presently, metal complexes occupy an important place in inorganic medicinal chemistry. Metal complexes provide versatile platforms for the design and development of metal-containing drugs (metallodrugs). The kinetic properties (ligand exchange rates) and thermodynamic properties (metal to ligand bond enthalpies, redox potentials, etc.) of metal complexes can be tuned and optimized by varying the central metal ion/atom and their oxidation states. It is quite noteworthy that not only the metal center but also ligand structures determine the bioactivities of metal complexes (Romero-Canelon & Sadler, 2013; Wani et al., 2016). Cisplatin (Fig. 5.1) is the most famous metallodrug used for treating various cancers (Ott & Gust, 2007). It was Peyrone in 1844 who first synthesized cisplatin. However, its anticancer effects were discovered much later by Rosenberg et al., in 1965, through their studies of the effect of alternating current on the growth of *Escherichia coli* (Bertini, Gray, Lippard, & Valentine, 1994; Kaim & Schwederski, 1993; Rosenberg, Van Camp, & Krigas, 1965). After exhaustive experimentation by the research group of Prof. Rosenberg, cisplatin was approved for treating testicular and ovarian cancers. Presently, cisplatin is the most commonly used drug for the treatment of testicular, ovarian, bladder, cervical, head and neck, and small cell and non-small cell lung cancers (Weiss & Christian, 1993). The fruitful, serendipitous discovery of cisplatin was well-publicized and thus served as the necessary impetus for the design and development of new metallodrugs. However, later it was exposed that cisplatin lacked tumor tissue selectivity and exhibited several kinds of adverse side effects, including ototoxicity (loss of balance/hearing), neurotoxicity, renal impairment, and others. Thus it was worthwhile for medical inorganic chemists to look for metallodrugs beyond cisplatin. As a result, second- and third-generation cisplatin analogs were developed with better anticancer properties and reduced side effects (Kostova, 2006). The second-generation cisplatin analogs, such as carboplatin, oxaliplatin, and nedaplatin (Fig. 5.1), were thus developed. Carboplatin was successful in the treatment of ovarian carcinoma, lung, and head and neck cancers. Oxaliplatin in combination therapy with

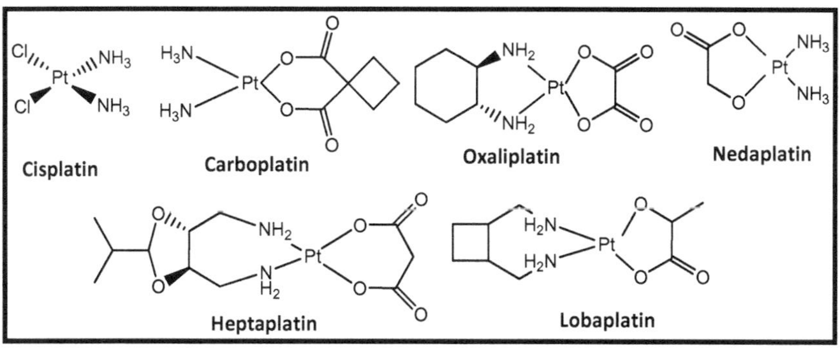

FIG. 5.1 Chemical structures of cisplatin and its second- and third-generation analogs.

5-fluorouracil (5FU) and folinic acid was successful in the treatment of adjuvant and metastatic colorectal cancers. Nedaplatin was used for the treatment of non-small cell lung cancer, small cell lung cancer, esophageal, and head and neck cancers. Heptaplatin and lobaplatin (Fig. 5.1) were the third-generation cisplatin analogs. Heptaplatin was successfully used in the treatment of gastric cancer and chronic myelogenous leukemia. Lobaplatin was effective for the treatment of inoperable metastatic breast cancer and single cell lung cancer. Then, the reports of treatment failures and the development of resistance to cisplatin analogs further expanded the development of metallodrugs. This resulted in the design and development of a large pool of platinum complexes with diverse structures and exciting anticancer properties. Currently, various platinum metallodrugs are in different drug development stages.

In response to some persistent drawbacks associated with platinum-based anticancer metallodrugs, anticancer metallodrug development was further extended to nonplatinum complexes. Consequently, a large number of structurally diverse anticancer metallodrugs based on metal ions/atoms other than platinum were developed. Metal ions belonging to the first transition series (scandium, titanium, vanadium, chromium, manganese, iron, cobalt, nickel, copper, zinc), second transition series (yttrium, zirconium, nobelium, molybdenum, technetium, ruthenium, rhodium, palladium, silver, cadmium), third transition series (tungsten, rhenium, osmium, iridium, gold), boron family (gallium), carbon family (tin), nitrogen family (antimony, bismuth), lanthanides (lanthanum, cerium, neodymium, samarium, europium, gadolinium, terbium, dysprosium) and actinides (uranium) (Fig. 5.2) have been frequently used for anticancer metallodrug design rationales (Mjos & Orvig, 2014). Out of the various metals/metal ions used for anticancer metallodrugs, some complexes based on ruthenium, copper, gold, palladium, iron, and tin have shown exciting anticancer properties.

Most of the patients with metastatic cancers have quite low 5-year survival rates, and the development of new anticancer drugs consumes a great deal of money and time. Thus it is better to develop strategies wherein multiple survival pathways are targeted, so that better results at a more reasonable cost can be obtained. Considering the complexity of the treatment and management of cancer, it is very difficult to treat cancer effectively by using a single anticancer drug in a chemotherapeutic regimen (monotherapy). However, cancer can be effectively treated with combination therapy by administering two or more than two anticancer drugs together. Combined administration of anticancer drugs is efficacious to the treatment procedure, as the combined drugs ensure targeting of the key multiple pathways in a typically synergistic or additive manner. Combination anticancer chemotherapy leads to the reduction of

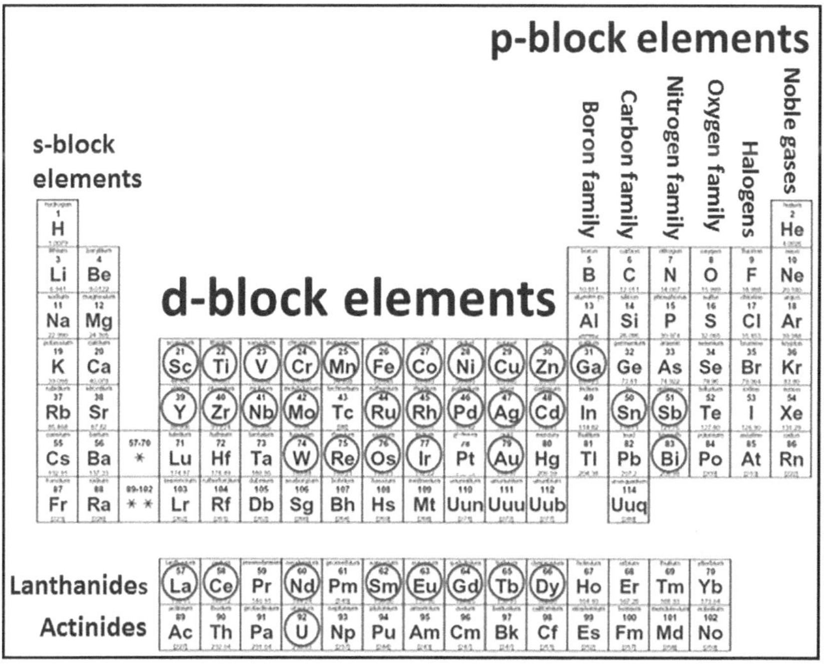

FIG. 5.2 Anticancer metallodrugs of nonplatinum metals [encircled in green (*gray in print version*)] have now been developed.

drug resistance, enhancement of anticancer effects by reduction of tumor growth, reduction in the metastatic potential of cancer cells, apprehension of the mitotically active cells, reduction of the populations of cancer stem cells, and the induction of apoptosis.

A comprehensive review of the research literature available on the use of metallodrugs in combination therapy of cancer indicates a progressive increase in interest in this field during the last two decades. The increased interest may be due to the increased demand for new and more effective anticancer drugs. There are no reviews/book chapters or books available that focus on metallodrug-based combination chemotherapy for cancer treatment. Therefore this chapter has been written to present the advances in the development of strategies for the treatment of cancer using metallodrugs in combination therapy. This chapter describes the state-of-the-art advances made in the metallodrug-driven combination therapy for the treatment of cancer. It discusses combination therapy with both platinum and nonplatinum metallodrugs. Finally, future outlooks for research and development in metallodrug-driven combination cancer therapy have been outlined.

5.2 Inception of combination therapy in cancer treatment

Combination therapy is also known as polytherapy. It is a therapeutic procedure wherein more than one drug or medication, or modality, is used for carrying out treatment of a disease. Cancer combination therapy utilizes two or more than two anticancer drugs for specifically targeting cancer-inducing or cell-sustaining pathways, and thus it forms a keystone of cancer therapy (Blagosklonny, 2004; Yap, Omlin, & de Bono, 2013). In 1965, Emil Frei, James F. Holland, and Emil J. Freireich were the first to propose combination chemotherapy for the treatment of acute leukemia (Frei et al., 1965). They used a combination of methotrexate, 6-mercaptopurine, vincristine, and prednisone (formally known as the POMP regimen) for the treatment of acute lymphocytic leukemia in pediatric patients. The combination drug regimen succeeded in reducing tumor burden and prolonging remission (Frei et al., 1965). The success of the POMP procedure provided the necessary impetus for research in combination therapies targeting different biological pathways in a quest to create a synergistic or additive effect. These synergistic or additive effects by virtue of combination chemotherapy have been achieved in several studies. Quinn and coworkers found that the combination chemotherapy utilizing sabutoclax (a pan-Bcl-2 inhibitor) and minocycline (an antibiotic that has previously displayed anticancer effects) worked synergistically on the intrinsic apoptotic pathway. Besides, selective toxicity and reduction in tumor growth both in vitro and in vivo were reported against pancreatic ductal adenocarcinoma (Quinn et al., 2015). Thus it is possible to generate synergistic or potentiation effects by using a combination of anticancer drugs with the ability to target different biological pathways, which consequently could yield significant anticancer results.

Monotherapy is still a routinely used treatment modality for different cancer types. However, it is often considered to be inefficient in comparison to combination therapy. Monotherapy is nonselective in nature and thus targets actively proliferating cells. This leads to the death of both healthy and cancerous cells. It is well known that chemotherapy can exhibit toxic effects and risks to the patient. Chemotherapy can also strongly weaken the immune systems of patients by affecting bone marrow cells and increasing susceptibility to host diseases (LeBaron, Zeltzer, LeBaron, Scott, & Zeltzer, 1988; Partridge, Burstein, & Winer, 2001). Though toxicity can be expected with combination therapy as well when one of the combined drugs used is cytotoxic, the toxicity is significantly less since different biological pathways are targeted. Eventually, the drug combination works in a synergistic or additive manner, thereby reducing the dosage of each individual drug (Albain et al., 2008; Mokhtari et al., 2013). Moreover, it is believed that combination therapy may lower toxic effects of the chemotherapeutic drugs on normal cells while at the same time producing cytotoxic effects on cancer cells. This occurs when one of the drugs in the combination regimen is antagonistic (in terms of cytotoxicity to normal cells) to another drug, which ultimately results in the protection of normal cells from cytotoxic effects (Blagosklonny, 2005). Prof. Blagosklonny demonstrated an increase in the therapeutic index of cancer therapy by involving combination therapy. A caspase-inhibitor, Z-DEVD-fmk, in combination therapy with an apoptosis-inducing agent was used. Z-DEVD-fmk-resistant cancer cells expressing p-glycoprotein pumped out Z-DEVD-fmk; nevertheless, the apoptosis-inducing agent induced apoptosis in the cancer cells. However, most of the normal cells unable to express p-glycoprotein were directly affected by Z-DEVD-fmk and resumed apoptosis (Blagosklonny, 2005).

Presently, combination therapy of cancer is in full bloom and a number of new and old anticancer drugs are continuously being screened in different possible drug combination regimens for the safe and effective treatment of cancer.

5.3 Metallodrugs in combination therapy for cancer treatment

Metallodrugs, both platinum and nonplatinum, have been extensively explored and used in anticancer chemotherapy regimens. Platinum-based anticancer drugs are presently more predominant in the market, though nonplatinum ones based on ruthenium and a copper metal center have also found considerable success as alternative therapeutic agents.

5.3.1 Platinum metallodrugs

Although most of the clinically used anticancer drugs are organic compounds and very few are metallodrugs, platinum metallodrugs such as cisplatin are the most commonly used against several types of cancers (Komeda & Casini, 2012). Cisplatin is the most famous anticancer metallodrug and is one of the oldest anticancer drugs. Even though cisplatin was first synthesized in 1844, it was only approved by the FDA in 1978 and was the first metal-based medical agent to enter into worldwide clinical use to treat cancer. Besides cisplatin, other FDA-approved platinum metallodrugs are carboplatin and oxaliplatin. Most of these platinum drugs involve activation by aquation inside cells due to varying Cl^- concentration, thereby binding DNA and arresting the cell cycle and transcription and causing apoptosis (Wang & Lippard, 2005). Cisplatin and other platinum metallodrugs have gained much attention in cancer treatment, but due to their high cytotoxicity against the human host cells, they are recommended to be used in combination with other drugs. In addition, the problem of acquired or intrinsic drug resistance against the platinum drugs is also a challenge, which could be answered with combination therapy, where two or more drugs with different drug targets are combined. Combination therapies for different cancers were trialed and proved to be superior to monotherapies (Chan & Coward, 2013).

Cisplatin has also been approved by the FDA to be used in combination with fluorouracil and folinic acid to target metastatic testicular, ovarian, and bladder cancers (Monneret, 2011). The other important combination therapy approved by the FDA to treat ovarian cancer is the combination of the second-generation platinum drug, carboplatin, with other drugs (Monneret, 2011). Oxaliplatin, a second-generation metallodrug, has also been licensed to be used with other chemotherapeutic drugs against colon and lung cancers (Chan & Coward, 2013). Another second-generation platinum drug, nedaplatin, was approved in Japan and has been extensively used in combination therapy to treat urological cancers (Uehara, Yamate, Torii, & Maruyama, 2011). In the United States, combination therapy of VP-16 and cisplatin to treat small-cell lung cancer has been the first-line therapy since the 1980s (Evans, Shepherd, Feld, et al., 1985). Besides combining two drugs to treat small-cell lung cancer, a phase trial combining three drugs, namely etoposide, paclitaxel, and topotecan, was also carried out; however, potential adverse effects from combining these drugs have also been noted (Miller, Bogart, Watson, et al., 2005). A doublet combination regimen of either cisplatin-based or 5FU-based chemotherapy is generally used for treatment of patients with advanced gastric cancer (AGC). However, there is still no consensus on the best regimen for treating AGC. Although various new chemotherapeutic agents, including oral 5FU, taxanes, and irinotecan, have been identified as improving the outcomes for AGC when used as a single agent or in combination with nonplatinum chemotherapy, it is still unclear whether nonplatinum-based chemotherapy is a viable treatment option for patients with AGC (Kang, Kim, Kwon, Ho, & Wansik, 2014). In a separate study, the importance of combining metallodrugs with nanomaterials to make more effective formulations to treat drug-resistant cancers have been reported (Ellahioui, Prashar, & Gómez-Ruiz, 2017). In the same study, metal-based drugs and combinations have been predicted as future anticancer drug chemotherapy.

The other challenge with the use of platinum drugs is delayed delivery, and therefore combining these drugs by encapsulating with different nanoparticles not only enhances their delivery but also increases the anticancer property. Platinum drugs, due to their toxicity and lack of target specificity, are strong candidates for nanocarrier delivery and a variety of different types of nanocarriers have been explored. Nanocarriers such as gold and lipoic based nanocarriers have been studied to enhance loading of platinum drugs and to enhance them to the specific targets and thereby decrease the related cytotoxicity (Caballero et al., 2019). In a recent study, the molecular mechanism of action of encapsulation of platinum anticancer drugs within a protein nanocage was reported as a novel strategy to improve the efficacy and selectivity of these drugs toward cancer cells (Monti, Ferraro, & Merlino, 2019).

Although FDA-approved drug combinations have been tested in clinical trials on genetically diverse patient populations, patient-to-patient variability remains a concern. In preclinical trials combination of drugs can reveal synergistic or additive effects; however, in the clinical population these results vary. Even in drug combinations, a patient can have a better response and receive more benefit from a single drug to which his cancer responds more strongly, without any benefit from other drug/s (Palmer & Sorger, 2017). Therefore, the long-term goal of cancer research remains the identification of drug combinations that act in an additive or synergistic fashion on human cancers at individual levels and that can thus contribute to improve patient benefits. Computational modeling, clinical data, and clinical trials guided by biomarkers that aim to predict which patients will respond to which drugs and combinations can lead the future research in this field. While treating patients, shifting from one therapy to another based on the disease's progression can also lead to improvement in the combination therapies.

5.3.2 Nonplatinum metallodrugs

The prevalence of platinum resistance in the clinic, both intrinsic and acquired, is a growing clinical concern, especially because platinum drugs are currently being used in over half of all chemotherapy regimens (Dilruba & Kalayda, 2016; Makovec, 2019; Ndagi, Mhlongo, & Soliman, 2017). This has initiated the development of nonplatinum chemotherapeutic agents based on different metals with better cellular selectivity, target specificity, and a different mechanism of action. Given the chemical similarity of the platinum group of metals (Pt, Pd, Rh, Ir, Ru, and Os), and the successes of platinum therapies, interest in ruthenium (II/III) complexes has increased considerably. Although complexes based on different metals have been developed to fight platinum resistance, ruthenium-based metal complexes are believed to be the most efficient. However, despite widespread efforts, only a few ruthenium complexes (RM175, RAPTA-C, NAMI-A, KP1019/NKP1339, and TLD1433) have reached clinical trials and are being developed as chemotherapeutic agents (see Fig. 5.3 for structures). Octahedral Ru(II) complex TLD1433 has potential as a photosensitizer for photodynamic therapy (Fong et al., 2015).

Berger and colleagues studied the combinations of ruthenium complexes and first-line anticancer drugs. They found that the clinical drug ruthenium complex KP1339 combined with multikinase inhibitor sorafenib was more effective in hepatoma therapy (Hep3B) than KP1339 or sorafenib alone (Heffeter et al., 2013). Specifically, the mean survival of patients was

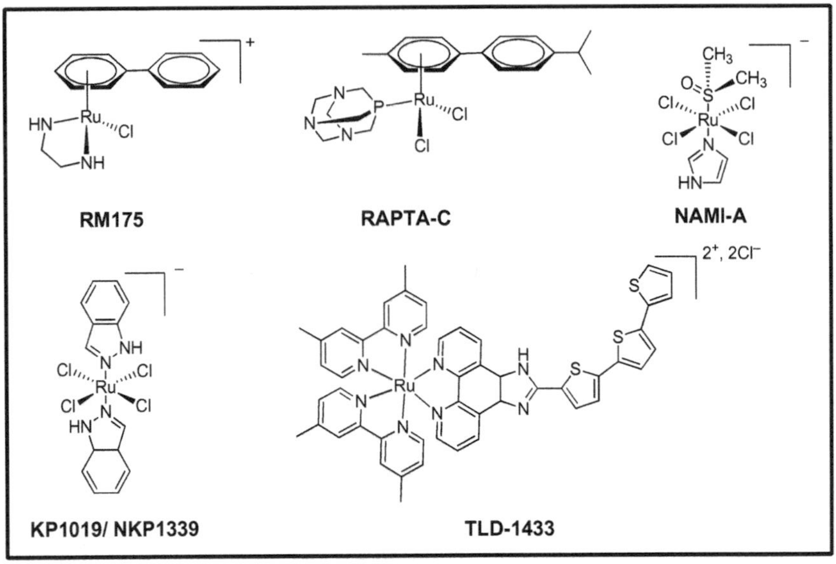

FIG. 5.3 Structures of ruthenium-based metal complexes in clinical trials.

extended by 3.9-fold by the combination, whereas KP1339 and sorafenib alone extended it by 2.4- and 1.9-fold, respectively. This combination also effectively inhibited the sorafenib-resistant tumor cells. Detailed studies also found that the combination substantially increased their intracellular accumulation and thereby interfered with the DNA synthesis process, rendering the cells unable to perform effective mitosis and enhancing apoptosis induction.

In clinical studies, NAMI-A combined with gemcitabine inhibited the activity of non-small cell lung cancer cells efficiently and reduced tolerance compared with the use of gemcitabine alone, but the combination of both had significant side effects, such as neutropenia, anemia, and renal impairment (Leijen et al., 2015). High-throughput screening has been focused on as a potential tool to identify promising drug combinations. (Bergamo, Riedel, Dyson, & Sava, 2015). NAMI-A and doxorubicin were shown to have a potent synergistic antitumor efficacy. NAMI-A effectively increased the accumulation of doxorubicin in breast carcinoma. This combination increased inhibition of tumor metastasis by 70%, as revealed by in vivo studies of mouse MCA mammary carcinoma. Promising synergistic effects were also observed in a lung metastasis preclinical tumor mice model using this combination, albeit with some noticeable side effects at maximum doses (Marien, Hillen, Vanderhoydonc, Swinnen, & Vande Velde, 2017). Studies on the combination of ovarian carcinoma chemotherapeutic doxorubicin and a ruthenium complex RAPTA-C have demonstrated that this combination significantly promoted the apoptosis of A2780 ovarian carcinoma cells compared to either single drug alone (Weiss et al., 2015). Normalization of tumor vasculature induced by apoptosis reduces vascular extravasations and provides adequate oxygen for oxygen-dependent phototherapy, achieving synergism (Goel et al., 2011). These studies provide valid evidence for the interaction between antiangiogenesis and antitumor effects. A ruthenium complex was combined with a second-line antitumor agent ketoconazole (KTZ) in hormone-refractory cancer therapy to form a $RuCl_2(KTZ)_2$ complex, which showed a favorable synergistic effect (Bozic et al., 2013). The combination of these two agents in a C8161 melanoma cell line significantly enhanced the expression of caspase-3 and promoted tumor cell apoptosis.

The combination of ruthenium complexes and photoacoustic (PA) imaging technology has significantly improved the accuracy and effectiveness of tumor diagnosis and therapy (Chen et al., 2014). In the therapy of tumors using drug combinations, ruthenium complexes have shown favorable efficacy. Combination of RAPTA-C and the epidermal growth factor receptor (EGFR) inhibitor erlotinib has been found to show strong synergistic inhibition of cell viability in human endothelial (ECRF24 and HUVEC) and human ovarian carcinoma (A2780 and A2780cisR) cells. In addition, erlotinib significantly enhanced the cellular uptake of RAPTA-C relative to treatment with RAPTA-C alone in human ovarian carcinoma cells, but not in endothelial cells. This drug combination has been found to induce the formation of chromosome bridges that persist after mitotic exit and delay abscission in A2780 and A2780cisR, therefore suggesting initiation of cellular senescence (Berndsen et al., 2017). Although a number of copper and gold metal complexes have shown promising anticancer activities, their combination effects have not yet been fully studied and there is no such combination approved by FDA yet.

5.4 Conclusions and future outlooks

To target various cancer pathways, multiple therapeutic designs have been developed during the last few decades. Some of the most exciting and effective results have been from combination therapy with regards to anticancer effects. The superiority of combination therapy rests on the ability of the combined drugs to target multiple pathways, which in effect reduces drug resistance, as cancer cells are normally unable to adapt to the simultaneous cytotoxic effects of two or more drugs (Zimmermann, Lehar, & Keith, 2007). The rapid proliferation rates of cancer cells are mainly due to the dysregulation of various pathways and the disrupted homeostatic environments. In addition, various types of cancers have mutations in tumor suppressor genes (such as p53) whose normal function is the activation of cell cycle arrest when DNA is damaged. Nevertheless, mutated tumor suppressor genes contribute to damaged DNA accumulation and cell cycle arrest inhibition, which leads to increased proliferation rates of cancer cells and a more aggressive cancer (Muller & Vousden, 2013). Also, upregulated production of autocrine growth factors or an upregulated autocrine loop in cancer cells further contributes to their growth (Marek et al., 2009; Stiehl et al., 2012). Thus it seems obvious that cancer control and the probability of cancer cells becoming increasingly malignant and incurable can be achieved by targeting the various pathways via multiple drug combinations. Combination therapy provides a choice to reduce the side effects, as in some cases the dose requirement of the administered drugs can be reduced, which is not possible with monotherapy treatment procedures (Delbaldo et al., 2004; Parhi, Mohanty, & Sahoo, 2012). An interesting benefit of combination therapy is that the multiple administered drugs can target the heterogeneous nature of tumors, which increases the chance of killing all cancer cell populations, including the cancer stem cells that generally contribute to drug resistance and cancer recurrence after remission in later years (Eyler & Rich, 2008; Wang, Narayanaswamy, Ren, & Torchilin, 2016; Yuan et al., 2013).

Despite the large number of advantages in combination therapy, some studies have shown certain drawbacks for combination therapy. The multiple drugs administered together can synergistically or additively create therapeutic benefits, but can similarly produce unwanted side effects (Delbaldo et al., 2004). This can complicate the identification of the drug responsible for the side effects and makes it challenging to ascertain which drug's dose should be lowered. For combined drugs with similar mechanisms of action and similar side-effect profiles, the accumulation of side effects can create more severe clinical symptoms and toxicities that can severely impact and alter the patient's life expectancy. Evidently, complications due to combination therapies can create additional problems for patients in terms of financial and general well-being. Thus it is advisable that drug interaction should be carefully understood when a combination therapy regimen for the treatment of cancer is developed. This must bring maximum efficacy and the least amount of toxicity for the combined drug regimen (Riechelmann et al., 2007).

This chapter demonstrates that metallodrugs currently in different developmental stages and those in the market being used in combination therapies warrant prospective therapeutic profiles. In the context of economic benefits, metallodrugs designed for cancer treatment have demonstrated promising results in clinical settings and have gained attention from oncologists all over the world. Metallodrugs in combination therapies have demonstrated enhanced efficacy and efficiency. Conspicuously, repositioning of metallodrugs passes toxicity and safety profiles and shows enhanced efficacy when used in combination treatments in an additive or synergistic manner, along with remarkable reduction of drug resistance.

References

Albain, K. S., Nag, S. M., Calderillo-Ruiz, G., Jordaan, J. P., Llombart, A. C., Pluzanska, A., et al. (2008). Gemcitabine plus paclitaxel versus paclitaxel monotherapy in patients with metastatic breast cancer and prior anthracycline treatment. *Journal of Clinical Oncology, 26*, 3950–3957.

Alderden, R. A., Hall, M. D., & Hambley, T. W. (2006). The discovery and development of cisplatin. *Journal of Chemical Education, 83*(5), 728–734.

Ali, I., Wani, W. A., Haque, A., & Saleem, K. (2013). Glutamic acid and its derivatives: Candidates for rational design of anticancer drugs. *Future Medicinal Chemistry, 5*(8), 961–978.

Ali, I., Wani, W. A., & Saleem, K. (2011). Cancer scenario in India with future perspectives. *Cancer Therapy, 8*, 56.

Ali, I., Wani, W. A., Saleem, K., & Haque, A. (2013). Platinum compounds: A hope for future cancer chemotherapy. *Anti-Cancer Agents in Medicinal Chemistry, 13*(2), 296–306.

Bergamo, A., Riedel, T., Dyson, P. J., & Sava, G. (2015). Preclinical combination therapy of the investigational drug NAMI-AC with doxorubicin for mammary cancer. *Investigational New Drugs, 33*, 53–63.

Berndsen, R. H., Weiss, A., Kulsoom Abdul, U., Wong, T. J., Meraldi, P., Grifoen, A. W., et al. (2017). Combination of ruthenium(II)-arene complex [Ru(η^6-*p*-cymene) Cl_2(pta)] (RAPTA-C) and the epidermal growth factor receptor inhibitor erlotinib results in efcient angiostatic and antitumor activity. *Science Reporter, 7*, 43005.

Bertini, I., Gray, H. B., Lippard, S. J., & Valentine, J. S. (1994). *Bioinorganic chemistry*. Sausalito, CA: University Science Books.

Blagosklonny, M. V. (2004). Analysis of FDA approved anticancer drugs reveals the future of cancer therapy. *Cell Cycle, 3*, 1035–1042.

Blagosklonny, M. V. (2005). Overcoming limitations of natural anticancer drugs by combining with artificial agents. *Trends in Pharmacological Sciences, 26*, 77–81.

Bozic, I., Reiter, J. G., Allen, B., Antal, T., Chatterjee, K., Shah, P., et al. (2013). Evolutionary dynamics of cancer in response to targeted combination therapy. *eLife, 2*, e00747.

Caballero, A. B., Cardo, L., Claire, S., Craig, J. S., Hodges, N. J., Vladyka, A., et al. (2019). Assisted delivery of anti-tumour platinum drugs using DNA-coiling gold nanoparticles bearing lumophores and intercalators: Towards a new generation of multimodal nanocarriers with enhanced action. *Chemical Science, 10*, 9244–9256.

Chan, B. A., & Coward, J. I. (2013). Chemotherapy advances in small-cell lung cancer. *Journal of Thoracic Disease, 5*(Suppl. 5), S565–S578.

Chen, Y., Lei, W., Jiang, G., Hou, Y., Li, C., Zhang, B., et al. (2014). Fusion of photodynamic therapy and photoactivated chemotherapy: A novel Ru(II) arene complex with dual activities of photobinding and photocleavage toward DNA. *Dalton Transactions, 43*, 15375–15384.

Delbaldo, C., Michiels, S., Syz, N., Soria, J. C., Le Chevalier, T., & Pignon, J. P. (2004). Benefits of adding a drug to a single-agent or a 2-agent chemotherapy regimen in advanced non-small-cell lung cancer: A meta-analysis. *Journal of the American Medical Association, 292*, 470–484.

Dilruba, S., & Kalayda, G. V. (2016). Platinum-based drugs: Past, present and future. *Cancer Chemotherapy and Pharmacology, 77*, 1103–1124.

Ellahioui, Y., Prashar, S., & Gómez-Ruiz, S. (2017). Anticancer applications and recent investigations of metallodrugs based on gallium, tin and titanium. *Inorganics, 5*(1), 4.

Evans, W. K., Shepherd, F. A., Feld, R., et al. (1985). VP-16 and cisplatin as first-line therapy for small-cell lung cancer. *Journal of Clinical Oncology, 3*, 1471–1477.

Eyler, C. E., & Rich, J. N. (2008). Survival of the fittest: Cancer stem cells in therapeutic resistance and angiogenesis. *Journal of Clinical Oncology, 26*, 2839–2845.

Fong, J., Kasimova, K., Arenas, Y., Kaspler, P., Lazic, S., Mandel, A., et al. (2015). A novel class of ruthenium-based photosensitizers effectively kills in vitro cancer cells and in vivo tumors. *Photochemical & Photobiological Sciences, 14*, 2014–2023.

Frei, E., Karon, M., Levin, R. H., Freireich, E. J., Taylor, R. J., Hananian, J., et al. (1965). The effectiveness of combinations of antileukemic agents in inducing and maintaining remission in children with acute leukemia. *Blood*, *26*, 642–656.

Goel, S., Duda, D. G., Xu, L., Munn, L. L., Boucher, Y., Fukumura, D., et al. (2011). Normalization of the vasculature for treatment of cancer and other diseases. *Physiological Reviews*, *91*, 1071–1121.

Heffeter, P., Atil, B., Kryeziu, K., Groza, D., Koellensperger, G., Korner, W., et al. (2013). The ruthenium compound KP1339 potentiates the anticancer activity of sorafenib in vitro and in vivo. *European Journal of Cancer*, *49*, 3366–3375.

Iqbal, P. F., Malik, M. A., & Wani, W. A. (2018). Serendipity of cisplatin, and the emergence of metallodrugs in cancer chemotherapy. *World Journal of Pharmacy and Pharmaceutical Sciences*, *7*(4), 654–664.

Kaim, W., & Schwederski, B. (1993). *Bioinorganic chemistry: Inorganic element in the chemistry of life*. Stuttgart: Wiley.

Kang, B. W., Kim, J. G., Kwon, O.-K., Ho, Y. C., & Wansik, Y. (2014). Non-platinum-based chemotherapy for treatment of advanced gastric cancer: 5-fluorouracil, taxanes, and irinotecan. *World Journal of Gastroenterology*, *20*(18), 5396–5402.

Komeda, S., & Casini, A. (2012). Next-generation anticancer metallodrugs. *Current Topics in Medicinal Chemistry*, *12*(3), 219–235. https://doi.org/10.2174/156802612799078964.

Kostova, I. (2006). Platinum complexes as anticancer agents. *Recent Patents on Anti-Cancer Drug Discovery*, *1*(1), 1–22.

LeBaron, S., Zeltzer, L. K., LeBaron, C., Scott, S. E., & Zeltzer, P. M. (1988). Chemotherapy side effects in pediatric oncology patients: Drugs, age, and sex as risk factors. *Medical and Pediatric Oncology*, *16*, 263–268.

Leijen, S., Burgers, S. A., Baas, P., Pluim, D., Tibben, M., van Werkhoven, E., et al. (2015). Phase I/II study with ruthenium compound NAMI-A and gemcitabine in patients with non-small cell lung cancer after first line therapy. *Investigational New Drugs*, *33*, 201–214.

Makovec, T. (2019). Cisplatin and beyond: Molecular mechanisms of action and drug resistance development in cancer chemotherapy. *Radiology and Oncology*, *53*, 148–158.

Malvezzi, M., Carioli, G., Bertuccio, P., Boffetta, P., Levi, F., La Vecchia, C., et al. (2019). European cancer mortality predictions for the year 2019 with focus on breast cancer. *Annals of Oncology*, *30*(5), 781–787.

Marek, L., Ware, K. E., Fritzsche, A., Hercule, P., Helton, W. R., Smith, J. E., et al. (2009). Fibroblast growth factor (FGF) and FGF receptor-mediated autocrine signaling in non-small-cell lung cancer cells. *Molecular Pharmacology*, *75*, 196–207.

Marien, E., Hillen, A., Vanderhoydonc, F., Swinnen, J. V., & Vande Velde, G. (2017). Longitudinal microcomputed tomography-derived biomarkers for lung metastasis detection in a syngeneic mouse model: Added value to bioluminescence imaging. *Laboratory Investigation*, *97*, 24–33.

Miller, A. A., Bogart, J. A., Watson, D. M., et al. (2005). Phase II trial of paclitaxel-topotecan-etoposide (PTE) followed by consolidation chemoradiotherapy for limited stage small cell lung cancer (LS-SCLC): CALGB 30002. *Journal of Clinical Oncology*, *23*(16 Suppl), 662s.

Mjos, K. D., & Orvig, C. (2014). Metallodrugs in medicinal inorganic chemistry. *Chemical Reviews*, *114*(8), 4540–4563.

Mokhtari, R. B., Kumar, S., Islam, S. S., Yazdanpanah, M., Adeli, K., Cutz, E., et al. (2013). Combination of carbonic anhydrase inhibitor, acetazolamide, and sulforaphane, reduces the viability and growth of bronchial carcinoid cell lines. *BMC Cancer*, *13*, 378.

Monneret, C. (2011). Platinum anticancer drugs. From serendipity to rational design. *Annales Pharmaceutiques Françaises*, *69*(6), 286–295.

Monti, D. M., Ferraro, G., & Merlino, A. (2019). Ferritin-based anticancer metallodrug delivery: Crystallographic, analytical and cytotoxicity studies. *Nanomedicine*, *20*, 101997.

Muller, P. A., & Vousden, K. H. (2013). p53 mutations in cancer. *Nature Cell Biology*, *15*, 2–8.

Murthy, N. S., & Mathew, A. (2004). Cancer epidemiology, prevention and control. *Current Science*, *86*, 518–527.

Nandakumar, A. (1990–1996). *Consolidated report of the population based cancer registries*. New Delhi, India: National Cancer Registry Programme, Indian Council of Medical Research.

Ndagi, U., Mhlongo, N., & Soliman, M. E. (2017). Metal complexes in cancer therapy—An update from drug design perspective. *Drug Design, Development and Therapy*, *11*, 599–616.

Orvig, C., & Abrams, M. J. (1999). Medicinal inorganic chemistry: Introduction. *Chemical Reviews*, *99*(9), 2201–2203.

Ott, I., & Gust, R. (2007). Non-platinum metal complexes as anti-cancer drugs. *Archiv der Pharmazie—Chemistry in Life Sciences*, *340*(3), 117–126.

Palmer, A. C., & Sorger, P. K. (2017). Combination cancer therapy can confer benefit via patient-to-patient variability without drug additivity or synergy. *Cell*, *171*(7), 1678–1691.e13.

Parhi, P., Mohanty, C., & Sahoo, S. K. (2012). Nanotechnology-based combinational drug delivery: An emerging approach for cancer therapy. *Drug Discovery Today*, *17*, 1044–1052.

Partridge, A. H., Burstein, H. J., & Winer, E. P. (2001). Side effects of chemotherapy and combined chemohormonal therapy in women with early-stage breast cancer. *Journal of the National Cancer Institute. Monographs*, (30), 135–142.

Quinn, B. A., Dash, R., Sarkar, S., Azab, B., Bhoopathi, P., Das, S. K., et al. (2015). Pancreatic cancer combination therapy using a BH3 mimetic and a synthetic tetracycline. *Cancer Research*, *75*, 2305–2315.

Rao, D. N., & Ganesh, B. (2008). Estimate of cancer incidence in India in 1991. *Indian Journal of Cancer*, *35*, 10–18.

Reichert, J. M., & Wenger, J. B. (2008). Development trends for new cancer therapeutics and vaccines. *Drug Discovery Today*, *13*, 30–37.

Riechelmann, R. P., Tannock, I. F., Wang, L., Saad, E. D., Taback, N. A., & Krzyzanowska, M. K. (2007). Potential drug interactions and duplicate prescriptions among cancer patients. *Journal of the National Cancer Institute*, *99*, 592–600.

Romero-Canelon, I., & Sadler, P. J. (2013). Next-generation metal anticancer complexes: Multitargeting via redox modulation. *Inorganic Chemistry*, *52*(21), 12276–12291.

Rosenberg, B., Van Camp, L., & Krigas, T. (1965). Inhibition of cell division in *Escherichia coli* by electrolysis products from a platinum electrode. *Nature*, *205*, 698–699.

Saleem, K., Wani, W. A., Haque, A., Milhotra, A., & Ali, I. (2013). Nanodrugs: Magic bullets in cancer chemotherapy. In K. Zaman, & Atta-u-Rahman (Eds.), *Topics in anti-cancer research* (pp. 437–494). USA: Bentham Science Publishers.

Siegel, R., Miller, K. D., & Jemal, A. (2018). Cancer statistics, 2019. *CA: A Cancer Journal for Clinicians*, *69*(1), 7–34.

Stiehl, D. P., Bordoli, M. R., Abreu-Rodríguez, I., Wollenick, K., Schraml, P., Gradin, K., et al. (2012). Non-canonical HIF-2alpha function drives autonomous breast cancer cell growth via an AREG-EGFR/ErbB4 autocrine loop. *Oncogene*, *31*, 2283–2297.

Uehara, T., Yamate, J., Torii, M., & Maruyama, T. (2011). Comparative nephrotoxicity of cisplatin and nedaplatin: Mechanisms and histopathological characteristics. *Journal of Toxicologic Pathology*, *24*(2), 87–94.

Wang, D., & Lippard, S. J. (2005). Cellular processing of platinum anticancer drugs. *Nature Reviews. Drug Discovery*, *4*, 307. Berners-Price, S. J.; Appleton, T. G. In Platinum-based drugs in cancer therapy; Kelland, L., Farrell, N. P., Eds.; Humana Press Inc.: Totowa, NJ, 2000; pp 3 – 35.

Wang, T., Narayanaswamy, R., Ren, H., & Torchilin, V. P. (2016). Combination therapy targeting both cancer stem-like cells and bulk tumor cells for improved efficacy of breast cancer treatment. *Cancer Biology & Therapy*, *17*, 698–707.

Wani, W. A., Prashar, S., Shreaz, S., & Gómez-Ruiz, S. (2016). Nanostructured materials functionalized with metal complexes: In search of alternatives for administering anticancer metallodrugs. *Coordination Chemistry Reviews*, *312*, 67–98.

Weiss, A., Bonvin, D., Berndsen, R. H., Scherrer, E., Wong, T. J., Dyson, P. J., et al. (2015). Angiostatic treatment prior to chemo- or photodynamic therapy improves anti-tumor efficacy. *Scientific Reports*, *5*, 8990.

Weiss, R. B., & Christian, M. C. (1993). New cisplatin analogues in development. A review. *Drugs*, *46*(3), 360–377.

Wynder, E. L., Covey, L. S., & Mabuchi, K. (1974). Current smoking habits by selected background variables: Their effect on future disease trends. *American Journal of Epidemiology*, *100*, 168–177.

Yap, T. A., Omlin, A., & de Bono, J. S. (2013). Development of therapeutic combinations targeting major cancer signaling pathways. *Journal of Clinical Oncology*, *31*, 1592–1605.

Yuan, S., Wang, F., Chen, G., Zhang, H., Feng, L., Wang, L., et al. (2013). Effective elimination of cancer stem cells by a novel drug combination strategy. *Stem Cells*, *31*, 23–34.

Zimmermann, G. R., Lehar, J., & Keith, C. T. (2007). Multi-target therapeutics: When the whole is greater than the sum of the parts. *Drug Discovery Today*, *12*, 34–42.

Zou, W. (2005). Immunosuppressive networks in the tumour environment and their therapeutic relevance. *Nature Reviews. Cancer*, *5*, 263–274.

CHAPTER

6

Combination antituberculosis therapy: Opportunities and challenges to combat drug-resistant tuberculosis

Sudarkodi Sukumar[a], *Md. Zafaryab*[b,c], *Md. Khurshid Alam Khan*[a], *Krishnan Hajela*[c], *Mohammad Nasiruddin*[d,e]

[a]School of Life Sciences, BS Abdur Rahman Crescent Institute of Science and Technology, Chennai, Tamil Nadu, India, [b]Department of Zoology, Kirori Mal College, University of Delhi, New Delhi, Delhi, India, [c]School of Life Sciences, Devi Ahilya Vishwavidyalaya, Indore, Madhya Pradesh, India, [d]Clinical Genomics and Molecular Diagnostics, Neuberg Anand Reference Laboratory, Bangalore, Karnataka, India, [e]Molecular Oncology and Transplant Immunology, MedGenome Labs Pvt Ltd., Bangalore, Karnataka, India

Abbreviations

AMI	amikacin
BSL3	biosafety level 3
CAP	capreomycin
DOTS	directly observed therapy
DR	drug resistant
DST	drug susceptibility testing
EMB	ethambutol
FQ	fluoroquinolone
HIV	human immunodeficiency virus
INH	isoniazid
KAN	kanamycin
LTBI	latent tuberculosis infections
MDR	multidrug resistant

https://doi.org/10.1016/B978-0-12-820576-1.00006-0

Mtb	*Mycobacterium tuberculosis*
PAS	*p*-amino salicylic acid
PZA	pyrazinamide
Rif	rifampicin
RR	rifampicin resistant
TAC	thioacetazone
TB	tuberculosis
TDR	totally drug resistant
WHO	World Health Organization
XDR	extensively drug resistant

6.1 Introduction

Tuberculosis (TB) is an ancient disease of humankind caused by *Mycobacterium tuberculosis* (Mtb), which primarily infects the lungs as pulmonary tuberculosis. TB is also capable of infecting other organs and this is known as extrapulmonary TB. The early predecessor of Mtb was estimated to exist in Africa around 3 million years ago (Gutierrez et al., 2005). Though TB is curable and preventable, still we are struggling to completely eradicate this ancient disease. According to the WHO's Global Tuberculosis Report of 2018, in the year 2017, an estimated 10 million people were infected with TB and 1.3 million HIV-negative people lost their lives to TB (WHO, 2018). In 2016 alone, multidrug resistant TB was estimated to cause 240,000 deaths across the globe (WHO, 2017a). Overall, TB is ranked as the ninth leading cause of death in the world and it overtook HIV as the major cause of death due to an infectious agent (WHO, 2017a). Besides these shocking statistics, the fact that 53 million people were cured of TB between 2000 and 2016 gives a slight ray of hope to end the TB epidemic by 2030.

Combination therapy is an important treatment strategy to combat TB; notably, TB was a forerunner in the development of combination drug regimens for other infectious diseases caused by HIV (Cihlar & Fordyce, 2016) and Hepatitis C (Naggie & Muir, 2017). However, the alarming increase in the number of drug-resistant TB cases is constantly causing the credibility of anti-TB combination drug therapy to be questioned. The aim of this chapter is to understand the circumstances under which the current anti-TB combination drug regimen emerged and to analyze the challenges ahead for improving the treatment regimen in the drug-resistant scenario. This chapter is focused only on the treatment methodologies and the associated challenges of adult pulmonary tuberculosis.

6.1.1 History of tuberculosis treatment

Tuberculosis treatment has had an interesting history, from strange potions made of animal and plant parts to isolation of patients in sanatoriums in the hopes of providing clean air and rest so as to eliminate the disease (Barberis, Bragazzi, Galluzzo, & Martini, 2017). The discovery of *M. tuberculosis* (Mtb), the etiological agent of TB, by Robert Koch in 1882 steered TB treatment towards addressing the actual disease cause. Mtb has a unique cell wall structure and massive metabolic capabilities. As a result, Mtb adopts a persistent and highly resistant phenotype in order to escape the host immune attack and/or inhibition by drugs, and multiple drugs are needed to inhibit its growth (Sarathy et al., 2018). These characteristic features make Mtb the most successful pathogen.

In general, the slow-growing nature and the lowered metabolic activities during the prolonged persistent behavior of Mtb make it a difficult target for therapeutics (Gupta, Kumar, Singh, Bisht, & Sharma, 2018). Streptomycin is an aminoglycoside antibiotic that, after its discovery in 1944, was used to examine antimycobacterial activity in vitro (Schatz & Waksman, 1944) and in vivo (Feldman & Hinshaw, 1944). This was followed up with several promising clinical trials, but the trials also revealed resistance development in Mtb against streptomycin monotherapy (Ferguson, Bristow, & Bull, 1949; Willis, 1948). During the same time period, several other studies demonstrated the antimycobacterial activity of *p*-amino salicylic acid (PAS) and these studies also eventually observed resistance development to PAS monotherapy (Lehmann, 1964). This later led to the historic clinical trial investigation by the British Medical Research Council on elucidating the efficiency of streptomycin and PAS combination anti-TB therapy (Medical Research Council, 1950). Thus began the odyssey of using more than one drug to combat resistant TB.

The first curative regimen included the hydrazide of isonicotinic acid, also referred to as isoniazid (INH), along with streptomycin and PAS (Kumagai et al., 1956; Seventh Report to the Medical Research Council, 1955). Since its discovery in 1952, INH has been the hallmark of TB chemotherapy, as it is safer and cheaper, and is the most potent drug to kill actively growing Mtb. INH is a prodrug activated by the catalase peroxidase (KatG) and inhibits enoyl acyl carrier protein reductase (InhA), which in turn inhibits mycolic acid biosynthesis (Lei, Wei, & Tu, 2000). Mycolic acid is the major hydrophobic lipid component of the Mtb cell envelope and is a crucial virulence factor in the progression of TB (Marrakchi, Laneelle, & Daffe, 2014). Apart from this action mechanism, several other concepts are employed to explain the complex mechanistic actions of INH (Unissa, Subbian, Hanna, & Selvakumar, 2016). INH monotherapy also faced resistance development and thus triple therapy (INH, PAS, streptomycin) became a set standard in TB treatment for the next 15 years.

In spite of the triple therapy success, the 18- to 24-months duration of the treatment resulted in several toxic side effects and an increase in the number of resistant cases. This in turn led to a huge proliferation in anti-TB drug-discovery studies (Christopher & Kerntzas, 2017). The development of the modern anti-TB regimen accelerated following the discovery of pyrazinamide (PZA, 1952), ethambutol (EMB, 1961), and rifampicin (Rif, 1963). The WHO has classified these four drugs, including INH, as the first-line anti-TB drugs (refer to Table 6.1 for anti-TB drug classification). PZA, through its active form, pyrazinoic acid, is known to disrupt membrane energetics and transport in Mtb (Zhang, Wade, Scorpio, Zhang, & Sun, 2003). Unlike INH, PZA targets nonreplicating tubercle bacilli and has a potent sterilizing effect, thereby enabling shorter treatment duration (Zhang, Shi, Zhang, & Mitchison, 2014). On the other hand, EMB is effective against actively replicating tubercle bacilli and is an inhibitor of arabinosyl transferases. The inhibition of arabinosyl transferases disrupts the arabinan synthesis, and thereby affects the complex formation between mycolic acids and the 5′-hydroxyl group of D-arabinose, which in synergy with other chemotherapeutic drugs affects the cell wall permeability of Mtb (Takayama & Kilburn, 1989). EMB replaced PAS and shortened the treatment duration to 18 months (Murray, Schraufnagel, & Hopewell, 2015). Further, the introduction of rifampicin reduced the treatment duration significantly, to 9 months (British Thoracic Association, 1982). Rif is the most commonly used rifamycin, which is a term used to collectively refer to rifampicin, rifabutin, and rifapentine. Rif inhibits the beta subunit of RNA polymerases, resulting in impaired RNA synthesis (Campbell et al., 2001) and offers a sterilizing effect along with PZA in TB treatment, thus controlling relapse of TB (Jindani, Dore, & Mitchison, 2003).

TABLE 6.1 Anti-TB drug classification (WHO, 2014).

Classification	Route of administration	Drugs
First line	Oral	Group 1: Isoniazid, rifampicin, pyrazinamide, ethambutol, rifapentine, or rifabutin
Second line	Injectable	Group 2: aminoglycosides and polypeptides: streptomycin, kanamycin, amikacin, capreomycin
	Oral/injectable	Group 3: fluoroquinolones: levofloxacin, moxifloxacin, ofloxacin, gatifloxacin
	Oral	Group 4: *p*-aminosalicylic acid, *p*-amino salicylate sodium cycloserine, terizidone, ethionamide, prothionamide
Third line	Oral	Group 5: bedaquiline, delamanid, linezolid, clofazimine, amoxicillin/clavulanate, imipenem/cilastatin, meropenem, high-dose isoniazid, thioacetazone, clarithromycin

Finally, the current TB regimen for drug-susceptible pulmonary tuberculosis is at least for 6 months. It is recommended in two phases: the first initiation phase comprises the four drugs INH, EMB, Rif, and PZA for 2 months; the second continuation phase is for 4 months with INH and Rif (WHO, 2010). This TB regimen achieves the three important goals of TB treatment: (i) rapid elimination of tubercle bacilli; (ii) elimination of the persistent bacilli to prevent relapse; (iii) prevention of drug-resistance emergence. The initiation phase sets up a balance between the bactericidal action and the sterilizing effects of the first-line drugs, which in turn ensures the minimization of bacterial load. Since PZA is highly effective in the first 2 months, it is restricted only to the initiation phase. The prolonged treatment in the continuation phase ensures relapse-free TB.

Thus the long history of TB treatment has clearly proven the fact that Mtb is capable of developing resistance against all first-line drugs in the monotherapy setting. This was an important lesson, which ensured the compulsory use of combination drugs to treat TB. The administration of the standard TB regimen under directly observed therapy (DOT) has resulted in a more than 80% cure rate among drug-susceptible TB patients (WHO, 2006). Despite the success of DOT, this standard regimen poses several challenges in the form of drug intolerance, toxicity issues, and interaction with other treatment procedures, such as antiretroviral therapy in the case of HIV coinfection with TB, which ultimately led to incomplete treatment adherence. Treatment nonadherence is widely believed to be one of the main reasons for treatment failure, which ultimately led to the epidemic of drug-resistant TB.

6.1.2 Evolution of combination therapy for drug-resistant TB

6.1.2.1 Multidrug-resistant TB management

In 1952, the visionary microbiologist Renee Dubos predicted that bacteria would develop resistance against antibiotics as a consequence of random mutations and natural selection (Moberg, 1996). Mtb is no exception to Dubos's foresight, as Mtb has undergone tremendous evolution leading to different types of drug resistance, such as multidrug resistance (MDR), extensive-drug resistance (XDR), and rifampicin resistance (RR). Mtb strains resistant to more than one drug of the triple therapy candidates INH, streptomycin, and PAS were reported in Great Britain in 1956, leading to the popularization of the term multidrug resistance. Despite such sporadic incidences of multidrug-resistant TB cases, it was largely neglected due to the overwhelming success of the short course DOT therapy.

In 1992, outbreaks of drug-resistant TB cases among HIV-infected people in New York hospitals (Frieden et al., 1993) forced the WHO to declare TB a global emergency in 1994 (WHO, 1994). Since then, MDR has remained the major focus of TB treatment. Nearly 25 years later, in the year 2016, there were an estimated 490,000 people with MDR-TB and the majority of these cases were reported to be in India, China, and the Russian Federation (WHO, 2017a).

There are several causes of drug resistance, including the taxing treatment regimen, magnifying effects of the DOT program, and community-based transmissions. MDR Mtb strains are resistant to both INH and Rif, the two most effective first-line TB drugs, and hence this created the need for other new drugs to tackle them. Random genetic mutation is the primary trigger for drug resistance in Mtb and not horizontal gene transfer, as in other bacteria (Smith, Wolff, & Nguyen, 2013). Chromosomal mutations occur at a frequency of 10^{-6} to 10^{-8} mycobacterial replications, with a probability of 10^{-18} to 10^{-20} of developing drug resistance (Dye & Williams, 2010). The resistance frequency is the highest for INH among all anti-TB drugs (Banerjee et al., 1994) and hence INH resistance has been mainly responsible for the increase in treatment failure and relapse.

Generally, there are three mechanisms by which antibiotic resistance arises: inactivation of the drug-activating enzymes, increased efflux of antibiotics from the bacterial cell, or modification of the drug target (Walsh, 2000). In the case of INH resistance, 40% of all resistant strains harbor the Ser315Thr mutation in KatG, the activating enzyme of INH (Marttila et al., 1998). On the other hand, Rif resistance is caused due to missense mutations in the gene *rpoB*, which codes for the beta subunit of the RNA polymerase enzyme, with the most common changes occurring in Ser531Leu, His526Tyr, and Asp516Val (Herrera, Jimenez, Valverde, Garcia-Aranda, & Saez-Nieto, 2003). According to the latest global surveillance data of the WHO, 19% of previously treated cases of TB were either RR or MDR. The most disturbing fact is that treatment success is less than 50% for MDR-TB, especially in the high burden countries like Indian and China (WHO, 2018).

The first obstacle in treating MDR-TB is the identification of such cases. The standard method used for detecting DR-TB is drug susceptibility testing (DST) using Mtb cultures. This method is time-consuming, extending up to several weeks on account of the slow growth rate of Mtb, which ultimately delays the appropriate treatment administration. Thus there is an unmet need to develop rapid tests for precise identification of DR-TB. Currently, the Xpert MTB/RIF assay is capable of rapidly detecting only RR (Herrera et al., 2003). However, due to a few false positive results of this assay, DST is still recommended for confirmatory results. Thus, DST is crucial in determining the personalized treatment measures to control MDR-TB. But in low-income countries, DST itself is inaccessible and expensive.

On the outset, there are several factors to be considered before choosing drugs for DR-TB, such as the past TB treatment history of the patient, the prevalent drug-resistance pattern in that location, the underlying medical status of the person, and finally the possible side effects of the drugs. The list of second-line drugs categorized by WHO is the only hope left for MDR-TB (refer Table 6.1). The second-line drugs in general are less effective, more toxic, and costlier than first-line drugs (Shehzad, Rehman, Ul-Islam, Khattak, & Lee, 2013). The intensive phase of MDR therapy is for 8 months with at least four of the active second-line drugs, plus any first-line drug given to the patient, and should be continued for 18 to 24 months after obtaining culture negatives (WHO, 2014). If a patient has undergone previous MDR treatment, then the current MDR regimen may be extended up to 28 months.

The second-line drugs are further subdivided into three groups. Group 2 includes injectables of aminoglycosides and polypeptides; group 3 comprises fluoroquinolone, whereas group 4 includes the more toxic oral agents, as listed in Table 6.1. It is interesting to note that the first antimycobacterial, streptomycin, is now classified as a second-line drug because it is an injectable, placing increased burden on patients by requiring many hospital visits for its administration. Streptomycin is recommended for MDR-TB only if the Mtb strain is found to be susceptible. Similar to streptomycin (Moazed & Noller, 1987) are the other second-line aminoglycosides like kanamycin (KAN) and amikacin (AMI); the cyclic peptides, such as capreomycin (CAP) and viomycin, also inhibit bacterial protein synthesis (Palomino & Martin, 2014). The mechanism of action of group 3 fluoroquinolones (FQs) is to inhibit DNA topoisomerases II (Fabrega, Madurga, Giralt, & Vila, 2009). Moxifloxacin and levofloxacin are the actively used FQs in MDR treatment. PAS, once a first-line drug, is now classified under group 4 and is used in MDR treatment. The mode of action of PAS is by inhibiting the enzymatic activity of dihydrofolate reductase of Mtb (Zheng et al., 2013). Cycloserine, another group 4 drug, is also used in MDR regimens and it possibly interferes with the peptidoglycan synthesis of Mtb (Zhang, 2005). Thioacetazone (TAC), also known as thiosemicarbazone, inhibits the cyclopropane mycolic acid synthases of Mtb (Alahari et al., 2007). It has been used in TB treatment since 1960 and is a rather old drug. TAC is always used in combination with other first-line drugs or against DR-TB, as it is less effective and is mainly used for preventing other first-line drug resistance (Falzon, Hill, Pal, Suwankesawong, & Jaramillo, 2014). Due to its severe side effects, even causing death among HIV coinfected TB patients, it is not widely used (Elliott & Foster, 1996). Ethionamide is an important second-line drug similar to INH and also inhibits mycolic acid synthesis (Tsukamura, Tsukamura, Mizuno, & Nakano, 1964).

Finally, the third-line drugs, also referred to as group 5, consist of oral agents like clofazimine, linezolid, amoxicillin plus clavulanate, imipenem plus cilastatin, and clarithromycin, whose role in TB treatment is incompletely defined. In general, group 5 and group 4 drugs are used as part of an MDR-TB treatment regimen only when active second-line drugs are unavailable, on a case-by-case basis. Linezolid is the first oxazolidinone to be approved for tuberculosis treatment and it acts by inhibiting protein synthesis (Richter, Rusch-Gerdes, & Hillemann, 2007). Clofazimine is recommended for DR-TB; it was traditionally used to treat leprosy through an unknown mechanism of action (Dookie, Rambaran, Padayatchi, Mahomed, & Naidoo, 2018). Other than MDR strains, there are polyresistant Mtb strains that are resistant to different combinations of first-line drugs, excluding the INH and Rif combination. Currently, polydrug-resistant strains are managed with other first-line drugs to which they are susceptible, and with some second-line drugs like FQ and KAN/AMI/CAP, for up to 18 months (WHO, 2014).

In summary, treating MDR-TB is complicated, for several reasons: (i) the second- and third-line drugs are more expensive than the first-line drugs and hence there is limited access in resource-deprived nations; (ii) DST procedures are not standardized for the second-line drugs in such countries, which further delays the treatment initiation; (iii) even after early treatment recommendations, the serious toxicity leads to treatment withdrawals and can even prove fatal in immune-compromised individuals.

6.1.2.2 *Extensive drug-resistant TB management*

Extensive drug-resistant TB (XDR-TB) is another form of MDR-TB, seen in strains with resistance to FQ and any of the second-line injectables such as CAP or KAN or AMI. The XDR-TB condition was first described in 2005 at the US Center for Disease Control and Prevention (CDC) (Shah, Wright, & Drobniewski, 2005). Following this, there were several reports of XDR-TB and of significance was the outbreak among HIV coinfected people in South Africa in 2006 (Moll et al., 2006). The main reason behind the emergence of XDR-TB is incomplete treatment adherence by MDR-TB patients, which has enabled those strains to accumulate additional resistance to other antibiotics. There is a high risk of developing XDR-TB in patients with inappropriately treated active TB. The risk further increases if the infected people are living in poorly ventilated, crowded residences and hospitals. According to the WHO 2016 update, 123 member countries have reported XDR-TB and a matter of concern is that many XDR-TB patients remain undetected. Further, 9% of the MDR-TB cases over the course of their life develop XDR-TB (WHO, 2013). Diagnosing and detecting XDR-TB is difficult, even in developed countries, extending up to several weeks. Usually a failed MDR treatment regimen is a strong predictor of underlying XDR-TB (WHO, 2014). An expert should advise personalized treatment with the existing second-line drugs after considering the severity of the disease and the immune status of the patient, as mortality could be higher in HIV-infected patients. Thus there is no standard regimen to treat XDR-TB and six or more active drugs may be used for treatment (Falzon et al., 2011). In many cases, treatment success is not guaranteed, but the chances can be improved with early diagnosis and appropriate treatment regimens.

6.1.2.3 *Totally drug-resistant TB management*

Globally, controlling MDR- and XDR-TB is a huge challenge and success remains elusive. Now, totally drug-resistant (TDR) TB has emerged, defined as Mtb strains that are resistant to all first- and second-line drugs. It was first reported in Italy in 2007 (Migliori, De Iaco, Besozzi, Centis, & Cirillo, 2007) shortly after the CDC description of MDR-TB. Soon there were reports of such cases in Iran (Velayati et al., 2009) and India as well (Udwadia, Amale, Ajbani, & Rodrigues, 2012). The definition of TDR is not completely approved by the WHO, due to reported inaccuracy in DST results for second-line drugs and the associated misinterpretations in the resistance definition for existing drugs (Centers for Disease Control and Prevention, 2006). Currently, it is proposed to use new drugs on compassionate grounds in the extreme cases resembling TDR-TB (WHO, 2012).

6.2 Challenges and opportunities for TB combination therapy

6.2.1 Complexity of Mtb biology

The difficult challenge of achieving a more effective combination therapy for TB is vested in the complicated physiology of Mtb (Delogu, Sali, & Fadda, 2013). Deeper understanding of the biology of Mtb was achieved after the publishing of its genome in 2000 by Cole et al. (1998). Mtb exists as a heterogeneous population in the host. Once Mtb infects a human host, the professional phagocytes in the lung engulf them. Usually our innate immune system is capable of eliminating the Mtb strains; however, any compromise of the immune functions could result in active tuberculosis. Meanwhile, Mtb takes shelter in all the neighboring cells and eventually attains dormancy in caseating granulomas, the hallmark of TB infection (Orme & Basaraba, 2014). Latent TB infections (LTBIs) do not have visible symptoms, unlike active TB, but in unfortunate circumstances, such as the weakening of the immune system, could lead to active TB. In the dormant state, Mtb is metabolically inactive and, further, the lipid rich cell envelopes of Mtb makes it difficult for drugs to permeate and eliminate them (Gengenbacher & Kaufmann, 2012).

Thus the multiple physiological phenotypes of Mtb have always been a crucial factor in deciding the course of a combination therapy. The standard regimen recommended for active tuberculosis is for 6 months duration, in order to kill every possible Mtb strain that would be hidden in the different host cells. With the current DR-TB epidemic, it is necessary to discover new drugs and assess their efficiency in eliminating the varied physiological states of Mtb. The slow growth rate of Mtb has always been a huge obstacle, not just in TB diagnosis (Nema, 2012) and DST, but also as a serious hampering factor in conducting research studies on Mtb. The current Mtb culturing methodology takes 4–6 weeks and a DST study may extend up to 8 weeks (Samra, Kaufman, Bechor, & Bahar, 2000). Further, Mtb is a contagious respiratory pathogen and hence it has to be cultured in strictly protected conditions, such as a biosafety level 3 (BSL3) facility, which is expensive and is a discouraging factor in the progress of TB research.

In this scenario, for some time many researchers have resorted to the use of surrogate models of Mtb, such as the rapid grower *Mycobacterium smegmatis* and the bovine TB-causing *Mycobacterium bovis*. However, discussions on the credibility of using such surrogate models are never-ending (Altaf, Miller, Bellows, & O'Toole, 2010; Reyrat & Kahn, 2001). Although the contributions from these models toward TB understanding can never be disregarded, there are several ongoing efforts to discover better models to study TB (Gupta & Bhakta, 2012; Phelan et al., 2015). Also, scientists involved in drug-discovery research have developed various innovative screening techniques to overcome the inherent limitations of Mtb culture systems (Zuniga, Early, & Parish, 2015). On the other hand, the complications are multiplied in replicating the dormant state of Mtb in laboratory conditions, because understanding the Mtb interaction with the host in its state of dormancy is central for identifying novel drug targets (Gengenbacher & Kaufmann, 2012). Recent works on optimizing shortened culture periods for Mtb look promising and might provide solutions to the previously mentioned challenges put forth by the complex physiology of Mtb (Ghodbane, Raoult, & Drancourt, 2014).

6.2.2 Host-pathogen coevolution: The new paradigm

One other upcoming challenge is the emergence of varied strains of Mtb across the different regions of the world. These genotypically and phenotypically varied Mtb strains are grouped into seven lineages (Gagneux, 2018). Numerous studies have deciphered the differences among these lineages in growth patterns (Banuls, Sanou, Anh, & Godreuil, 2015), profiles of virulence factors (Portevin et al., 2014; Reiling et al., 2013), drug susceptibility, and, importantly, their differential immune responses in the host (Coscolla, 2017; Tientcheu et al., 2017). Mtb H37Rv and Erdman strains are the conventional model strains in tuberculosis research. But studies based on these strains that belong to Lineage 4 may not be representative of the physiological status of the other six lineages. Hence, it is crucial to identify reference strains in all the seven lineages to enable the meaningful translation of research findings to clinical settings. Accurate interpretation of such lineage-based studies could be pivotal in shaping the personalized treatment regimens for the Mtb strains from different lineages (Manson et al., 2017; Panwalkar, Chauhan, & Desikan, 2017).

Apart from the pathogen evolution into different lineages, there is a parallel coevolution in the interacting host as well (Gagneux, 2012) and, further, the host cells are instrumental in shaping the genetic diversity of Mtb (Copin et al., 2016). Thus this newfound challenge is a blessing in disguise, as it helps in achieving the future goals of TB treatment by enabling the determination of novel drug targets in TB. Moreover, with an increase in the understanding of host-centric roles in the pathogenesis of TB, several studies have explored new host-centric adjuvant treatment approaches to combat Mtb. Thus host-centric therapeutics represents an exciting venture in TB drug discovery, with immense potential to be considered in designing novel anti-TB combination therapeutic strategies.

6.2.3 Systems biology solutions for tuberculosis

Noteworthy advancements in genomics and proteomics technology have influenced the research on many human diseases and TB is not an exception to such an influence. This is evident in the number of publications involving proteomics/genomics in TB, which has nearly doubled since the last decade. Though this can be considered as a positive trend in our efforts to obtain an improved understanding of TB pathogenesis, on the other hand these studies can lead to information overload. It is indispensable to ensure best utilization of these datasets in deciphering novel therapeutic targets for TB.

Computational biologists have come to the rescue to make biological sense of these data. Several computational models have analyzed the global as well as the local metabolic states of Mtb (Rienksma, Schaap, Martins dos Santos, & Suarez-Diez, 2018). Most importantly, systems biology investigations of chemo-genomics data are enabling design of new combination drug regimens (Chandrasekaran et al., 2016). Even the establishment of integrated platforms by different research groups to facilitate analysis and visualization of the omics data is promising (Kapopoulou, Lew, & Cole, 2011; Reddy et al., 2009). However, such platforms are underutilized when compared to cBioPortal and other cancer-related analysis platforms. Thus more interdisciplinary science collaboration should be initiated to enrich data exchange and valid interpretations in the TB research community to aid the drug discovery pipelines.

6.2.4 Global initiatives to accelerate TB drug discovery pipelines

The World Health Organization (WHO), through its End TB strategy, aims at not just controlling TB but eliminating it as a public health concern worldwide by 2050. Research and development (R&D) are important pillars of the End TB strategy. To achieve this, WHO calls for fundamental research focusing on new drug regimens, efficient diagnostic kits, and safe vaccines. Further, WHO has developed a global TB strategy to ensure a targeted and effective approach spanning different directions in the world. At this juncture, joint efforts by scientists from interdisciplinary fields would definitely be a step in the right direction, as envisioned by the WHO. With this viewpoint, it is encouraging to witness several united efforts across the globe that are specifically targeted on identifying new drug candidates for TB (Table 6.2). One such global initiative is the Open Source Drug Discovery pipeline (OSDD), which has successfully identified 10 new drug targets in TB with research inputs from a wide network of students, clinicians, and researchers. The Working Group on New TB drugs along with the StopTB partnership is playing a crucial role in streamlining the clinical phase trials of over 20 drugs, including newly discovered hopeful drugs such as delamanid and bedaquiline (refer to Table 6.3 for the WHO list of drugs in clinical development). Thus these

TABLE 6.2 Global initiatives to accelerate TB drug discovery.

Global tuberculosis initiatives	URL
Open Source Drug Discovery	http://www.osdd.net
StopTB Partnership	http://www.stoptb.org
Working Group on New TB Drugs	http://www.newtbdrugs.org
TB Alliance	http://www.tballiance.org
More Medicines for TB	http://www.mm4tb.org
The Tuberculosis TB Consortium	http://www.tbtrialsnetwork.org

TABLE 6.3 Antibacterial products in clinical development for Mtb, as of January 2018 reported by WHO (2017c).

Phase I	Phase II	Phase III
GSK-3036656 OPC-167832 Q203 TBA-7371	Delpazolid (LC601-0371) Levofloxacin SQ109 Sutezolid/PNU-100480 PBTZ-169	Bedaquiline Pretomanid (PA-824)

global initiatives should be truly global and involve the nations with high TB burden in order to realize the goal of achieving a shorter duration, cheaper, and more effective combination drug regimen, to eliminate all forms of TB.

6.3 Conclusion

WHO is putting its best foot forward to achieve the 2035 End TB strategy of reducing TB deaths by 95% and decreasing new TB cases by 90% (WHO, 2015). WHO, along with all the concerned countries, must unite in their fight against TB against all odds. The lack of effective prophylactic measures, insufficient treatment regimens, limited resources in heavily burdened nations, ever-increasing DR-TB cases, gaps in funding policies worldwide, and the immature control measures of certain countries with TB epidemics are all part of an exhausting list of factors leading to the global pandemic of TB. Combination therapy is a well-proven treatment strategy for TB and, hence, it is a strong weapon that needs to be sharpened further, not just to treat TB but to completely eliminate TB from the world. This will become a reality if there is a concerted effort on both the global level and national level to ensure proper TB disease management, improved research funding, and fruitful collaborations between industry and academia.

References

Alahari, A., Trivelli, X., Guérardel, Y., Dover, L. G., Besra, G. S., Sacchettini, J. C., et al. (2007). Thiacetazone, an antitubercular drug that inhibits cyclopropanation of cell wall mycolic acids in mycobacteria. *PLoS One*, *2*(12).

Altaf, M., Miller, C. H., Bellows, D. S., & O'Toole, R. (2010). Evaluation of the *Mycobacterium smegmatis* and BCG models for the discovery of *Mycobacterium tuberculosis* inhibitors. *Tuberculosis (Edinburgh, Scotland)*, *90*(6), 333–337.

Banerjee, A., Dubnau, E., Quemard, A., Balasubramanian, V., Um, K. S., Wilson, T., et al. (1994). inhA, a gene encoding a target for isoniazid and ethionamide in *Mycobacterium tuberculosis*. *Science*, *263*(5144), 227–230.

Banuls, A. L., Sanou, A., Anh, N. T., & Godreuil, S. (2015). *Mycobacterium tuberculosis*: Ecology and evolution of a human bacterium. *Journal of Medical Microbiology*, *64*(11), 1261–1269.

Barberis, I., Bragazzi, N. L., Galluzzo, L., & Martini, M. (2017). The history of tuberculosis: From the first historical records to the isolation of Koch's bacillus. *Journal of Preventive Medicine and Hygiene*, *58*(1), E9–E12.

British Thoracic Association. (1982). A controlled trial of six months chemotherapy in pulmonary tuberculosis. Second report: Results during the 24 months after the end of chemotherapy. British Thoracic Association. *The American Review of Respiratory Disease*, *126*(3), 460–462.

Campbell, E. A., Korzheva, N., Mustaev, A., Murakami, K., Nair, S., Goldfarb, A., et al. (2001). Structural mechanism for rifampicin inhibition of bacterial RNA polymerase. *Cell*, *104*(6), 901–912.

Centers for Disease Control and Prevention. (2006). Emergence of *Mycobacterium tuberculosis* with extensive resistance to second-line drugs worldwide. *MMWR. Morbidity and Mortality Weekly Report*, *55*, 301–305.

Chandrasekaran, S., Cokol-Cakmak, M., Sahin, N., Yilancioglu, K., Kazan, H., Collins, J. J., et al. (2016). Chemogenomics and orthology-based design of antibiotic combination therapies. *Molecular Systems Biology*, *12*(5), 872.

Christopher, A., & Kerntzas, W. R. J. (2017). Origins of combination therapy for tuberculosis: Lessons for future antimicrobial development and application. *American Society of Microbiology*, *8*(2). mBio:01586-01516.

Cihlar, T., & Fordyce, M. (2016). Current status and prospects of HIV treatment. *Current Opinion in Virology*, *18*, 50–56.

Cole, S. T., Brosch, R., Parkhill, J., Garnier, T., Churcher, C., Harris, D., et al. (1998). Deciphering the biology of *Mycobacterium tuberculosis* from the complete genome sequence. *Nature*, *393*(6685), 537–544.

Copin, R., Wang, X., Louie, E., Escuyer, V., Coscolla, M., Gagneux, S., et al. (2016). Within host evolution selects for a dominant genotype of *Mycobacterium tuberculosis* while T cells increase pathogen genetic diversity. *PLoS Pathogens*, *12*, e1006111.

Coscolla, M. (2017). Biological and epidemiological consequences of MTBC diversity. *Advances in Experimental Medicine and Biology*, *1019*, 95–116.

Delogu, G., Sali, M., & Fadda, G. (2013). The biology of *Mycobacterium tuberculosis* infection. *Mediterranean Journal of Hematology and Infectious Diseases*, *5*(1), e2013070.

Dookie, N., Rambaran, S., Padayatchi, N., Mahomed, S., & Naidoo, K. (2018). Evolution of drug resistance in *Mycobacterium tuberculosis*: A review on the molecular determinants of resistance and implications for personalized care. *The Journal of Antimicrobial Chemotherapy*, *73*(5), 1138–1151.

Dye, C., & Williams, B. G. (2010). The population dynamics and control of tuberculosis. *Science*, *328*(5980), 856–861.

Elliott, A. M., & Foster, S. D. (1996). Thiacetazone: Time to call a halt? Considerations on the use of thiacetazone in African populations with a high prevalence of human immunodeficiency virus infection. *Tubercle and Lung Disease*, *77*(1), 27–29.

Fabrega, A., Madurga, S., Giralt, E., & Vila, J. (2009). Mechanism of action of and resistance to quinolones. *Microbial Biotechnology*, *2*(1), 40–61.

Falzon, D., Hill, G., Pal, S. N., Suwankesawong, W., & Jaramillo, E. (2014). Pharmacovigilance and tuberculosis: Applying the lessons of thioacetazone. *Bulletin of the World Health Organization*, *92*(12), 918–919.

Falzon, D., Jaramillo, E., Schunemann, H. J., Arentz, M., Bauer, M., Bayona, J., et al. (2011). WHO guidelines for the programmatic management of drug-resistant tuberculosis: 2011 update. *The European Respiratory Journal*, *38*(3), 516–528.

Feldman, W. H., & Hinshaw, H. (1944). Effects of streptomycin on experimental tuberculosis in Guinea pigs: A preliminary report. *Proceedings of the Staff Meetings. Mayo Clinic*, *19*, 593–599.

Ferguson, J. B., Bristow, V. G., & Bull, P. R. (1949). Streptomycin in tuberculosis. *Medical Journal of Australia*, *2*(6), 192–194.

Frieden, T. R., Sterling, T., Pablos-Mendez, A., Kilburn, J. O., Cauthen, G. M., & Dooley, S. W. (1993). The emergence of drug-resistant tuberculosis in New York City. *The New England Journal of Medicine*, *328*(8), 521–526.

Gagneux, S. (2012). Host-pathogen coevolution in human tuberculosis. *Philosophical Transactions of the Royal Society of London. Series B, Biological Sciences*, *367*(1590), 850–859.

Gagneux, S. (2018). Ecology and evolution of *Mycobacterium tuberculosis*. *Nature Reviews. Microbiology*, *16*(4), 202–213.

Gengenbacher, M., & Kaufmann, S. H. E. (2012). *Mycobacterium tuberculosis*: Success through dormancy. *FEMS Microbiology Reviews*, *36*, 3.

Ghodbane, R., Raoult, D., & Drancourt, M. (2014). Dramatic reduction of culture time of *Mycobacterium tuberculosis*. *Scientific Reports*, *4*, 4236.

Gupta, A., & Bhakta, S. (2012). An integrated surrogate model for screening of drugs against *Mycobacterium tuberculosis*. *The Journal of Antimicrobial Chemotherapy*, *67*(6), 1380–1391.

Gupta, V. K., Kumar, M., Singh, D., Bisht, D., & Sharma, S. (2018). Drug targets in dormant *Mycobacterium tuberculosis*: Can the conquest against tuberculosis become a reality? *Infectious Diseases*, *50*(2).

Gutierrez, M. C., Brisse, S., Brosch, R., Fabre, M., Omais, B., Marmiesse, M., et al. (2005). Ancient origin and gene mosaicism of the progenitor of *Mycobacterium tuberculosis*. *PLoS Pathogens*, *1*(1), e5.

Herrera, L., Jimenez, S., Valverde, A., Garcia-Aranda, M. A., & Saez-Nieto, J. A. (2003). Molecular analysis of rifampicin-resistant *Mycobacterium tuberculosis* isolated in Spain (1996-2001). Description of new mutations in the rpoB gene and review of the literature. *International Journal of Antimicrobial Agents*, *21*(5), 403–408.

Jindani, A., Dore, C. J., & Mitchison, D. A. (2003). Bactericidal and sterilizing activities of antituberculosis drugs during the first 14 days. *American Journal of Respiratory and Critical Care Medicine*, *167*(10), 1348–1354.

Kapopoulou, A., Lew, J. M., & Cole, S. T. (2011). The MycoBrowser portal: A comprehensive and manually annotated resource for mycobacterial genomes. *Tuberculosis (Edinburgh, Scotland)*, *91*(1), 8–13.

Kumagai, T., Oka, S., Suzuki, C., Kurobane, T., Kanno, I., Sato, S., et al. (1956). Treatment of pulmonary tuberculosis on a combined regimen of para-amino-salicylic acid, streptomycin and isonicotinic acid hydrazide. *The Science Reports of the Research Institutes, Tohoku University. Series C, Medicine*, *6*(4), 401–432.

Lehmann, J. (1964). Twenty years afterward historical notes on the discovery of the antituberculosis effect of paraaminosalicylic acid (PAS) and the first clinical trials. *The American Review of Respiratory Disease*, *90*, 953–956.

Lei, B., Wei, C. J., & Tu, S. C. (2000). Action mechanism of antitubercular isoniazid. Activation by *Mycobacterium tuberculosis* KatG, isolation, and characterization of inha inhibitor. *The Journal of Biological Chemistry*, *275*(4), 2520–2526.

Manson, A. L., Abeel, T., Galagan, J. E., Sundaramurthi, J. C., Salazar, A., Gehrmann, T., et al. (2017). *Mycobacterium tuberculosis* whole genome sequences from Southern India suggest novel resistance mechanisms and the need for region-specific diagnostics. *Clinical Infectious Diseases*, *64*(11), 1494–1501.

Marrakchi, H., Laneelle, M. A., & Daffe, M. (2014). Mycolic acids: Structures, biosynthesis, and beyond. *Chemistry & Biology*, *21*(1), 67–85.

Marttila, H. J., Soini, H., Eerola, E., Vyshnevskaya, E., Vyshnevskiy, B., Otten, T. F., et al. (1998). A Ser315Thr substitution in KatG is predominant in genetically heterogeneous multidrug-resistant *Mycobacterium tuberculosis* isolates originating from the St. Petersburg area in Russia. *Antimicrobial Agents and Chemotherapy*, *42*(9), 2443–2445.

Medical Research Council. (1950). Treatment of pulmonary tuberculosis with streptomycin and para-aminosalicylic acid; A Medical Research Council investigation. *British Medical Journal*, *2*(4688), 1073–1085.

Migliori, G. B., De Iaco, G., Besozzi, G., Centis, R., & Cirillo, D. M. (2007). First tuberculosis cases in Italy resistant to all tested drugs. *Euro Surveillance*, *12*(5), E070517.1.

Moazed, D., & Noller, H. F. (1987). Interaction of antibiotics with functional sites in 16S ribosomal RNA. *Nature*, *327*(6121), 389–394.

Moberg, C. L. (1996). Rene Dubos: A harbinger of microbial resistance to antibiotics. *Microbial Drug Resistance*, *2*(3), 287–297.

Moll, A., Gandhi, N., Pawinski, R., Andrews, J., Zeller, K., Sturm, A. W., et al. (2006). Identification of a multidrug-resistant tuberculosis cluster as a cause of death among HIV co-infected patients in rural South Africa. In *13th Conference on Retroviruses and Opportunistic Infections; 7 February 2006; Denver, Colorado, United States of America. 2006.*

Murray, J. F., Schraufnagel, D. E., & Hopewell, P. C. (2015). Treatment of tuberculosis. A historical perspective. *Annals of the American Thoracic Society*, *12*(12), 1749–1759.

Naggie, S., & Muir, A. J. (2017). Oral combination therapies for hepatitis C virus infection: Successes, challenges, and unmet needs. *Annual Review of Medicine*, *68*, 345–358.

Nema, V. (2012). Tuberculosis diagnostics: Challenges and opportunities. *Lung India*, *29*(3), 259–266.

Orme, I. M., & Basaraba, R. J. (2014). The formation of the granuloma in tuberculosis infection. *Seminars in Immunology*, *26*(6), 601–609.

Palomino, J. C., & Martin, A. (2014). Drug resistance mechanisms in *Mycobacterium tuberculosis*. *Antibiotics (Basel)*, *3*(3), 317–340.

Panwalkar, N., Chauhan, D. S., & Desikan, P. (2017). Spoligotype defined lineages of *Mycobacterium tuberculosis* and drug resistance: Merely a casual correlation? *Indian Journal of Medical Microbiology*, *35*(1), 27–32.

Phelan, J., Maitra, A., McNerney, R., Nair, M., Gupta, A., Coll, F., et al. (2015). The draft genome of *Mycobacterium aurum*, a potential model organism for investigating drugs against *Mycobacterium tuberculosis* and *Mycobacterium leprae*. *International Journal of Mycobacteriology*, *4*(3), 207–216.

Portevin, D., Sukumar, S., Coscolla, M., Shui, G., Li, B., Guan, X. L., et al. (2014). Lipidomics and genomics of *Mycobacterium tuberculosis* reveal lineage-specific trends in mycolic acid biosynthesis. *Microbiology*, *3*(6), 823–835.

Reddy, T. B., Riley, R., Wymore, F., Montgomery, P., DeCaprio, D., Engels, R., et al. (2009). TB database: An integrated platform for tuberculosis research. *Nucleic Acids Research*, *37*(Database issue), D499–D508.

Reiling, N., Homolka, S., Walter, K., Brandenburg, J., Niwinski, L., Ernst, M., et al. (2013). Clade-specific virulence patterns of *Mycobacterium tuberculosis* complex strains in human primary macrophages and aerogenically infected mice. *MBio*, *4*(4).

Reyrat, J. M., & Kahn, D. (2001). *Mycobacterium smegmatis*: An absurd model for tuberculosis? *Trends in Microbiology*, *9*(10), 472–474.

Richter, E., Rusch-Gerdes, S., & Hillemann, D. (2007). First linezolid-resistant clinical isolates of *Mycobacterium tuberculosis*. *Antimicrobial Agents and Chemotherapy*, *51*(4), 1534–1536.

Rienksma, R. A., Schaap, P. J., Martins dos Santos, V. A. P., & Suarez-Diez, M. (2018). Modeling the metabolic state of *Mycobacterium tuberculosis* upon infection. *Frontiers in Cellular and Infection Microbiology*, *8*, 264.

Samra, Z., Kaufman, L., Bechor, J., & Bahar, J. (2000). Comparative study of three culture systems for optimal recovery of mycobacteria from different clinical specimens. *European Journal of Clinical Microbiology & Infectious Diseases*, *19*(10), 750–754.

Sarathy, J. P., Via, L. E., Weiner, D., Blanc, L., Boshoff, H., Eugenin, E. A., et al. (2018). Extreme drug tolerance of *Mycobacterium tuberculosis* in Caseum. *Antimicrobial Agents and Chemotherapy*, *62*(2).

Schatz, A., & Waksman, S. (1944). Effect of streptomycin and other antibiotic substances upon *Mycobacterium tuberculosis* and related organisms. *Experimental Biology and Medicine*, *57*, 244–248.

Seventh Report to the Medical Research Council. (1955). Various combinations of isoniazid with streptomycin or with P.A.S. in the treatment of pulmonary tuberculosis. *British Medical Journal*, *1*(4911), 435–445.

Shah, N. S., Wright, A., & Drobniewski, F. (2005). Extreme drug resistance in tuberculosis (XDR-TB): Global survey of supranational reference laboratories for *Mycobacterium tuberculosis* with resistance to second-line drugs. *The International Journal of Tuberculosis and Lung Disease*, *9*, S77.

Shehzad, A., Rehman, G., Ul-Islam, M., Khattak, W. A., & Lee, Y. S. (2013). Challenges in the development of drugs for the treatment of tuberculosis. *The Brazilian Journal of Infectious Diseases*, *17*(1), 74–81.

Smith, T., Wolff, K. A., & Nguyen, L. (2013). Molecular biology of drug resistance in *Mycobacterium tuberculosis*. *Current Topics in Microbiology and Immunology*, *374*, 53–80.

Takayama, K., & Kilburn, J. O. (1989). Inhibition of synthesis of arabinogalactan by ethambutol in *Mycobacterium smegmatis*. *Antimicrobial Agents and Chemotherapy*, *33*(9), 1493–1499.
Tientcheu, L. D., Koch, A., Ndengane, M., Andoseh, G., Kampmann, B., & Wilkinson, R. J. (2017). Immunological consequences of strain variation within the *Mycobacterium tuberculosis* complex. *European Journal of Immunology*, *47*(3), 432–445.
Tsukamura, M., Tsukamura, S., Mizuno, S., & Nakano, E. (1964). The mechanism of action of ethionamide. *The American Review of Respiratory Disease*, *89*, 933–935.
Udwadia, Z. F., Amale, R. A., Ajbani, K. K., & Rodrigues, C. (2012). Totally drug-resistant tuberculosis in India. *Clinical Infectious Diseases*, *54*(4), 579–581.
Unissa, A. N., Subbian, S., Hanna, L. E., & Selvakumar, N. (2016). Overview on mechanisms of isoniazid action and resistance in *Mycobacterium tuberculosis*. *Infection, Genetics and Evolution*, *45*, 474–492.
Velayati, A. A., Masjedi, M. R., Farnia, P., Tabarsi, P., Ghanavi, J., ZiaZarifi, A. H., et al. (2009). Emergence of new forms of totally drug-resistant tuberculosis bacilli: Super extensively drug-resistant tuberculosis or totally drug-resistant strains in Iran. *Chest*, *136*(2), 420–425.
Walsh, C. (2000). Molecular mechanisms that confer antibacterial drug resistance. *Nature*, *405*, 775–781.
WHO. (1994). *TB: A global emergency, WHO report, 1994*. Geneva: WHO [WHO/TB/94.177].
WHO. (2006). *Global tuberculosis control: Surveillance, planning, financing: WHO report 2006*. Geneva: WHO [WHO/HTM/TB/2006.362].
WHO (Ed.), (2010). *Treatment of tuberculosis: Guidelines* (4th ed.). Geneva: WHO [WHO/HTM/TB/2009.420].
WHO. (2012). *"Totally Drug-Resistant TB": A WHO consultation onthediagnostic definition and treatment options*. Dates: 21–22 March 2012 Geneva: WHO/HQ.
WHO. (2013). *Multidrug-resistant tuberculosis (MDR-TB) 2013 update*. Geneva: WHO [PDF) (Retrieved 14 June 2013].
WHO. (2014). *Companion handbook to the WHO guidelines for the programmatic management of drug-resistant tuberculosis*. Geneva: WHO [WHO/HTM/TB/2014.11].
WHO. (2015). *The END TB strategy*. Geneva: WHO [WHO/HTM/TB/2015.19].
WHO. (2017a). *Global tuberculosis report 2017*. Geneva: WHO [licence: CC BY-NC-SA 3.0 IGO].
WHO. (2017c). *Antibacterial agents in clinical development: An analysis of the antibacterial clinical development pipeline, including tuberculosis*. Geneva: WHO [WHO/EMP/IAU/2017.12, licence: CC BY-NC-SA 3.0 IGO].
WHO. (2018). *Global tuberculosis report 2018*. Geneva: WHO [licence: CC BY-NC-SA 3.0 IGO].
Willis, H. S. (1948). Streptomycin in tuberculosis. *Radiology*, *50*, 6.
Zhang, Y. (2005). The magic bullets and tuberculosis drug targets. *Annual Review of Pharmacology and Toxicology*, *45*, 529–564.
Zhang, Y., Shi, W., Zhang, W., & Mitchison, D. (2014). Mechanisms of pyrazinamide action and resistance. *Microbiology Spectrum*, *2*(4), MGM2-0023-2013.
Zhang, Y., Wade, M. M., Scorpio, A., Zhang, H., & Sun, Z. (2003). Mode of action of pyrazinamide: Disruption of *Mycobacterium tuberculosis* membrane transport and energetics by pyrazinoic acid. *The Journal of Antimicrobial Chemotherapy*, *52*(5), 790–795.
Zheng, J., Rubin, E., Bifani, P., Mathys, V., Lim, V., Au, M., et al. (2013). Para-aminosalicylic acid is a prodrug targeting dihydrofolate reductase in *Mycobacterium tuberculosis*. *The Journal of Biological Chemistry*, *288*(32), 23447–23456.
Zuniga, E. S., Early, J., & Parish, T. (2015). The future for early-stage tuberculosis drug discovery. *Future Microbiology*, *10*(2), 217–229.

Further reading

WHO. (2017b). *Guidelines for treatment of drug-susceptible tuberculosis and patient care (2017 update)*. Geneva: WHO [WHO/HTM/TB/2017.05].

CHAPTER

7

Synergistic effect of drugs against multiple drug-resistant swine pathogen *Streptococcus suis*

Shama Khan

Department of Clinical Microbiology and Infectious Diseases, School of Pathology, University of Witwatersrand, Johannesburg, South Africa

Abbreviations

DNA	deoxyribonucleic acid
EMA	European Medicines Agency
ESBLs	extended spectrum beta lactamases
FDA	Food and Drug Administration
MDR	multidrug resistant
MIC	minimum inhibitory concentration
PBPs	penicillin binding proteins
PCR	polymerase chain reaction
PRRSV	respiratory syndrome virus
SIV	swine influenza virus
S. suis	*Streptococcus suis*

7.1 Introduction

Streptococcus suis (*S. suis*) is a worldwide leading pathogen in swine and an anaerobic Gram-positive coccus that causes substantial commercial losses in the porcine industry. *S. suis* is also a well-known evolving zoonotic agent, after the first cases of infection in humans were testified to in 1986 (Yu et al., 2018) in Denmark and other northern European countries. Over the last few years, the number of *S. suis* cases recorded in the literature has risen drastically (Wertheim et al., 2009). *S. suis* can cause arthritis, meningitis, and sepsis in pigs, while the major clinical disorder in humans is meningitis, subsequently followed by endocarditis, pneumonia, septicemia, septic shock, and arthritis (Yu et al., 2018).

https://doi.org/10.1016/B978-0-12-820576-1.00007-2

The most common problems found in humans infected by meningitis syndrome are vestibular dysfunction and deafness. In bacterial meningitis, deafness was the most prevalent consequence discovered (34%), followed by multiple deficiencies (20%), with more incidents focused in the African and Southeast Asian regions (Rayanakorn, Goh, Lee, Khan, & Saokaew, 2018). This pathogen causes infection in pigs by invading their upper respiratory tract. However, in humans, the septicity mostly follows exposure to infected animals or eating or handling contaminated pork-derived products (Huong et al., 2014). Thus diseases in humans usually develop in butchers, pig breeders, veterinarians, and pork processing workers (Yu et al., 2018). The infection from this pathogen is possibly inevitable (Huong et al., 2014). Currently, more than 30 serotypes of *S. suis* have been identified based on capsular polysaccharides. The most extremely prevalent and virulent one out of all serotypes is type 2, usually isolated from infected humans and swine (Devi et al., 2017).

7.2 Biological landscapes

S. suis is a facultative anaerobic, Gram-positive, ovoid or coccoid that occurs as single cells, short chains, and in pairs; 35 serotypes have been isolated from type 1–34 and 1–2, based on their capsular polysaccharides. The most commonly reported serotype globally is type 2, which is related to infections in humans and swine. All human diseases initiated by *S. suis* are recognized as serotype 2, excluding two human cases triggered by type 1 and only a single case of septicemia affected by type 14 (Kerdsin et al., 2011). Serotype 2 of the *S. suis* pathogen is highly resistant to numerous environmental factors. This pathogen can survive at 60°C for 10 min, at 50°C for 2 h, and at 10°C for 6 weeks in carcasses. It can persist for 1 month at 0°C in filth and for more than 2 months in manure, while at 25°C, this pathogen can persist in dust for 24 h and in feces for 8 days. Conversely, serotype 2 can be simply exterminated with 5% bleach at 1:799 dilution (Lun, Wang, Chen, Li, & Zhu, 2007). The *S. suis* genome has been successfully sequenced and comprises 20,074,917 bp with G+C content of 41.3%. The roles of one-third of the genes are still not known; various genes that could play a functional role in the *S. suis* pathogenesis have been discovered, comprising virulence-associated proteins, polysaccharide production, iron-restriction factors, capsular transport, suilysin, many enzymes, IgG binding proteins, and the arginine deiminase system (Zhu et al., 2018). Research on various virulence aspects of *S. suis* infections has been initiated to explore the pathogenesis system of this microorganism. The virulence of this pathogen varies between serotypes and among various strains of the identical serotype. Most of the research on *S. suis* virulence has focused on serotype 2. Many virulence features have been identified, involving muraminidase-release protein, suilysin, capsule, adhesins, and extracellular protein factors (Segura, Fittipaldi, Calzas, & Gottschalk, 2017).

7.3 Epidemiology, transmission, and clinical features of S. *suis* in swine and human

The infections caused by *S. suis* are multifactorial, with adverse environmental circumstances promoting the advancement of disease (Shen et al., 2014). The upper respiratory tract of swine, especially the nasal cavities and tonsils, is a reservoir niche for *S. suis* and other pathogenic commensals and microorganisms. In this nasal cavity commensals can behave as harmless microbes that genetically occupy the breathing sheath and can help various other microbes to produce irrefutable disease (Vötsch, Willenborg, Weldearegay, & Valentin-Weigand, 2018).

S. suis is one of the most critical bacterial swine pathogens, with approximately 100% of porcine farmsteads globally having carrier animals. Diffusion of *S. suis* within organisms is expressed mostly over the respiratory route. Meningitis and septicemia are the most prominent features of the numerous manifestations of the diseases; however, arthritis, endocarditis, and pneumonia can also be perceived. In severe incidents of septicity, swine are frequently found deceased with no predictive indications of infection. Corroboration of the disease is obligatory and must be attained by description and isolation of the pathogen (Goyette-Desjardins, Auger, Xu, Segura, & Gottschalk, 2014).

The carrier rate of *S. suis* infection in swine is almost 100%, although the duration of disease differs from time to time. The death of animals can reach up to 20% due to the lack of on-time treatment. The early clinical symptoms include high fever that may initially occur without other symptoms, later followed by septicemia, which may remain for many days if untreated. Commonly, an unstable fever and varying degrees of appetite, shifting lameness, and depression appear (Gottschalk, 2019). Meningitis is the most noticeable symptom, on which the initial diagnosis is generally based. In the early stages of meningitis, pigs may squint their eyes and also clutch their ears back. Other symptoms comprise incoordination, adoption of unusual stances, and depression, which soon proceeds to paddling, nystagmus, convulsions, and inability to stand. Other signs of endocarditis and polyarthritis are also discovered in older pigs, with sudden mortality rate showing wasting, dyspnea, and cyanosis. Symptoms of other respiratory diseases may also be noticed in a few outbreaks; however, the function of *S. suis* as a main cause of pneumonia remains unclear in the absence of other pathogens (Gottschalk, 2019).

The first report of human disease by *S. suis* occurred in Denmark; since then an increasing number of cases has been discovered in many other countries (Wertheim et al., 2009). *S. suis* can cause infection in humans from consuming undercooked or raw pork products. Humans also get infection via skin exposure from touching infected piglets or diseased swine products (Vötsch et al., 2018). In southern Vietnam, a large sequence of 151 *S. suis* meningitis cases was recounted over a period of 10 years. While most accounts contain sporadic cases of infection, an epidemic of *S. suis* infection resulted in China, which involved 215 cases and 38 deaths, accentuating the significance of *S. suis* as a developing zoonosis. In humans, serotype 2 of *S. suis* infecting pigs is the most common cause of this disease (Wertheim et al., 2009). In Asian countries, people are affected specifically due to prevalent dishes, for example uncooked blood pudding, *tietcanh*. The swine industry is growing rapidly because the society is unaware of the risks of disease. For *S. suis*, synergistic actions with various bacterial pathogens like *Mycoplasma hyopneumoniae* and *Pasteurella multocida*, as well as others such as the porcine reproductive and respiratory syndrome virus (PRRSV), swine influenza virus (SIV), and porcine circovirus type 2, may increase the risk of intrusive infections. PRRSV and SIV are known principal pathogens, as they create the path for *S. suis* infections prominent in serious respiratory indications and severe pneumonia (Tan, Zhang, Chen, & Zhou, 2019; Vötsch et al., 2018).

In humans, *S. suis* instigates a severe infection by affecting multiple organs; meningitis is the most prevalent scientific indication. The signs/symptoms of *S. suis* meningitis are typically related to those of other microbial pyogenic meningitis that involve vomiting, fever, meningeal signs, and headache. The period of sickness before hospitalization is usually 2–5 days. One noticeable symptom is subjective loss of hearing that may be disclosed by up to 50% of the patients immediately or a few days later. Roughly 31% of patients are also diagnosed with extensive skin problems comprising purpura, ecchymoses, petechiae, and skin necrosis. At a later stage in the disease, gangrene of the toes and fingers may also be noticed in a small group of patients. Other indications of *S. suis* infection involve subacute and acute endocarditis, brainstem ophthalmoplegia, acute pyogenic arthritis, epidural abscess, and endophthalmitis and uveitis (Wertheim et al., 2009).

7.4 Mechanism and emergence of drug resistance

A major health problem disturbing humans globally is the emergence of drug resistance among the most problematic bacterial pathogens. Multidrug resistant (MDR) organisms have appeared in hospitals and are also recognized in communal surroundings, signifying that reservoirs of antibiotic-resistant pathogens exist outside the hospital. Diseases caused by MDR bacteria are accompanied by increased mortality associated with those diseases caused by vulnerable pathogens and they convey a significant financial loss of approximately 20 billion dollars each year in the United States. Thus improving our knowledge of the genetic and biochemical basis of resistance is most important for the development of approaches to restrain the spread and emergence of resistance and discover advanced therapeutic methods against multidrug-resistant pathogens (Munita & Arias, 2016).

The overconsumption of antibiotics in veterinary medication and humans has encouraged the spread and emergence of antimicrobial-resistant microorganisms (Schmithausen et al., 2018). The concept of antimicrobial compounds was initially offered in 1908 by Paul Ehrlich, who suggested consumption of a few chemicals that he eventually referred to as "magic bullets" to kill microbes precisely, with the least possible detriment to humans and animals. He and his partners developed and successfully used an arsenic compound, named Salversan, on humans to cure syphilis, a contagious disease spread through sexual interaction. This represented the first time such a compound was used in the treatment of microbial disease. Researchers and scientists worked conscientiously to discover new antibiotics after the first antibiotic, penicillin, was found in the 1920s, and introduced clinically in the 1940s. This led to the development of various new antibiotics, such as gentamicin, chloramphenicol, and tetracycline and since the 1950s major infectious diseases have been treated using antibiotics in humans (Ashraf, Mustafa, Rehman, Khalid Bashir, & Adnan Ashraf, 2019).

Various aspects of the resulting antimicrobial resistance have been published in the literature, since the introduction of antimicrobial agents in therapeutic practice. Most accounts principally define the emergence of pathogens that are resistant to specific antimicrobial drugs, the incidences of antibiotic resistance, and investigations into accumulating minimum inhibitory concentration (MIC) standards for many drugs. As mentioned previously, one of the most significant factors behind the spread and emergence of antibiotic resistance is overuse and overexposure to antibiotics (Cantón & Morosini, 2011).

After the clinical implementation of antibiotics, an incident was discovered in *Staphylococcus aureus* by means of which this microbe was no longer susceptible to penicillin. It generated the penicillinase enzyme, which could simply collapse the beta lactam ring of penicillin. This ring is essential for binding bacteria (penicillin binding proteins; PBPs) and hence for penicillin's bacterial killing capability. This reaction was called antibiotic/antimicrobial resistance in pathogens. Since then, various types of microbes/bacteria are increasingly becoming resistant to many types of antibiotics. This effect of antibiotic resistance in bacteria is an insistently constant progression, escalating with every day. Through evolution of various molecular techniques, different methods for the advancement of antibiotic resistance have been perceived over the past 50 years. Currently, the most prevalent risk is from the extended spectrum beta lactamase (ESBL) generating Gram-negative bacteria. Currently, one of the major emerging concerns is resistance against multiple drugs, for example, cephalosporins and fluoroquinolones. Colistin is one of the old antibiotics that was extensively used to cure *Escherichia coli* infections, have been withdrawn so that it may be used in the therapy of multiple diseases caused by the MDR Gram-negative bacterial species (Ashraf et al., 2019). Table 7.1 shows some of the mechanisms of the bacterial resistance against multiple drugs.

TABLE 7.1 Mechanisms of antibacterial drug resistance.

Antibiotic class	Member	Target	Resistance mechanism
Aminoglycosides	Ribosomes	Cell membrane	Inactivation by bacterial enzymes
Beta lactams	Penicillin Amoxicillin Ampicillin Cloxacillin Cefuroxine Cefalexin Ceftiofur Cefquinome	Cell wall (inhibit the synthesis of peptidoglycan)	Efflux, hydrolysis altered drug target
Cationic polypeptides	Colistin, polymyxin B	Cell membrane	Increased efflux of the drug
Fluoroquinolones	Enrofloxacin Marbofloxacin Orbifloxacin Ibafloxacin Pradofloxacin Lincomycin	DNA replication (inhibit the DNA gyrase)	Altered drug target, drug efflux, inactivation of the drug by the bacterial enzymes
Lincosamides	Clindamycin Tylosin	Ribosome	Drug efflux, inactivation of the drug, altered drug target
Macrolides	Tilmicosin Tiamulin Erythromycin Azithromycin Clarithromycin Spiramycin Tulathromycin Nitrofurazone	Ribosome	Glycosylation, hydrolysis, phosphorylation, efflux, altered drug target

TABLE 7.1 Mechanisms of antibacterial drug resistance—cont'd

Antibiotic class	Member	Target	Resistance mechanism
Nitrofurans	Nitrofurantoin	DNA	Modification of the target in bacteria, changes in porins
Polypeptide	Bacitracin	Cell wall	Efflux
Pleuromutilins	Valnemulin	Ribosome	Altered drug target, efflux of the drug
Phenicols	Florfenicol Chloramphenicol	Ribosome	Drug efflux, inactivation of the drug by enzymes, altered drug target
Pyrimidines	Trimethoprim	Folic acid synthesis	Modification of the target in bacteria, enhanced drug efflux
Sulfonamides	Sulfamethoxazole	Folic acid synthesis	Altered drug target, increase efflux of the drug
Streptogramins	Virginiamycin	Ribosome	Efflux, change in drug target, drug inactivation by bacterial enzymes
Tetracycline	Tetracycline Oxytetracycline Chlortetracycline Doxycycline Minocycline	Ribosome	Drug efflux, change in binding site, drug modification by bacterial enzymes

7.5 Diagnosis and multidrug treatment

The identification of *S. suis* infection is commonly based on medical symptoms, animal age, abrasions, and history. Corroboration of the diagnosis can be attained by isolating the infective pathogen and also by assessing the microscopic lesions in affected tissues. After isolation, biochemical description of *S. suis* isolates retrieved from infected animals is applicable with minimum biochemical tests (Dutkiewicz et al., 2017). To confirm and implement preventive measures, serotyping is required and other multiplex PCR checks to easily serotype *S. suis* strains have also been developed. Validated and specific serologic tests for *S. suis* antibody detection are still not available. For epidemiological studies, genetic characterization comprising multilocus typing is carried out in some laboratories. Testing of *S. suis* virulent strains from nasal cavities and tonsils cannot be achieved, as the global virulence factors are not known. These locations are extremely infected and conventional bacterial isolation has very low sensitivity. Virulent strains isolated from tonsils must be identified through *S. suis* specific PCR tests, because other biochemical tests are not adequate to properly distinguish this bacterial class from other streptococci that usually exist in the upper respiratory tract of swine (Gottschalk, 2019).

In the last few years, new cases of *S. suis* has been reported in humans, with two major epidemics in 1998 and 2005 in China. A vaccine for streptococcosis on porcine farmsteads is not yet possible and efficient antimicrobial drugs are still required to restrict *S. suis* infection. The most common drug being used to treat and control the infection is penicillin. It has been reported that the penicillin strains are resistant to other β-lactam antibiotics; thus potential and novel antibiotics or more accurate therapeutic approaches against *S. suis* are immediately required (Zhou et al., 2017).

A worldwide public health problem has been identified after the upsurge in the level of antibiotic resistance reported in the past decade (Zhang et al., 2015). This is considered to be a worldwide human health concern and the global occurrence of multidrug-resistant phenotypes is triggering high interest (Hernandez-Garcia et al., 2017). The use of antimicrobial drugs in veterinary applications is being closely observed for therapeutic use, to decrease the high-risk antimicrobial resistance in humans. Multidrug-resistant characteristics of *S. suis* have been identified in human and swine cases. Cephalosporins and penicillin are the most commonly used therapeutic drugs against *S. suis* infection, although resistance to these drugs has been disclosed in China and Europe (Zhang et al., 2015). Resistance to tetracycline in *S. suis* was discovered in several European countries, North America, and in some Asian countries, varying from 87% to 100% in swine (Seitz, Valentin-Weigand, & Willenborg, 2016). Excessive resistance to trimethoprim, amphenicols, macrolide, lincosamide, and aminoglyside β-lactams has also been reported. The bacteria can spread in different species, such as rabbits, horses, cattle, cats, dogs, and birds, or may be commensal and prevalent in nature. The situation with *S. suis* is triggering alarm in the human health field and the porcine industry (Yu et al., 2018). Therapies using combined drugs have specific benefits over a single drug in terms of their synergistic effects, including smaller dosage and delay of antibiotic resistance (Zhao et al., 2019).

Tulathromycin is a prolonged effective, semisynthetic macrolide antibiotic drug, manufactured for the therapy and regulation of bacterial respiratory diseases in swine and cattle. There are also a few antimicrobial drugs approved by the European Medicines Agency (EMA) and the US Food and Drug Administration (FDA) for *Actinobacillus pleuropneumoniae*, *Pasteurella multocida*, and *Mannheimia haemolytica*. Tulathromycin is also an approved drug to cure interdigital necrobacillosis caused by *Porphyromonas levii*; however, currently this drug has not been approved for therapy of *S. suis* infections (Zhou et al., 2017).

The main goal of using antimicrobial drugs in clinical applications is to increase therapeutic efficiency while decreasing bacterial resistance. *S. suis* is considered to be a zoonotic and opportunistic pathogen; thus in the authentication of combining penicillin with aminoglycosides it should be considered, particularly for human antimicrobial drugs. A synergism of aminoglycosides and pleuromutilin drugs could only be relevant for animal use, because retapamulin is the single pleuromutilin drug permitted by the FDA against *S. aureus* infection in humans. Two combinations of drugs, apramycin with ampicillin and spectinomycin with tiamulin, have been identified as significantly strengthened and also as having high synergistic antimicrobial activity against multiresistant *S. suis* strains from swine (Yu et al., 2018).

Over the past few years, different levels of antimicrobial resistance have been noticed in various counties and serotypes. *S. suis* isolated from diseased or medically healthy swines was frequently unaffected by regularly used antimicrobial drugs or vaccines. Random use of antibiotics leads to the expansion of *S. suis* resistance to these drugs globally (Devi et al., 2017).

7.6 Conclusion

S. suis infections are recognized as economically and universally troublesome in the porcine industry. Additionally, *S. suis* is a zoonotic agent that affects humans in interaction with diseased pork-derived products or swine. During the past 40 years, various sporadic cases of *S. suis* infections have been reported in humans. A large outbreak in 2005 in humans was reported in China. Various problems such as rapid disease progression, higher mortality rate, and shorter incubation time gained attention from the general media and scientific communities (Gottschalk, Xu, Calzas, & Segura, 2010). *S. suis* is a developing pathogen in humans that was initially designated as a major source of septicemia in swine in the 1950s. Meningitis and septicemia are human diseases caused by *S. suis*, which was discovered in Denmark in the late 1960s. In the last decade, the number of cases described in the literature has increased significantly, mainly in China; it persists as an extremely sporadic disease in the United States (Dejace, Bagley, & Wood, 2017).

Most of the *S. suis* human infection cases reported are in Asian countries, where the infection can be considered prevalent. This is largely due to the large number of swine in these countries, intake of uncooked or slightly cooked pork products, and slaughtering practices without precautionary actions (Huong et al., 2014). *S. suis* infections are a potential risk for both animal and human health, with high resistance to various antimicrobial agents. Globally, huge percentages of *S. suis* resistance to lincosamides, tetracyclines, and macrolides have been reported in both human and swine isolates. Furthermore, *S. suis* resistant to other antibiotics such as trimethoprim-sulfamethoxazoles, β-lactams, florfenicol/chloramphenicols, fluoroquinolones, and aminoglycosides have also been often reported (Gurung et al., 2015).

Presently, human vaccination is not available; however, easy preventive methods like washing hands after handling uncooked pork products, proper cooking of pork, and wearing gloves while slaughtering or processing pork can prevent most of the cases. Tourists must be alert that eating habits in some nations may pose an uncertainty of *S. suis* infection, as well as other infectious diseases. Controlled research to investigate the risk factors accompanying *S. suis* infection is immediately needed, as it is the major cause of adult meningitis in Vietnam, Thailand, and China. Greater attention from both microbiologists and clinicians in required to completely acknowledge the significance of *S. suis* as a human pathogen (Wertheim et al., 2009).

7.7 Future perspective

S. suis is gaining more attention from the scientific community and human health officials as a zoonotic pathogen. The present situation with human *S. suis* infections should be reassessed, and comprehensive approaches implemented in terms of epidemiology. It is recommended that attempts be made to advance community attention to decrease occupational infections by academic education and commercialization. In the future, support is needed to investigate structural biology-related pathogenesis, which could provide a strong basis to design and develop small molecule inhibitors targeting bacterial virulence factors (Feng et al., 2014).

Infections of *S. suis* are not unusual. The small number of cases identified is mainly due to lack of awareness and underdiagnosis of the disease. The pathogen is frequently misdiagnosed in laboratories, causing delayed or incomplete therapy. It is necessary that patients with indicative *S. suis* signs with affecting risk features get suitable attention while waiting for test reports, in spite of negative results, either due to previous antibiotic administration or misidentification. Advancement of new protocols for screening could be beneficial to support treatment decisions. The identification should not be challenging, if relevant diagnosis is implemented. Urgent therapy, with antimicrobial drugs or penicillin that the organism is susceptible to, before problems appear, specifically hearing loss, could be important in controlling long-term morbidity and mortality rates. The preferred way to control the disease is to prevent transmission in the absence of vaccination. General public awareness comprising a food safety campaign could be useful to improve public knowledge of the infection, particularly in situations where there is a strong association between eating uncooked pork and conventional values (Rayanakorn et al., 2018). Also, comprehensive knowledge of the mechanisms involved in antibiotic resistance of the pathogen is of utmost importance for developing and designing innovative approaches to counter the resistance risks. We are in urgent need of new and better antibiotic drugs, but we must understand that the pathogen will react to them and resistance will appear. Hence, there should be continuous, steady and resilient efforts to develop antibiotics and research the development of antibiotic resistance. This will likely be a long-haul "battle" against an enemy with a strong capability to acclimate and survive (Munita & Arias, 2016).

References

Ashraf, M., Mustafa, B.-E., Rehman, S.-U., Khalid Bashir, M., & Adnan Ashraf, M. (2019). Emergence of antimicrobial resistance, causes, molecular mechanisms, and prevention strategies: a bovine perspective. In *Bovine science—A key to sustainable development.* https://doi.org/10.5772/intechopen.79757.

Cantón, R., & Morosini, M.-I. (2011). Emergence and spread of antibiotic resistance following exposure to antibiotics. *FEMS Microbiology Reviews, 35*(5), 977–991. https://doi.org/10.1111/j.1574-6976.2011.00295.x.

Dejace, J., Bagley, P., & Wood, E. (2017). *Streptococcus suis* meningitis can require a prolonged treatment course. *International Journal of Infectious Diseases, 65,* 34–36. https://doi.org/10.1016/j.ijid.2017.09.011.

Devi, M., Dutta, J. B., Rajkhowa, S., Kalita, D., Saikia, G. K., Das, B. C., ... Mahato, G. (2017). Prevalence of multiple drug resistant *Streptococcus suis* in and around Guwahati, India. *Veterinary World, 10*(5), 556–561. https://doi.org/10.14202/vetworld.2017.556-561.

Dutkiewicz, J., Sroka, J., Zając, V., Wasiński, B., Cisak, E., Sawczyn, A., ... Wójcik-Fatla, A. (2017). *Streptococcus suis*: A re-emerging pathogen associated with occupational exposure to pigs or pork products. Part I—Epidemiology. *Annals of Agricultural and Environmental Medicine, 24*(4), 683–695. https://doi.org/10.26444/aaem/79813.

Feng, Y., Zhang, H., Wu, Z., Wang, S., Cao, M., Hu, D., & Wang, C. (2014). *Streptococcus suis* infection: An emerging/reemerging challenge of bacterial infectious diseases? *Virulence, 5*(4), 477–497. https://doi.org/10.4161/viru.28595.

Gottschalk, M. (2019). *Streptococcus suis* infection—Generalized conditions. In *Veterinary manual.* Retrieved from https://www.msdvetmanual.com/generalized-conditions/streptococcal-infections-in-pigs/streptococcus-suis-infection. (Accessed June 17, 2019).

Gottschalk, M., Xu, J., Calzas, C., & Segura, M. (2010). *Streptococcus suis*: A new emerging or an old neglected zoonotic pathogen? *Future Microbiology, 5*(3), 371–391. https://doi.org/10.2217/fmb.10.2.

Goyette-Desjardins, G., Auger, J.-P., Xu, J., Segura, M., & Gottschalk, M. (2014). *Streptococcus suis*, an important pig pathogen and emerging zoonotic agent—An update on the worldwide distribution based on serotyping and sequence typing. *Emerging Microbes & Infections, 3*(6), e45. https://doi.org/10.1038/emi.2014.45.

Gurung, M., Tamang, M. D., Moon, D. C., Kim, S.-R., Jeong, J.-H., Jang, G.-C., ... Lim, S.-K. (2015). Molecular basis of resistance to selected antimicrobial agents in the emerging zoonotic pathogen *Streptococcus suis*. *Journal of Clinical Microbiology, 53*(7), 2332–2336. https://doi.org/10.1128/JCM.00123-15.

Hernandez-Garcia, J., Wang, J., Restif, O., Holmes, M. A., Mather, A. E., Weinert, L. A., ... on behalf of the BRADP1T Consortium. (2017). Patterns of antimicrobial resistance in *Streptococcus suis* isolates from pigs with or without streptococcal disease in England between 2009 and 2014. *Veterinary Microbiology, 207,* 117–124. https://doi.org/10.1016/j.vetmic.2017.06.002.

Huong, V. T. L., Ha, N., Huy, N. T., Horby, P., Nghia, H. D. T., Thiem, V. D., ... Hirayama, K. (2014). Epidemiology, clinical manifestations, and outcomes of *Streptococcus suis* infection in humans. *Emerging Infectious Diseases, 20*(7). https://doi.org/10.3201/eid2007.131594.

Kerdsin, A., Dejsirilert, S., Puangpatra, P., Sripakdee, S., Chumla, K., Boonkerd, N., ... Oishi, K. (2011). Genotypic profile of *Streptococcus suis* serotype 2 and clinical features of infection in humans, Thailand. *Emerging Infectious Diseases*, *17*(5), 835–842. https://doi.org/10.3201/eid1705.100754.

Lun, Z.-R., Wang, Q.-P., Chen, X.-G., Li, A.-X., & Zhu, X.-Q. (2007). *Streptococcus suis*: An emerging zoonotic pathogen. *The Lancet Infectious Diseases*, *7*(3), 201–209. https://doi.org/10.1016/S1473.

Munita, J. M., & Arias, C. A. (2016). Mechanisms of antibiotic resistance. *Microbiology Spectrum*, *4*(2). https://doi.org/10.1128/microbiolspec.VMBF-0016-2015.

Rayanakorn, A., Goh, B.-H., Lee, L.-H., Khan, T. M., & Saokaew, S. (2018). Risk factors for *Streptococcus suis* infection: A systematic review and meta-analysis. *Scientific Reports*, *8*(1), 13358. https://doi.org/10.1038/s41598-018-31598-w.

Schmithausen, R., Schulze-Geisthoevel, S., Heinemann, C., Bierbaum, G., Exner, M., Petersen, B., ... Steinhoff-Wagner, J. (2018). Reservoirs and transmission pathways of resistant indicator bacteria in the biotope pig stable and along the food chain: A review from a one health perspective. *Sustainability*, *10*(11), 3967. https://doi.org/10.3390/su10113967.

Segura, M., Fittipaldi, N., Calzas, C., & Gottschalk, M. (2017). Critical *Streptococcus suis* virulence factors: Are they all really critical? *Trends in Microbiology*, *25*(7), 585–599. https://doi.org/10.1016/j.tim.2017.02.005.

Seitz, M., Valentin-Weigand, P., & Willenborg, J. (2016). Use of antibiotics and antimicrobial resistance in veterinary medicine as exemplified by the swine pathogen *Streptococcus suis*. In Vol. 398. *Current topics in microbiology and immunology* (pp. 103–121). https://doi.org/10.1007/82_2016_506.

Shen, C., Li, M., Zhang, W., Yi, Y., Wang, Y., Hou, Q., ... Lu, C. (2014). Modeling transmission dynamics of *Streptococcus suis* with stage structure and sensitivity analysis. *Discrete Dynamics in Nature and Society*, *2014*, 1–10. https://doi.org/10.1155/2014/432602.

Tan, C., Zhang, A., Chen, H., & Zhou, R. (2019). Recent proceedings on prevalence and pathogenesis of *Streptococcus suis*. *Current Issues in Molecular Biology*, *32*, 473–520. https://doi.org/10.21775/cimb.032.473.

Vötsch, D., Willenborg, M., Weldearegay, Y. B., & Valentin-Weigand, P. (2018). *Streptococcus suis*—The "two faces" of a pathobiont in the porcine respiratory tract. *Frontiers in Microbiology*, *9*, 480. https://doi.org/10.3389/fmicb.2018.00480.

Wertheim, H. F. L., Nghia, H. D. T., Taylor, W., Schultsz, C., Taylor, W., & Schultsz, C. (2009). *Streptococcus suis*: An emerging human pathogen. *Clinical Infectious Diseases*, *48*(5), 617–625. https://doi.org/10.1086/596763.

Yu, Y., Fang, J.-T., Zheng, M., Zhang, Q., Walsh, T. R., Liao, X.-P., ... Liu, Y.-H. (2018). Combination therapy strategies against multiple-resistant *Streptococcus suis*. *Frontiers in Pharmacology*, *9*, 489. https://doi.org/10.3389/fphar.2018.00489.

Zhang, C., Zhang, Z., Song, L., Fan, X., Wen, F., Xu, S., & Ning, Y. (2015). Antimicrobial resistance profile and genotypic characteristics of *Streptococcus suis* capsular type 2 isolated from clinical carrier sows and diseased pigs in China. *BioMed Research International*, *2015*, 1–9. https://doi.org/10.1155/2015/284303.

Zhao, F., Yang, N., Wang, X., Mao, R., Hao, Y., Li, Z., ... Wang, J. (2019). In vitro/vivo mechanism of action of MP1102 with low/nonresistance against *Streptococcus suis* type 2 strain CVCC 3928. *Frontiers in Cellular and Infection Microbiology*, *9*, 48. https://doi.org/10.3389/fcimb.2019.00048.

Zhou, Y.-F., Peng, H.-M., Bu, M.-X., Liu, Y.-H., Sun, J., & Liao, X.-P. (2017). Pharmacodynamic evaluation and PK/PD-based dose prediction of tulathromycin: A potential new indication for *Streptococcus suis* infection. *Frontiers in Pharmacology*, *8*, 684. https://doi.org/10.3389/fphar.2017.00684.

Zhu, H., Ni, Y., Zhou, J., Yu, Z., Mao, A., Wang, D., & He, K. (2018). Complete genome sequence of *Streptococcus suis* serotype 2 virulent strain SS2-1. *Genome Announcements*, *6*(10). https://doi.org/10.1128/genomeA.00067-18.

CHAPTER

8

Combination therapy and multidrug resistance in malaria parasite

Abdul Hafiz[a], *Mahmood A. Alam*[b], *Othman A. Alghamdi*[c], *Arif Mohammed*[c]

[a]Faculty of Medicine, Umm Al-Qura University, Makkah, Kingdom of Saudi Arabia, [b]Lord Kelvin/Adam Smith (LKAS) fellow, Wellcome Centre for Integrative Parasitology, Institute of Infection, Immunity and Inflammation, University of Glasgow, Glasgow, Scotland, [c]Department of Biological Sciences, Faculty of Science, University of Jeddah, Jeddah, Kingdom of Saudi Arabia

Abbreviations

ACTs artemisinin-based combination therapies
KELCH13 Kelch propeller protein
MDR multidrug resistance
PfCRT *P. falciparum* chloroquine resistance transporter
PfMDR1 *P. falciparum* multidrug resistance transporter-1
WHO World Health Organization

8.1 Introduction

Malaria caused about 219 million clinical cases and 435,000 deaths worldwide in the year 2017. About 92% of global malaria cases and 93% of global malaria-related deaths are reported from the World Health Organization's African region. Most of the malaria fatalities are in children under 5 years of age, accounting for 61% of all malaria deaths (World Health Organization, 2018a). Malaria disease is caused by a protozoan parasite belonging to the genus *Plasmodium*. Plasmodium species infect

Combination Therapy Against Multidrug Resistance
https://doi.org/10.1016/B978-0-12-820576-1.00008-4

host blood cells and cause malaria in mammals, birds, and reptiles. There are six Plasmodium species that are known to infect humans: *Plasmodium falciparum*, *P. vivax*, P. malariae, *P. ovale* curtisi, *P. ovale* wallikeri (Joste et al., 2018), and *P. knowlesi* (Pattyn & Verdoolaege-Van Loo, 1972). Globally, *P. falciparum* is the most prevalent and most deadly human malaria parasite (World Health Organization, 2018a).

The malaria parasite needs two consecutive hosts to complete its life cycle. The asexual stage of the human malaria parasite is completed in the human host, while the sexual stage is completed in the mosquito host (vector). Malaria control programs use drugs in human hosts to treat malaria and control malaria transmission. Vector control programs are used to check malaria transmission (Zaw & Lin, 2019). In addition, vaccines are also being developed against malaria, with little success so far. In this chapter, we will discuss use of drugs against malaria, the issue of drug resistance in the malaria parasite, and combination therapy approaches to prevent drug resistance in malaria.

8.1.1 Malaria life cycle

The malaria life cycle in human host begins with a bite of the malaria-infected female Anopheles mosquito. If malaria sporozoites are present in the salivary glands of the mosquito, it can infect the human with malaria when it bites the human host for a blood meal. Sporozoites are released into the human bloodstream during the process of mosquito biting. In order to stop human blood from clotting, the mosquito releases its saliva at the site of its biting on the human skin. The sporozoites thus enter the human bloodstream with the saliva and circulate through the blood until they reach the human liver. In the human liver, each sporozoite traverses through several hepatocytes before establishing itself inside a single hepatocyte. The parasites multiply and proliferate in the hepatocytes by a process called liver schizogony. Tens of thousands of merozoites are formed in the process of liver schizogony leading to rupture of hepatocytes. Upon hepatocyte rupture, merozoites are released into the bloodstream in the form of bunches covered with vesicles. Once in the bloodstream, these vesicles are thought to burst and merozoites are released into the blood. The merozoites readily invade erythrocytes to start the blood stage of the asexual life cycle in human host (Cowman, Healer, Marapana, & Marsh, 2016; Meibalan & Marti, 2017; Phillips et al., 2017).

The invasion of erythrocytes by Plasmodium merozoite is a quick, dynamic, and multistep process that includes attachment of the merozoite to the erythrocyte, tight junction formation between the merozoite apical end and the erythrocyte surface, and finally the active penetration of the merozoite into the erythrocyte. This process of attachment contains initial merozoite contact with the erythrocyte and apical reorientation of the merozoite to direct the apical end of the merozoite toward the erythrocyte surface. The formation of a tight junction between the merozoite and the erythrocyte surface involves specific receptor ligand interactions between the merozoite cell surface ligand and the erythrocyte cell surface receptors. The merozoite enters the erythrocyte by a process known as penetration, which is an active process that requires merozoite cytoskeleton movements to push the merozoite inside the erythrocyte. Postinvasion, the merozoite grows through the ring and trophozoite stages. The trophozoite enters into the schizont stage as it divides to produce multiple merozoites. Up to 32 merozoites are released from a single schizont, once it bursts upon maturation. These blood merozoites reinvade fresh erythrocytes and the erythrocyte cycle continues (Baum, Richard, & Riglar, 2017; Vaughan & Kappe, 2017). Some of the merozoites differentiate into male and female gametocytes that sets off the sexual stage of the malaria life cycle. However, the sexual life cycle of Plasmodium cannot be completed in the human host, as the gametocytes require low temperature for further development and fertilization (Gaur & Chitnis, 2011; Ngotho et al., 2019).

8.1.2 Malaria disease

Malaria is considered to be a curable disease. All the clinical symptoms of malaria are manifested due to the blood stage of the Plasmodium life cycle in the human host (Ngotho et al., 2019). Malaria infection remains asymptomatic as long as the parasite stays in the human liver. However, a small percentage of malaria cases develop into severe, life-threatening disease for reasons not yet fully understood. Even in endemic areas where there is repeated exposure to the parasite, immunity to malaria develops slowly. Sterile immunity, where a person is protected from subsequent infections upon initial exposure to the pathogen, may never develop for malaria. However, adults living in the endemic areas may be protected from symptomatic infection, where the individual does not show any sign of malaria disease despite being infected with the malaria parasite (Mayor et al., 2007; Miller, Ackerman, Su, & Wellems, 2013). Such protection from the symptomatic disease is the result of acquired immunity upon continuous repeated infections.

Children below 5 years of age are particularly vulnerable to severe malaria in endemic regions, as their immune system has not been exposed to enough malaria. Most of the malaria-related deaths are reported in children below 5 years of age (Cohee & Laufer, 2017; Krzych, Zarling, & Pichugin, 2014). Another group vulnerable to severe malaria in endemic regions is pregnant women (Mfueni et al., 2018). Adult women who live in malaria endemic areas are protected against symptomatic malaria infection and severe disease, as are their male counterparts (van Eijk, Hill, Noor, Snow, & ter Kuile, 2015). However, when a woman gets pregnant, she becomes vulnerable to severe falciparum malaria (Fried & Duffy, 2017). Several factors contribute to this vulnerability of pregnant women toward severity of malaria (Fana, Bunza,

Anka, Imam, & Nataala, 2015). The best-characterized factor that contributes to the severity of malaria in pregnant women is the sequestration of *P. falciparum* mature stage infected erythrocytes into the blood vasculature of the placenta of pregnant women (Matteelli, Caligaris, Castelli, & Carosi, 1997). In fact, primigravida women are more susceptible to a severe manifestation of malaria as compared to multigravida women, suggesting acquisition of immunity against severe disease upon repeated infections during pregnancy (Autino, Corbett, Castelli, & Taramelli, 2012; Marinho et al., 2009; Shulman et al., 2001).

There are two main hypotheses proposed to explain the severe manifestation of malaria in some patients as compared to others (Garg, 2000; Shulman & Dorman, 2003). The cytokine theory suggests that it is the cytokines released during the course of malaria infection that determines the severe vs mild outcome of the malaria. The mechanical theory suggests that severe malaria is the result of sequestration of erythrocytes infected with mature stages of mostly *P. falciparum* (Warrell, 1992). Sequestration is the result of cytoadherence, a property of mostly *P. falciparum* infected erythrocytes by which the infected erythrocytes bind to the host microvascular endothelial cells and get stuck in the blood vasculature of the human host. Cytoadherence causes obstruction in blood flow, endothelial cell activation, and release of proinflammatory cytokines (Mayor et al., 2011). Evidence suggests that cytoadhcrcnce in certain organs such as brain and placenta lead to severe malaria (Barrera et al., 2015; Boudová et al., 2017; Frevert & Nacer, 2014; Milner et al., 2015; Molyneux, 1990; Ponsford et al., 2012).

8.1.2.1 Severe malaria

Clinical malaria is defined as fever or history of fever with presence of the asexual-stage malaria parasite in the blood. Severe malaria is described as clinical malaria plus presence of at least one of the severe malaria clinical features such as cerebral malaria, severe anemia, acidosis or respiratory distress, prostration, hypoglycemia, multiple seizures, circulatory collapse/shock renal impairment, or clinical jaundice. Cerebral malaria is defined as Blantyre Coma Score ≤ 2 in children and Glasgow coma score ≤ 9 in adults. Severe malarial anemia is defined as packed blood cell volume $< 15\%$ or hemoglobin < 5 g/dL in children and packed blood cell volume $< 20\%$ or hemoglobin < 5 g/dL in adults. Acidosis and respiratory distress syndrome in malaria is defined as lactate > 5 mM and/or chest in-drawing or deep breathing. Prostration is defined as inability to sit unaided in adults and in children who can sit by themselves by virtue of their age. In children who are too young to be able to sit by themselves, prostration is defined as inability to breastfeed. Hypoglycemia is defined as blood glucose < 2.2 mM. Multiple seizures are defined as ≥ 2 convulsions within 24 h. Circulatory collapse or shock is defined as systolic blood pressure

< 50 mmHg in children and < 80 mmHg in adults. Renal impairment is defined as serum creatinine levels > 265 μmol/L. Clinical jaundice is defined as serum bilirubin level > 2.5 to 3 mg/dL with presentation of yellow skin and sclera (Bartoloni & Zammarchi, 2012; van Spronsen, Schneider, & Atasige, 2012).

8.1.2.2 Nonsevere malaria

Nonsevere malaria is defined as clinical malaria without the presence of any of the previously mentioned severe malaria clinical features (van Spronsen et al., 2012).

8.2 Drug resistance in malaria

Drug resistance in pathogens is now a common phenomenon that limits the option for treatment of a disease. Drug resistance by the malaria parasite can be achieved by diverse mechanisms. For instance, the dihydrofolate inhibitors such as pyrimethamine, proguanil, sulfones, and sulfonamides act on enzymes in the folate pathway of the malaria parasite. The parasite acquires resistance to these drugs when the mutations in the target enzyme decrease the affinity of binding of the drug with the target enzyme (Heinberg & Kirkman, 2015; World Health Organization, 2014). Another mechanism by which the malaria parasite develops resistance to a drug is by decreasing the accumulation of the drug in the parasite. For instance, in the case of chloroquine resistance, there is increased efflux of the drug from the parasite by transporter proteins (Mohapatra et al., 2014; Sanchez et al., 2007).

Multidrug resistance (MDR) is a phenotype in which a pathogen shows resistance to multiple drugs. MDR can make the treatment of a disease precarious or unsuccessful sometimes. MDR was first reported for enteric bacteria such as *Escherichia coli*, Shigella and Salmonella around the 1950s (Ecker, Lehane, Clain, & Fidock, 2012). The spread of MDR through mobile genetic elements is not as common in parasites as it is in bacteria. This is the reason why MDR is less common in parasites as compared to bacteria (Watanabe, 1963). However, the limited availability of drugs against malaria makes the tackling of multidrug resistance more challenging in malaria.

There are three main factors that contribute to the emergence and propagation of drug resistance in the malaria parasite. The most common cause of emergence and spread of drug resistance is subtherapeutic doses of the drug. The subtherapeutic doses are a result of poor compliance and substandard quality drug supplies in poor countries (Goldberg, Siliciano, & Jacobs, 2012). The second important contributing factor for higher chances of drug resistance mutation is high parasitemia (Pattyn & Verdoolaege-Van Loo, 1972). Asymptomatic malaria infections have lower levels of parasitemia. In such infections, the chances of emergence of drug resistance are low (Thu, Phyo, Landier, Parker, & Nosten, 2017). The third factor that influences the emergence and spread of drug resistance in the malaria parasite is the ease of the genetic appearance of the drug resistance mutation. Some mutation phenotypes are easier to be manifested at the genetic level, such as single point mutation at a single genetic locus (Hastings, Kay, & Hodel, 2015). Atovaquone resistance is one such example, where a single point mutation in the cytochrome *b* gene of *P. falciparum* confers resistance to atovaquone (malERA Refresh Consultative Panel on Insecticide and Drug Resistance, 2017). On the other hand, if the drug resistance phenotype demands multiple genetic changes, then such phenotypes appear more slowly. For instance, the emergence of artemisinin resistance was mapped to multiple independent mutations in a gene encoding a Kelch propeller protein (KELCH13) plus nonsynonymous mutations in several genes encoding ferredoxin, apicoplast ribosomal protein S10, MDR protein 2 and *P. falciparum* chloroquine resistance transporter (PfCRT) (Staines et al., 2018).

8.2.1 Combination therapy for severe and nonsevere malaria

Spreading drug resistance is a serious cause of public health concerns. Unlike bacterial disease where the antibiotics options are quite numerous, antimalarial drugs are scarce. Thus preventing or delaying the development of resistance against antimalarial drugs is essential for the long-term control and elimination goals of malaria. To achieve these goals, it is recommended that all the episodes of malaria should be treated with at least two effective antimalarial drugs that have a different cellular target or mechanism of action. Under the WHO malaria control program, malaria endemic countries are required to adopt a national malaria treatment policy. Such a national policy describes specific drugs to be given for treatment of nonsevere malaria, severe malaria, malaria in infants, malaria in pregnancy, and malaria in lactating women. The policy also describes alternative drugs, in case the first-line treatment fails (Tilley, Straimer, Gnädig, Ralph, & Fidock, 2016; Zaw & Lin, 2019).

WHO recommends artemisinin-based combination therapies (ACTs) for effective treatment of malaria and to delay the development of resistance. ACT combines at least two drugs, one of which is a rapidly acting artemisinin derivative, while the other partner may be another type of antimalarial drug with slow antimalarial action. The artemisinin component is meant to quickly clear the asexual and sexual stages of the malaria parasite, while the slow acting partner drug clears any remaining parasites that may not have been killed by the action of the artemisinin derivative. Such combinations effectively eliminate any scope of development of artemisinin resistance and provide posttreatment prophylaxis as well.

In addition, the quick killing action of the artemisinin derivative helps in controlling transmission of malaria.

8.2.1.1 Combination therapy for nonsevere falciparum malaria

WHO currently recommends ACTs for use against nonsevere and severe *P. falciparum* malaria. Artemisinin derivatives such as artesunate, artemether, and dihydroartemisinin are recommended for ACT. Partner drugs such as mefloquine, piperaquine, lumefantrine, amodiaquine, sulfadoxine/pyrimethamine, and chlorproguanil/dapsone are recommended to be used. To ensure both drugs are used, it is important that coformulated drugs in which the ACT is formulated as a single tablet are used (Tilley et al., 2016; World Health Organization, 2018b; Zaw & Lin, 2019).

The choice of ACT component drugs should be determined based on the presence of drug resistance prevalent in local *P. falciparum* strains. The ACT should be administered for 3 days. In cases where the infecting malaria species is not known or in cases of mixed infection, the same ACT therapy should be followed. However, women in the first trimester of their pregnancy should not be given this therapy regimen. Such women should be treated with quinine + clindamycin for 7 days (Tilley et al., 2016; World Health Organization, 2018b; Zaw & Lin, 2019).

8.2.1.2 Combination therapy for severe malaria

Infants, children, and all adults, including pregnant women in all trimesters and lactating women, with severe malaria should be administered intravenous or intramuscular artesunate for the first 24 h. After 24 h, if the patient can take oral medication, ACT should be followed for 3 days. Otherwise, artesunate should continue (Tilley et al., 2016; World Health Organization, 2018b; Zaw & Lin, 2019).

8.2.2 Drugs against malaria

Several antimalarial drugs are available for treatment and prevention of malaria. Most antimalarial drugs are known to eliminate the blood stage of the malaria parasite in the human host. Antimalarial drugs can be biologically classified into five classes based on their malaricidal mechanism of action, namely: tissue schizontocides, blood schizontocides, gametocytocides, sporontocides, and hypnozoitocides. Tissue schizontocides and hypnozoitocides are the drugs that act on the liver stage of the malaria parasite in human; pyrimethamine and primaquine fall under this class of drugs. Tissue schizontocides act on the active infective stage of the malaria parasite in the liver, while hypnozoitocides act on the dormant stage hypnozoites. Blood schizontocides, on the other hand, act on the blood stages of the parasite and examples include chloroquine, quinine, mefloquine, halofantrine, pyrimethamine, sulfadoxine, sulfones, and tetracyclines.

Gametocytocides kill the sexual stages of the human malaria parasite in blood, thus preventing malaria transmission. Primaquine is an example of a gametocytocyde that is effective against all species of malaria, whereas chloroquine and quinine have gametocytocidal activity against *P. vivax* and *P. malariae* only. Sporontocides are malaria transmission blocking drugs that prevent the development of oocysts in the mosquito examples; they include primaquine and chloroguanide (World Health Organization, 2018b; Zaw & Lin, 2019).

Chemically, antimalarial drugs can be classified into five broad classes: arylaminoalcohols, aminoquinolines, artemisinins, antifolates, and antibiotics.

8.2.2.1 Arylaminoalcohols

The arylaminoalcohol class of antimalarials includes quinine, mefloquine, and lumefantrine. Arylaminoalcohol drugs target the blood stage of the human malaria life cycle.

A. Quinine

Quinine was first introduced in Europe by the Spanish Countess of Cinchona in the 17th century (Blasco, Leroy, & Fidock, 2017). In 1820, the first antimalarial drug was developed with the isolation of quinine from the cinchona bark of South American origin. The South American cinchona bark contained four antimalarial alkaloids: quinine, quinidine, cinchonine, and cinchonidine. However, Javan cinchona bark, which later dominated the cinchona bark supply chain in drug markets, predominantly contains quinine (Permin, Norn, Kruse, & Kruse, 2016). Quinine is highly basic in nature and belongs to the aryl amino alcohol group of chemicals. Quinine is a schizonticidal drug. In 1944, quinine was synthesized chemically (Goss, 2014). Quinine with clindamycin is recommended to treat uncomplicated falciparum malaria in the first trimester of pregnancy (Seeman, 2007).

The mechanism of antimalarial action of quinine is not fully understood. However, quinine is known to interrupt the heme digestion and hemozoin formation in the parasite's digestive vacuole (Moore, Salman, & Davis, 2016). Heme digestion is absolutely essential for malaria parasite survival in the human host during the blood stage of the malaria life cycle. Resistance to quinine was first reported in 1910 in Brazil (Combrinck et al., 2013). The quinine resistance phenotype is a result of mutations in genes encoding transporter proteins; *P. falciparum* multidrug resistance transporter 1 (PfMDR1) (Peters, 1990), PfCRT (Cheruiyot et al., 2014), and *P. falciparum* sodium/proton exchanger 1 (PfNHE1) are examples of such transporter proteins that have been implicated in quinine resistance (Peters, 1990).

B. Mefloquine

Mefloquine is also an arylaminoalcohol derivative, developed in the 1970s to counter the quinine and chloroquine resistances. Mefloquine is not a widely used drug due to its neuropsychiatric side effects (Hrycyna et al., 2014). Mefloquine is also not recommended to be used during pregnancy. The antimalarial mechanisms of action of mefloquine have not been completely understood. However, it is known that it partially inhibits formation of hemozoin in the digestive vacuole. Mefloquine also inhibits plasmodium protein synthesis by interacting with ribosome (Eick-Cost, Hu, Rohrbeck, & Clark, 2017). Mefloquine resistance has been mapped to PfMDR1 polymorphism (Bohórquez, Juliano, Kim, & Meshnick, 2013; Wong et al., 2017). Mefloquine has been used in drug combination with artesunate in ACT.

8.2.2.2 *Aminoquinolines*

Aminoquinoline antimalarial drugs are of two types: eight aminoquinolines and four aminoquinolines. The four aminoquinoline antimalarial drugs are schizontocidal and gametocytocidal in nature. The four aminoquinolines include chloroquine, amodiaquine, and piperaquine. Primaquine, which is an eight aminoquinoline, acts against hypnozoites.

A. Chloroquine

Chloroquine is a synthetic derivative of quinine, which was first synthesized by German scientists as Resochin in 1934 (Reiling & Rohrbach, 2015). The first report of chloroquine-resistant *P. falciparum* appeared at the Thai-Cambodian border in the late 1950s (Jensen & Mehlhorn, 2009). *Plasmodium falciparum* chloroquine resistance transporter protein PfCRT, which is located at the digestive vacuole membrane, has been implicated in chloroquine resistance. Another *Plasmodium falciparum* protein, PfMDR1, has also been implicated in chloroquine resistance (Wongsrichanalai, Pickard, Wernsdorfer, & Meshnick, 2002). Due to widespread resistance to chloroquine, it is only used in monotherapy in areas where there is no report of resistance. Chloroquine is used in combination with primaquine for the treatment of mainly nonfalciparum malaria infections. Another eight aminoquinoline, piperaquine, is used in ACT with dihydroartemisinin in severe malaria infections.

B. Primaquine

Primaquine is an eight aminoquinoline that acts against hypnozoites, liver stage, and sexual and asexual blood stages of the malaria parasite (Asare et al., 2017). It generates reactive oxygen radicals and deregulates the electron transports in the respiratory metabolism. It is thought to obstruct the intracellular respiration of blood stage Plasmodium (Chu & White, 2016).

8.2.2.3 Artemisinins

Artemisinin and its derivatives belong to this class of antimalarial drugs, which were first isolated from the *Artemisia annua* plant (Balaji & Trivedi, 2013). Artemisinin and its derivatives are sesquiterpenic lactones, meaning they are sesquiterpenoids that contain a lactone ring (Li, Zhang, Gong, & Wang, 2018). These compounds have an endoperoxide bond that generates free radicals and reactive oxygen species for antimalarial activity (Gaur, Darokar, Ajayakumar, Shukla, & Bhakuni, 2014). Artemisinins are active against both the sexual and asexual stages of the human malaria parasite. Artemisinins disrupt the heme detoxification process in the malaria parasite (Gupta, Singh, Gupta, & Pandey, 2016).

Artemisinin derivatives inhibit the hemozoin polymerization, induce the depolarization of the parasite mitochondrial and plasma membranes, and target the *P. falciparum* endoplasmic reticulum ATPase 6 (PfATP6). Artemisinin resistances are associated with mutation in PfATP6 and overexpression of PfMDR1 (Gupta et al., 2016).

8.2.2.4 Antifolates

Tetrahydrofolate is a cofactor that is essential for the nucleic acid synthesis in Plasmodium. There are two main enzymes that are essentially required for tetrahydrofolate synthesis in the malaria parasite: dihydrofolate reductase (DHFR) and dihydropteroate synthase (DHPS). DHPS is absent in mammals, so it is an excellent target of antimalarial therapies. Sulfonamide drugs such as sulfadoxine and sulfalene act as inhibitor DHPS (Moles, Oliva, Sánchez-González, & Safont, 2010). Diaminopyrimidine drugs such as pyrimethamine proguanil and chlorproguanil inhibit Plasmodium DHFR (Oguike et al., 2016). Although DHFR is present in human, diaminopyrimidine drugs have lesser affinity for human DHFR. The first report of antifolate resistances in the malaria parasite emerged in the late 1980s. Studies have suggested that mutations in PfDHFR and PfDHPS are responsible for such resistance.

8.2.2.5 Antibiotics

There are several antibiotics that exhibit antimalarial properties, notably tetracyclines and clindamycin (Tarnchompoo et al., 2002). Tetracyclines have broad-spectrum antibiotic properties. The antimalarial potential of tetracycline was first realized in the 1950s (Puri & Dutta, 1982). Doxycycline is known to inhibit protein synthesis in mitochondria (Gaillard, Madamet, & Pradines, 2015). It also decreases de novo pyrimidine synthesis by inhibiting dihydroorotate dehydrogenase. Currently, doxycycline is given with quinine in areas where MDR is widespread.

8.3 Future perspectives

The pipeline for the development of novel antimalarials needs regular updates and should be ready to meet the challenges of the rise in drug-resistant forms of malaria. Most of the work and the resources needed for the discovery of novel antimalarials has largely been made possible due to partnerships developed between the pharmaceutical industry, academia, and with the support of government organizations or philanthropists. Various approaches have been used for the discovery of different chemical classes of compounds, which mainly comprise target-based screening, phenotypic screening, and hypothesis-based design. In terms of success, phenotypic screening has been more successful than the target-based approaches in antimalarial drug discovery. However, with the advancement of various molecular and genetic engineering technologies along with the increase in the knowledge of malaria parasite biology, the target-based approaches will play a bigger role in the discovery of novel antimalarials in the near future (Gamo et al., 2010).

The World Health Organization (WHO) has set an ambitious goal to decrease the malaria mortality rate by 90% for the coming 15 years. Despite the agenda set by WHO, there is an estimate that a child will die of malaria every 10 min by 2030, suggesting that even a greater effort should be made to eradicate malaria completely. With the proper use and availability of ACTs it appears that much of the malaria burden from poor countries can be reduced. The availability of two newly approved ACTs that are sweetened and taste-masked for pediatric use is also going to be very useful (World Health Organization, 2016). However, a major challenge has been posed due to the emergence of insecticide-resistant mosquitoes and MDR Plasmodium spp. Since the first report of emergence of the artemisinin resistance from the Thai-Cambodian border region, there is no report of its spread in African countries. Interestingly, the disease severity of wild-type and the resistant form is not much different; however, in the absence of other alternatives, the emergence of the resistant form cannot be effectively treated. Therefore, combination therapy provides a better way to deal with such problems (Zaw & Lin, 2019). Further, the fight against malaria will also depend on the availability of easy, noninvasive diagnostic tests for malaria. An easy detection method of drug-resistant strains is further going to enhance our capability to fight against this deadly disease (World Health Organization, 2013).

8.4 Conclusion

The presence of MDR organisms has created a real problem for world health and for policymakers. For the past several decades, there has been great demand for the development of new drugs to combat the several MDR pathogens. Malaria remains one of the deadly diseases in many parts of the world. Some progress has been made in combating the burden of malaria but still much research work is needed to develop an effective therapy against it. Combination therapy has provided new hope in the fight against malaria. It provides a new opportunity of mixing different drugs for an effective antimalarial response. The initial success of various drug combinations against different MDR pathogens also raises hope for a lasting drug combination therapy against malaria. Combination therapies based on artemisinin have been successful; however, new combination therapies must be developed quickly to face the challenge of new drug-resistant malaria pathogens.

References

Asare, K. K., Boampong, J. N., Duah, N. O., Afoakwah, R., Sehgal, R., & Quashie, N. B. (2017). Synergism between Pfcrt and Pfmdr1 genes could account for the slow recovery of chloroquine sensitive *Plasmodium falciparum* strains in Ghana after chloroquine withdrawal. *Journal of Infection and Public Health*, *10*(1), 110–119.

Autino, B., Corbett, Y., Castelli, F., & Taramelli, D. (2012). Pathogenesis of malaria in tissues and blood. *Mediterranean Journal of Hematology and Infectious Diseases*, *4*(1), e2012061.

Balaji, S. N., & Trivedi, V. (2013). Methemoglobin incites primaquine toxicity through single-electron oxidation and modification. *Journal of Basic and Clinical Physiology and Pharmacology*, *24*(2), 105–114.

Barrera, V., Hiscott, P. S., Craig, A. G., White, V. A., Milner, D. A., Beare, N. A., et al. (2015). Severity of retinopathy parallels the degree of parasite sequestration in the eyes and brains of malawian children with fatal cerebral malaria. *The Journal of Infectious Diseases*, *211*(12), 1977–1986.

Bartoloni, A., & Zammarchi, L. (2012). Clinical aspects of uncomplicated and severe malaria. *Mediterranean Journal of Hematology and Infectious Diseases*, *4*(1), e2012026.

Baum, J., Richard, D., & Riglar, D. T. (2017). Malaria parasite invasion: achieving superb resolution. *Cell Host & Microbe*, *21*(3), 294–296.

Blasco, B., Leroy, D., & Fidock, D. A. (2017). Antimalarial drug resistance: linking *Plasmodium falciparum* parasite biology to the clinic. *Nature Medicine*, *23*(8), 917–928.

Bohórquez, E. B., Juliano, J. J., Kim, H. S., & Meshnick, S. R. (2013). Mefloquine exposure induces cell cycle delay and reveals stage-specific expression of the pfmdr1 gene. *Antimicrobial Agents and Chemotherapy*, *57*(2), 833–839.

Boudová, S., Divala, T., Mungwira, R., Mawindo, P., Tomoka, T., & Laufer, M. K. (2017). Placental but not peripheral *Plasmodium falciparum* infection during pregnancy is associated with increased risk of malaria in infancy. *The Journal of Infectious Diseases*, *216*(6), 732–735.

Cheruiyot, J., Ingasia, L. A., Omondi, A. A., Juma, D. W., Opot, B. H., Ndegwa, J. M., et al. (2014). Polymorphisms in Pfmdr1, Pfcrt, and Pfnhe1 genes are associated with reduced in vitro activities of quinine in Plasmodium falciparum isolates from western Kenya. *Antimicrobial Agents and Chemotherapy*, *58*(7), 3737–3743.

Chu, C. S., & White, N. J. (2016). Management of relapsing *Plasmodium vivax* malaria. *Expert Review of Anti-Infective Therapy*, *14*(10), 885–900.

Cohee, L. M., & Laufer, M. K. (2017). Malaria in children. *Pediatric Clinics of North America*, *64*(4), 851–866.

Combrinck, J. M., Mabotha, T. E., Ncokazi, K. K., Ambele, M. A., Taylor, D., Smith, P. J., et al. (2013). Insights into the role of heme in the mechanism of action of antimalarials. *ACS Chemical Biology*, *8*(1), 133–137.

Cowman, A. F., Healer, J., Marapana, D., & Marsh, K. (2016). Malaria: biology and disease. *Cell*, *167*(3), 610–624.

Ecker, A., Lehane, A. M., Clain, J., & Fidock, D. A. (2012). PfCRT and its role in antimalarial drug resistance. *Trends in Parasitology*, *28*(11), 504–514.

Eick-Cost, A. A., Hu, Z., Rohrbeck, P., & Clark, L. L. (2017). Neuropsychiatric outcomes after mefloquine exposure among U.S. military service members. *The American Journal of Tropical Medicine and Hygiene*, *96*(1), 159–166.

Fana, S. A., Bunza, M. D., Anka, S. A., Imam, A. U., & Nataala, S. U. (2015). Prevalence and risk factors associated with malaria infection among pregnant women in a semi-urban community of north-western Nigeria. *Infectious Diseases of Poverty*, *4*, 24.

Frevert, U., & Nacer, A. (2014). Fatal cerebral malaria: a venous efflux problem. *Frontiers in Cellular and Infection Microbiology*, *4*, 155.

Fried, M., & Duffy, P. E. (2017). Malaria during pregnancy. *Cold Spring Harbor Perspectives in Medicine*, *7*(6), a025551.

Gaillard, T., Madamet, M., & Pradines, B. (2015). Tetracyclines in malaria. *Malaria Journal*, *14*, 445.

Gamo, F. J., Sanz, L. M., Vidal, J., de Cozar, C., Alvarez, E., Lavandera, J. L., et al. (2010). Thousands of chemical starting points for antimalarial lead identification. *Nature*, *465*(7296), 305–310.

Garg, R. K. (2000). Cerebral malaria. *The Journal of the Association of Physicians of India*, *48*(10), 1004–1013.

Gaur, D., & Chitnis, C. E. (2011). Molecular interactions and signaling mechanisms during erythrocyte invasion by malaria parasites. *Current Opinion in Microbiology*, *14*(4), 422–428.

Gaur, R., Darokar, M. P., Ajayakumar, P. V., Shukla, R. S., & Bhakuni, R. S. (2014). In vitro antimalarial studies of novel artemisinin biotransformed products and its derivatives. *Phytochemistry*, *107*, 135–140.

Goldberg, D. E., Siliciano, R. F., & Jacobs, W. R. (2012). Outwitting evolution: Fighting drug-resistant TB, malaria, and HIV. *Cell*, *148*(6), 1271–1283.

Goss, A. (2014). Building the world's supply of quinine: Dutch colonialism and the origins of a global pharmaceutical industry. *Endeavour*, *38*(1), 8–18.

Gupta, R., Singh, A., Gupta, M. M., & Pandey, R. (2016). Cumulative role of bioinoculants on growth, antioxidant potential and artemisinin content in *Artemisia annua* L. under organic field conditions. *World Journal of Microbiology and Biotechnology*, *32*(10), 167.

Hastings, I. M., Kay, K., & Hodel, E. M. (2015). How robust are malaria parasite clearance rates as indicators of drug effectiveness and resistance. *Antimicrobial Agents and Chemotherapy*, *59*(10), 6428–6436.

Heinberg, A., & Kirkman, L. (2015). The molecular basis of antifolate resistance in *Plasmodium falciparum*: looking beyond point mutations. *Annals of the New York Academy of Sciences*, *1342*, 10–18.

Hrycyna, C. A., Summers, R. L., Lehane, A. M., Pires, M. M., Namanja, H., Bohn, K., et al. (2014). Quinine dimers are potent inhibitors of the *Plasmodium falciparum* chloroquine resistance transporter and are active against quinoline-resistant *P. falciparum*. *ACS Chemical Biology*, *9*(3), 722–730.

Jensen, M., & Mehlhorn, H. (2009). Seventy-five years of Resochin in the fight against malaria. *Parasitology Research*, *105*(3), 609–627.

Joste, V., Kamaliddin, C., Kendjo, E., Hubert, V., Argy, N., & Houzé, S. (2018). Distinction of Plasmodium ovale wallikeri and Plasmodium ovale curtisi using quantitative polymerase chain reaction with high resolution melting revelation. *Scientific Reports*, *8*(1), 300.

Krzych, U., Zarling, S., & Pichugin, A. (2014). Memory T cells maintain protracted protection against malaria. *Immunology Letters*, *161*(2), 189–195.

Li, J., Zhang, C., Gong, M., & Wang, M. (2018). Combination of artemisinin-based natural compounds from *Artemisia annua* L. for the treatment of malaria: pharmacodynamic and pharmacokinetic studies. *Phytotherapy Research, 32*(7), 1415–1420.

malERA Refresh Consultative Panel on Insecticide and Drug Resistance. (2017). malERA: an updated research agenda for insecticide and drug resistance in malaria elimination and eradication. *PLoS Medicine, 14*(11), e1002450.

Marinho, C. R., Neres, R., Epiphanio, S., Gonçalves, L. A., Catarino, M. B., & Penha-Gonçalves, C. (2009). Recrudescent Plasmodium berghei from pregnant mice displays enhanced binding to the placenta and induces protection in multigravida. *PLoS One, 4*(5), e5630.

Matteelli, A., Caligaris, S., Castelli, F., & Carosi, G. (1997). The placenta and malaria. *Annals of Tropical Medicine and Parasitology, 91*(7), 803–810.

Mayor, A., Aponte, J. J., Fogg, C., Saúte, F., Greenwood, B., Dgedge, M., et al. (2007). The epidemiology of malaria in adults in a rural area of southern Mozambique. *Malaria Journal, 6*, 3.

Mayor, A., Hafiz, A., Bassat, Q., Rovira-Vallbona, E., Sanz, S., Machevo, S., et al. (2011). Association of severe malaria outcomes with platelet-mediated clumping and adhesion to a novel host receptor. *PLoS One, 6*(4), e19422.

Meibalan, E., & Marti, M. (2017). Biology of malaria transmission. *Cold Spring Harbor Perspectives in Medicine, 7*(3), a025452.

Mfueni, E., Devleesschauwer, B., Rosas-Aguirre, A., Van Malderen, C., Brandt, P. T., Ogutu, B., et al. (2018). True malaria prevalence in children under five: Bayesian estimation using data of malaria household surveys from three sub-Saharan countries. *Malaria Journal, 17*(1), 65.

Miller, L. H., Ackerman, H. C., Su, X. Z., & Wellems, T. E. (2013). Malaria biology and disease pathogenesis: insights for new treatments. *Nature Medicine, 19*(2), 156–167.

Milner, D. A., Lee, J. J., Frantzreb, C., Whitten, R. O., Kamiza, S., Carr, R. A., et al. (2015). Quantitative assessment of multiorgan sequestration of parasites in fatal pediatric cerebral malaria. *The Journal of Infectious Diseases, 212*(8), 1317–1321.

Mohapatra, P. K., Sarma, D. K., Prakash, A., Bora, K., Ahmed, M. A., Sarma, B., et al. (2014). Molecular evidence of increased resistance to anti-folate drugs in *Plasmodium falciparum* in North-East India: a signal for potential failure of artemisinin plus sulphadoxine-pyrimethamine combination therapy. *PLoS One, 9*(9), e105562.

Moles, P., Oliva, M., Sánchez-González, A., & Safont, V. S. (2010). A topological study of the decomposition of 6,7,8-trioxabicyclo[3.2.2]nonane induced by Fe(II): modeling the artemisinin reaction with heme. *The Journal of Physical Chemistry B, 114*(2), 1163–1173.

Molyneux, M. E. (1990). Cerebral malaria in children: clinical implications of cytoadherence. *The American Journal of Tropical Medicine and Hygiene, 43*(2 Pt. 2), 38–41.

Moore, B. R., Salman, S., & Davis, T. M. (2016). Treatment regimens for pregnant women with falciparum malaria. *Expert Review of Anti-Infective Therapy, 14*(8), 691–704.

Ngotho, P., Soares, A. B., Hentzschel, F., Achcar, F., Bertuccini, L., & Marti, M. (2019). Revisiting gametocyte biology in malaria parasites. *FEMS Microbiology Reviews, 43*(4), 401–414.

Oguike, M. C., Falade, C. O., Shu, E., Enato, I. G., Watila, I., Baba, E. S., et al. (2016). Molecular determinants of sulfadoxine-pyrimethamine resistance in *Plasmodium falciparum* in Nigeria and the regional emergence of dhps 431V. *International Journal for Parasitology: Drugs and Drug Resistance, 6*(3), 220–229.

Pattyn, S. R., & Verdoolaege-Van Loo, G. (1972). Absence of enhancing effect of trimethoprim on the activity of a sulfonamide against *Mycobacterium leprae*. *International Journal of Leprosy and Other Mycobacterial Diseases, 40*(3), 320.

Permin, H., Norn, S., Kruse, E., & Kruse, P. R. (2016). On the history of Cinchona bark in the treatment of Malaria. *Dansk Medicinhistorisk Årbog, 44*, 9–30.

Peters, W. (1990). Plasmodium: resistance to antimalarial drugs. *Annales de Parasitologie Humaine et Comparée, 65*(Suppl. 1), 103–106.

Phillips, M. A., Burrows, J. N., Manyando, C., van Huijsduijnen, R. H., Van Voorhis, W. C., & Wells, T. N. C. (2017). Malaria. *Nature Reviews. Disease Primers, 3*, 17050.

Ponsford, M. J., Medana, I. M., Prapansilp, P., Hien, T. T., Lee, S. J., Dondorp, A. M., et al. (2012). Sequestration and microvascular congestion are associated with coma in human cerebral malaria. *The Journal of Infectious Diseases, 205*(4), 663–671.

Puri, S. K., & Dutta, G. P. (1982). Antibiotics in the chemotherapy of malaria. *Progress in Drug Research, 26*, 167–205.

Reiling, S. J., & Rohrbach, P. (2015). Monitoring PfMDR1 transport in *Plasmodium falciparum. Malaria Journal, 14*, 270.

Sanchez, C. P., Rohrbach, P., McLean, J. E., Fidock, D. A., Stein, W. D., & Lanzer, M. (2007). Differences in trans-stimulated chloroquine efflux kinetics are linked to PfCRT in *Plasmodium falciparum. Molecular Microbiology, 64*(2), 407–420.

Seeman, J. I. (2007). The Woodward-Doering/Rabe-Kindler total synthesis of quinine: setting the record straight. *Angewandte Chemie (International Ed. in English), 46*(9), 1378–1413.

Shulman, C. E., & Dorman, E. K. (2003). Importance and prevention of malaria in pregnancy. *Transactions of the Royal Society of Tropical Medicine and Hygiene, 97*(1), 30–35.

Shulman, C. E., Marshall, T., Dorman, E. K., Bulmer, J. N., Cutts, F., Peshu, N., et al. (2001). Malaria in pregnancy: adverse effects on haemoglobin levels and birthweight in primigravidae and multigravidae. *Tropical Medicine & International Health, 6*(10), 770–778.

Staines, H. M., Burrow, R., Teo, B. H., Chis Ster, I., Kremsner, P. G., & Krishna, S. (2018). Clinical implications of Plasmodium resistance to atovaquone/proguanil: a systematic review and meta-analysis. *The Journal of Antimicrobial Chemotherapy, 73*(3), 581–595.

Tarnchompoo, B., Sirichaiwat, C., Phupong, W., Intaraudom, C., Sirawaraporn, W., Kamchonwongpaisan, S., et al. (2002). Development of 2,4-diaminopyrimidines as antimalarials based on inhibition of the S108N and C59R+S108N mutants of dihydrofolate reductase from pyrimethamine-resistant *Plasmodium falciparum. Journal of Medicinal Chemistry, 45*(6), 1244–1252.

Thu, A. M., Phyo, A. P., Landier, J., Parker, D. M., & Nosten, F. H. (2017). Combating multidrug-resistant *Plasmodium falciparum* malaria. *The FEBS Journal, 284*(16), 2569–2578.

Tilley, L., Straimer, J., Gnädig, N. F., Ralph, S. A., & Fidock, D. A. (2016). Artemisinin action and resistance in *Plasmodium falciparum. Trends in Parasitology, 32*(9), 682–696.

van Eijk, A. M., Hill, J., Noor, A. M., Snow, R. W., & ter Kuile, F. O. (2015). Prevalence of malaria infection in pregnant women compared with children for tracking malaria transmission in sub-Saharan Africa: a systematic review and meta-analysis. *The Lancet Global Health, 3*(10), e617–e628.

van Spronsen, J. H., Schneider, T. A., & Atasige, S. (2012). Placental malaria and the relationship to pregnancy outcome at Gushegu District Hospital, Northern Ghana. *Tropical Doctor, 42*(2), 80–84.

Vaughan, A. M., & Kappe, S. H. I. (2017). Malaria parasite liver infection and exoerythrocytic biology. *Cold Spring Harbor Perspectives in Medicine, 7*(6), a025486.

Warrell, D. A. (1992). Cerebral malaria. *Schweizerische Medizinische Wochenschrift, 122*(23), 879–886.

Watanabe, T. (1963). Infective heredity of multiple drug resistance in bacteria. *Bacteriological Reviews, 27*, 87–115.

Wong, W., Bai, X. C., Sleebs, B. E., Triglia, T., Brown, A., Thompson, J. K., et al. (2017). Mefloquine targets the *Plasmodium falciparum* 80S ribosome to inhibit protein synthesis. *Nature Microbiology, 2*, 17031.

Wongsrichanalai, C., Pickard, A. L., Wernsdorfer, W. H., & Meshnick, S. R. (2002). Epidemiology of drug-resistant malaria. *The Lancet Infectious Diseases, 2*(4), 209–218.

World Health Organization. (2013). *Emergency response to artemisinin resistance in the greater Mekong subregion. Regional framework for action 2013–2015.* WHO. http://www.who.int/malaria/publications/atoz/9789241505321/en/.

World Health Organization. (2016). *Global technical strategy for Malaria 2016–2030*. WHO. http://who.int/malaria/areas/global_technical_strategy/en/.

World Health Organization. (2018a). *World malaria report*. WHO. https://www.who.int/malaria/publications/world-malaria-report-2018/en/.

World Health Organization. (2018b). *Overview of malaria treatment*. WHO. https://www.who.int/malaria/areas/treatment/overview/en/.

World Health Organization. (2014). Severe malaria. *Tropical Medicine & International Health, 19*(Suppl. 1), 7–131.

Zaw, M. T., & Lin, Z. (2019). Human *Plasmodium knowlesi* infections in South-East Asian countries. *Journal of Microbiology, Immunology, and Infection, 52*(5), 679–684.

Further reading

Chukwudi, C. U. (2016). rRNA binding sites and the molecular mechanism of action of the tetracyclines. *Antimicrobial Agents and Chemotherapy, 60*(8), 4433–4441.

CHAPTER

9

Combination therapy as an effective tool for treatment of drug-resistant viral infections

Musa Marimani[a,*], *Aijaz Ahmad*[a,b], *Adriano Duse*[a,b]

[a]Department of Clinical Microbiology and Infectious Diseases, School of Pathology, Health Sciences, University of the Witwatersrand, Johannesburg, South Africa, [b]Division of Infection Control, Charlotte Maxeke Johannesburg Academic Hospital, National Health Laboratory Service (NHLS), Johannesburg, South Africa

9.1 Emergence, clinical impact, and mechanisms of microbial drug resistance

Antimicrobial resistance is the ability of microorganisms such as viruses, bacteria, fungi, and parasites to replicate and disseminate in the presence of antimicrobial agents that would normally retard their growth and pathogenicity. Microbial pathogens have evolved a number of innovative strategies to render antimicrobial therapies inefficient. These include alteration of enzyme catalytic activity (Latallo et al., 2017), activation of membrane impermeability and efflux systems (Geisinger & Isberg, 2017), and induction of genetic mutations in the therapeutic target (Zhao et al., 2018). In most cases, the presence of resistance genes drives the emergence of drug-resistant pathogens following exertion of immense selection pressure on the therapeutic molecule (Hupalo et al., 2016). The major strategy with which microbial pathogens acquire drug resistance is by inducing genetic mutations (Dong et al., 2015; Martinez & Baquero, 2000; Yang

*Musa Marimani wrote the book chapter with the assistance and guidance of Aijaz Ahmad and Adriano Duse.

https://doi.org/10.1016/B978-0-12-820576-1.00009-6

et al., 2015) and via horizontal gene transfer (Chen et al., 2016; Gilbert & Cordaux, 2017; Liu et al., 2010).

Incorporation of genetic mutations allows the pathogens to develop drug resistance toward antimicrobial therapy (Fig. 9.1), while horizontal gene transfer enables dissemination of resistant genes from a drug-resistant microbe to a susceptible one (Fig. 9.2). Importantly, transfer of genetic mutations between microorganisms condones acquisition of drug-resistant genes by other members of the microbial population. Furthermore, administration of antimicrobial molecules across different industries also

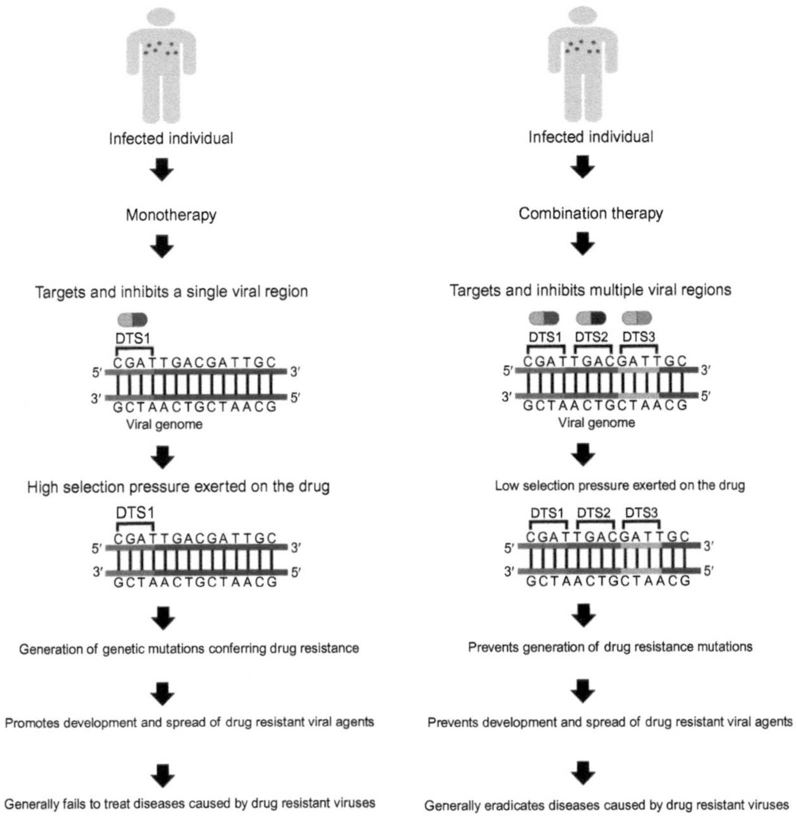

FIG. 9.1 Effect of mono- and combination therapy on development of drug-resistant viruses. Long-term monotherapy targeting a specific viral region results in high selection pressure being exerted on the antiviral agent. Mutations conferring drug resistance develop, leading to emergence of drug-resistant viral strains that can tolerate and survive in the presence of conventional antiviral molecules. Combination therapy consists of a cocktail of antimicrobial molecules targeting different viral regions simultaneously. Low selection pressure is exerted on each individual drug by viral pathogens and prevents development and spread of drug-resistant viral escape mutants. Relative to monotherapy, combination therapy is highly effective in the treatment of drug-resistant infections. *DTS*, drug target site.

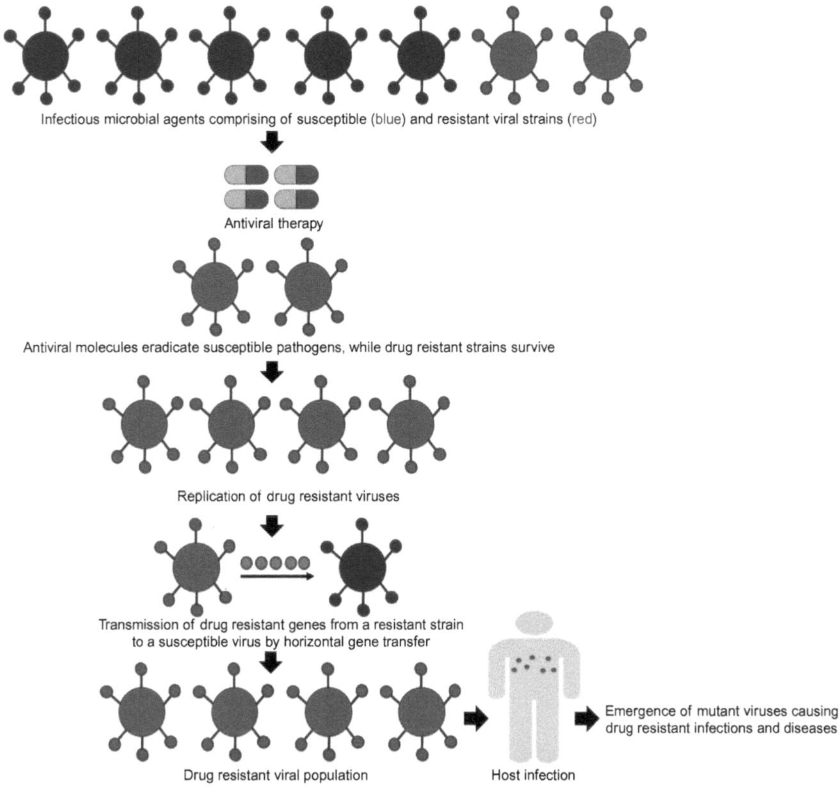

FIG. 9.2 Emergence and dissemination of drug-resistant viral strains. The microbial population is composed of susceptible and resistant strains. The susceptible viruses are killed, while the resistant strains survive and proliferate following treatment with antiviral agents. Horizontal gene transfer facilitates transmission of resistant genes from resistant viral strains to susceptible species. Host infection by drug-resistant strains leads to development of diseases that are untreatable by conventional antiviral therapies.

complicates and undermines stringent infection-control measures. For example, transmission of drug-resistant pathogens may occur from:

(a) an infected individual to a healthy individual,
(b) an infected person to healthy livestock,
(c) an infected livestock to healthy livestock, or
(d) an infected livestock to a healthy individual after ingestion of infected animal products or via zoonosis.

Additionally, overuse of a single antimicrobial drug (i.e., monotherapy) (Fig. 9.1) allows the development of drug escape mutants that replicate and propagate resistant traits to the next generation following microbial cell division (Fig. 9.2). This situation is further complicated by the long research process required to produce novel and effective therapeutic compounds to eradicate drug-resistant strains. The presence of drug-resistant microorganisms thus has devastating clinical implications in both developed and developing countries (Gulen et al., 2015). Specifically, antimicrobial resistance results in administration of more powerful and expensive antimicrobial therapies, lengthy recovery time and hospital stay, dissemination of drug-resistant strains to unsuspecting individuals, as well as high morbidity and mortality rates (Gulen et al., 2015). Numerous factors have been associated with the existence of nondrug-responsive pathogens and include, among others:

(a) Incorrect diagnosis,
(b) Inappropriate drug use,
(c) Incorrect prescription and/or overprescribing antimicrobial drugs, and
(d) Noncompliance with treatment plan.

This chapter will focus on various factors that critically contribute to development of drug resistance in viral pathogens of clinical significance, including hepatitis B, influenza A, herpes simplex and cytomegalovirus. Of paramount importance, the application of combination therapy to combat the devastating clinical impact of drug resistant pathogens is also examined.

9.2 Mechanism of drug resistance in viral pathogens of clinical significance

9.2.1 Drug resistance in hepatitis B virus

The hepatitis B virus (HBV) belongs to the *Hepadnaviridae* family and includes avian hepadnaviruses and orthohepadnaviruses. The HBV genome is composed of four overlapping open reading frames (ORFs). These include *surface* (*pre-S1*, *pre-S2*, and *S*), *polymerase* (*Pol*), *hepatitis B x* (*HBx*) and *precore/core* (Tuttleman, Pourcel, & Summers, 1986). These viral RNAs code for seven proteins including the three surface proteins (pres-S1, pre-S2, and S), polymerase, core, HBx, and hepatitis B core antigen (HBcAg). Depending on the host immune competence, HBV infection can induce an acute or chronic disease state. Acute HBV infection does not usually require medical attention as the pathogen is cleared by the host immune system. However, chronic HBV may result in dire medical complications such as cirrhosis (Perz et al., 2006) and hepatocellular carcinoma (Benvegnu et al., 1994). Globally, an estimated 257 million individuals were living with chronic HBV infection in 2015 (World Health Organization, 2017a, 2017b). The majority of these patients were located in the African and Western Pacific regions.

The virus may be transmitted:

(a) By being in contact with an infected person (horizontal transmission) (Tabor et al., 1985),
(b) By sexual contact with an infected individual (Hadchouel et al., 1985),
(c) By contact with contaminated body fluids or blood (parenteral transmission) (Scott et al., 1980), or
(d) From mother to child during gestation (perinatal transmission) (Zuckerman, 1984).

Clinically, an individual is defined as having the chronic form of HBV infection when the hepatitis B surface antigen (HBsAg), which indicates recent HBV infection, persists in the serum for more than 6 months. An effective vaccine to prevent HBV infection is available and generally provides about 90%–95% protection in immunized individuals (Margolis, 1998). This vaccine may also be taken in conjunction with hepatitis B immune globulin (HBIG) to prevent mother-to-child transmission. However, unvaccinated individuals remain at risk of acquiring chronic HBV infection. Antimicrobial therapies used to treat chronic HBV infection include nucleoside and nucleotide drugs such as adefovir, lamivudine, telbivudine, tenofovir, and entecavir, which inhibit the reverse transcription phase during viral replication (Doong et al., 1991). Immunotherapies such as PEGylated interferon alpha (IFN-α), PEGylated IFN-α 2a, IFN-α and interferon alpha beta (IFN-β), which activate the host immunity to combat infection, are also employed for chronic HBV treatment.

Drug resistance to HBV therapies is linked to virologic breakthrough, which is described as an increase in HBV DNA levels in patient serum by about 1.0 $\log_{10}$ (Lok, 2008). Subsequently, biochemical breakthrough, which is accompanied by a significant increase in hepatic aspartate transaminase and alanine aminotransferase levels, is activated (Gitlin, 1997; Kim et al., 2008). Interestingly, approximately 30%–50% of viral breakthroughs reported during clinical trials have been demonstrated to be effected by lack of medical compliance (Lok, 2008). The emergence of drug-resistant HBV strains has a negative impact on clinical and research milestones aimed at eliminating chronic HBV infection. Both nucleotide and nucleoside therapies target the reverse transcriptase enzyme responsible for converting viral RNA to DNA. The HBV pathogens that are resistant to nucleotide and nucleoside analogues have mutations in the reverse transcriptase gene, particularly within the B, C, and D domains. The presence of these mutations weakens the binding affinity between viral enzyme and antimicrobial drugs (Das et al., 2001). Additionally, these mutations also induce conformational changes that significantly alter substrate binding or structural changes that impede the therapeutic activity of the administered antimicrobial molecule (Sharon & Chu, 2008).

Adefovir-resistant strains harbor the N236T and A181V/T mutations (Angus et al., 2003; Hadziyannis et al., 2006). These mutations regulate HBV replication in infected host cells and trigger cross-resistance to adefovir and tenofovir (Yang et al., 2004). The M204I/V (Allen et al., 1998) and A181V/T (Villet et al., 2008) mutations confer resistance to lamivudine. Furthermore, M204I/V together with secondary mutations such as L801/V or L180M have also been shown to render HBV resistant to lamivudine therapy (Liaw et al., 2009). Viral strains carrying the M204I/V mutation display increased cross-resistance to telbivudine, but do not alter viral susceptibility to tenofovir and adefovir (Liaw et al., 2009; Yang et al., 2005). Drug resistance to entecavir is thought to occur via a 2-hit system. In this mechanism, lamivudine-resistant HBV mutants possessing M204I/V and L180M are selected in patients undergoing lamivudine or primary entecavir treatment (Kobashi et al., 2009). During the course of entecavir therapy, development of additional mutations such as M250V and I169T or S2021 and T184G drives entecavir resistance (Langley et al., 2007; Tenney et al., 2004, 2007). Resistance to emtricitabine is a result of M204I/V mutation alone or accompanied by V173L and L180M mutations, thus conferring cross-resistance to both telbivudine and lamivudine (Strasfeld & Chou, 2010).

9.2.2 Drug resistance in influenza A virus

Influenza A virus is an infectious airborne pathogen that causes respiratory diseases and seasonal flu outbreaks (Chen et al., 2013; Horimoto & Kawaoka, 2005). This virus belongs to the *Orthomyxoviridae* family and infects humans and other animals (Liu, Shi, & Gao, 2014; Wu et al., 2014). About 290,000–650,000 mortalities occur each year as a result of seasonal flu (World Health Organization, 2017c). The viral genome is composed of eight negative-sense single-stranded RNAs that code for viral proteins. These include polymerase (PB1, PB2, and PA), nucleoprotein (NP), transmembrane proteins (HA, NA, and M2), viral nuclear export protein (NEP), matrix1 protein (M1), and nonstructural protein 1 (NS1) (Jagger et al., 2012; Muramoto et al., 2013; Wise et al., 2009). The most essential viral protein is NP, which plays a crucial role during RNA binding (Elton et al., 1999a; Portela & Digard, 2002), oligomerization (Elton et al., 1999b), and viral replication (Lamb & Choppin, 1983). Furthermore, NP phosphorylation is critical for maintaining viral replication and polymerase catalytic activity (Cohen, 2000) in infected host cells, leading to prolonged intracellular survival (Li et al., 2018). The virus may be spread by:

(a) Inhalation of contaminated droplets,
(b) Being in contact with contaminated objects, or
(c) Being in contact with an infected individual.

Treatment of influenza A viral infection includes the use of rimantadine, amantadine, oseltamivir, and zanamivir. Viral pathogens resistant to the adamantane drug class of antiviral agents (e.g., amantadine and rimantadine) have been described. These antiviral therapies target the M2 proton channel, which regulates the pH across the viral membrane during viral entry, replication, and maturation in infected host cells. The presence of adamantane-resistant strains has a devastating clinical impact, as drug-resistant levels of about 90% have been reported (Bright et al., 2006; Nelson et al., 2009). A point mutation in serine-to-asparagine at amino acid 31 (S31N) of the M2 protein has been implicated as the major contributor in conferring resistance to adamantane drugs (Bright et al., 2006; Nelson et al., 2009). Additionally, it is also a leading mutation responsible for adamantane resistance in H1N1, H5N1, and H3N2 strains extracted from swine, birds, and humans (Furuse, Suzuki, & Oshitani, 2009; Krumbholz et al., 2009; Schmidtke et al., 2006). However, other amino acid mutations that play a role in inducing adamantane resistance include A30T/V, G34E, L26F, L38F, and V27A located in the M2 transmembrane region lining the channel pore. These mutations result in a significant increase in hydrophilicity and pore size of the channel and may also destabilize helix-to-helix assembly or decrease the pore size, thus rendering antiviral therapy ineffective (Cady et al., 2010; Pielak, Schnell, & Chou, 2009; Schnell & Chou, 2008; Thomaston & DeGrado, 2016). It has also been demonstrated that strains harboring drug-resistant genes weaken the transmembrane packing and markedly increase structural heterogeneity (Pielak et al., 2009).

9.2.3 Drug resistance in herpes simplex virus

The herpes simplex virus (HSV) is a viral agent that causes contagious sores, mainly around the mouth and genitals. The virus belongs to the *Herpesviridae* family, which consists of eight members and is classified into three subfamilies, α, β and γ, according to their respective biological and genetic properties (Gilbert, Bestman-Smith, & Boivin, 2002). The α-HSVs include HSV-1, HSV-2, and varicella zoster virus (VZV) (Burrel et al., 2013; Jones et al., 2014; Markus et al., 2011; Suazo et al., 2015). HSV-1 is mainly associated with perioral and oral infections, while HSV-2 is mainly correlated with genital infections. An estimated 3.7 billion and 417 million people globally have HSV-1 and HSV-2 infections, respectively (World Health Organization, 2017a, 2017b). Infection by HSV results in a wide variety of mild and chronic clinical manifestations such as keratitis, cold sores, encephalitis, and corneal blindness. These manifestations compromise the host immune response and dramatically increase the risk of acquiring and developing HIV infection (Jiang et al., 2016; Mesquita et al., 2017). Infection caused by HSV is prevalent in immunocompromised individuals, transplant recipients, pregnant women, and newborns. Viral dissemination is through:

(a) Kissing or sharing contaminated objects, or
(b) Oral secretions or sores on infected skin.

In contrast, HSV-2 is generally transmitted via sexual contact. The three groups of antiviral compounds used to treat HSV infections are acyclic guanosine (Drake et al., 2012; Gopal et al., 2013; Perti et al., 2013), acyclic nucleoside (Gokulgandhi et al., 2012; Momper et al., 2013; Vossen et al., 2014) and pyrophosphate drugs (Gilbert et al., 2011; Minces et al., 2014). These agents target and inhibit viral DNA replication and commonly used antiviral compounds are acyclovir, cidofovir, and foscarnet. Of these, acyclovir is the most effective agent and is employed for prophylaxis and treatment of HSV diseases. This antiviral agent along with other nucleoside drugs is phosphorylated to monophosphate in infected host cells by the enzyme thymidine kinase. Subsequently, cellular kinases catalyze the conversion of monophosphate to diphosphate and triphosphate active derivatives. The triphosphate derivative is also incorporated into replicating viral particles due to the absence of the 3′ hydroxyl group, leading to termination of viral DNA replication (De Clercq, 2008; De Clercq & Holy, 2005; Gilbert et al., 2002).

Long-term application of acyclovir and other anti-HSV drugs led to development of drug-resistant viral escape mutants capable of surviving therapeutic onslaught (Martinez et al., 2006; Safrin et al., 1991; Sims et al., 2007). These drug-resistant strains survive and flourish in infected host cells by:

(a) Producing decreased levels of thymidine kinase enzyme,
(b) Halting or decreasing activity of thymidine kinase, or
(c) Producing DNA polymerase and thymidine kinase with modified substrate specificity.

The mutations responsible for inducing acyclovir resistance are found in viral DNA polymerase protein (UL30) and activating/phosphorylation gene (*UL23*) in thymidine kinase (Breton et al., 1998; Crassard et al., 2000; Saint-Leger et al., 2001). Generally, genetic alterations in the *thymidine kinase* gene lead to production of deficient enzymes due to deletion or addition of nucleotides. An estimated 95% of acyclovir-resistant strains are as a result of thymidine kinase deficiency. Of great significance, even a single nucleotide change in DNA polymerase may render the virus resistant to various antiviral therapies (Bryan et al., 2008; Levin et al., 2003; Morfin et al., 1999). Consequently, many acyclovir resistant mutants also display resistance to penciclovir and this is activated by a single mutation in DNA polymerase (Jiang et al., 2016). Mutations in the *DNA polymerase* gene are infrequent and are mainly located in functional domains II and

III (Bestman-Smith & Boivin, 2003; Gilbert et al., 2002). Interestingly, specific genetic mutations in thymidine kinase emanating from penciclovir administration vary from those induced following acyclovir application. Nevertheless, cross-resistance to ganciclovir (Ljungman et al., 1990) and other antimicrobial compounds may occur in viral pathogens with thymidine kinase deficiency (Bacon et al., 2003). Encouragingly, drug-resistant strains with a deficient thymidine kinase remain susceptible to antiviral agents (e.g., cidofovir and foscarnet) that do not rely on viral-mediated phosphorylation, provided mutations in DNA polymerase are absent. In murine models, it has been illustrated that thymidine kinase-deficient mutants have the capacity to resurge from a latent state. This is achieved by reversion as a result of replication errors or ribosomal frameshifting that enables formation of drug-resistant HSV strains with modified thymidine kinase structure and activity (Griffiths et al., 2003, 2006). Although vidarabine displayed impressive anti-HSV activity in cultured cells (De Clercq, 1984), it was rendered clinically ineffective when exposed to acyclovir resistant mutants (Safrin et al., 1991).

9.2.4 Drug resistance to cytomegalovirus

Cytomegalovirus (CMV) is a viral agent that belongs to the *Herpesviridae* family and causes cytomegalovirus disease. This condition can affect different body parts and tissues. The most leading clinical manifestations are pneumonitis, hepatitis, nephritis, and infection of the central nervous system, respectively (Plosa et al., 2012). Infections of the central nervous system include encephalitis, which affects the brain and transverse myelitis, causing inflammation and disrupting normal functioning of the spinal cord. Transmission of CMV is achieved by being in contact with body fluids contaminated with viral particles such as urine, breast milk, saliva, and semen. Similarly to HSV, this viral pathogen mainly infects infants, pregnant women, and immunocompromised individuals (Belo et al., 2012; Micallef & Galea, 2018; Saliba et al., 2004). Specifically, high-risk groups for acquiring CMV infection include:

(a) Neonates infected before birth (congenital CMV infection),
(b) Newborns infected during or shortly after birth (perinatal infection), and
(c) Individuals with compromised immune systems, e.g., HIV patients and organ transplant recipients.

Ganciclovir is the gold standard for treating CMV infection; however, valganciclovir is the drug of choice for preventing diseases in solid-organ transplant recipients. This drug exerts its antiviral activity by blocking incorporation of deoxyguanosine triphosphate into DNA, thereby disrupting the activity of viral DNA polymerase. Other therapies employed for treating CMV diseases are cidofovir, foscarnet, brincidofovir and maribavir. Impressively, letermovir, which inhibits CMV terminase enzyme complex, has recently been approved as a prophylactic drug in adult CMV-seropositive patients undergoing allogeneic hematopoietic stem cell transplant. Brincidofovir, an oral form of cidofovir, and maribavir were tested during clinical trials and have been utilized to treat multidrug-resistant CMV strains (Lischka et al., 2010; Melendez & Razonable, 2015).

With regard to CMV infection, the vast majority of drug-resistance cases were reported in previously seronegative transplant patients (Elion, 1993; Gilbert et al., 2002; Ormrod, Scott, & Perry, 2000). More than 90% of ganciclovir-resistant strains possess mutations in the viral *UL97 kinase* gene (Chou et al., 2002). This leads to a significant decline in the phosphorylation of ganciclovir and does not alter enzyme function during viral replication (Chou et al., 2002; Prichard et al., 1999; Wolf et al., 2001). The CMV-induced UL97 mutations are located at codons 460, 520, and 590–607. The dominant nucleotide alterations found in ganciclovir-resistant strains are canonical mutations, namely A594V, C592G, C603W, H520Q, L595S, and M460V/I (Chou et al., 2002) and are found in > 80% of ganciclovir-resistant clinical isolates (Gilbert et al., 2002). These individual mutations cause ganciclovir resistance and display an IC_{50} ratio of 5–10, while C592G has the lowest resistance level with an approximate IC_{50} ratio of 2.5 (Chou et al., 2002). These reported IC_{50} ratios were obtained from recombinant phenotyping analysis pertaining to various UL97 mutations. Importantly, the majority of ganciclovir mutations are found in the viral DNA polymerase exonuclease domains and generally induce cross-resistance to cidofovir (Chou et al., 2003; Chou et al., 2008).

The UL54 genetic mutations in DNA polymerase allow resistance to all currently approved anti-CMV therapies. Resistance to foscarnet is activated by mutations located in DNA polymerase catalytic regions II, III, and IV. Similarly, mutations present in nonconserved loci can also confer resistance to foscarnet and other antimicrobial compounds such as ganciclovir and cidofovir (Chou et al., 1998, 2003; Gilbert et al., 2002). Additionally, a single nucleotide alteration in UL54 can trigger drug resistance to foscarnet, cidofovir, and ganciclovir (Chou et al., 2008; Chou, Miner, & Drew, 2000; Scott et al., 2007).

The UL97 mutation in viral kinase is also responsible for triggering drug resistance. This enzyme catalyzes phosphorylation of threonine and serine amino acids of viral proteins and also plays a critical role in efficient viral replication (Prichard, 2009). The kinase enzyme is also required for ganciclovir phosphorylation, which enables effective antiviral activity. Consequently, UL97 mutations that disrupt drug phosphorylation render the therapeutic molecule inefficient in neutralizing viral replication and subsequent proliferation. It has been reported that about 90% of ganciclovir resistance cases emanated from solid-organ transplant patients and led to symptom development and an extended treatment schedule (Kotton et al., 2010). In this cohort, high drug resistance levels have been observed in patients undergoing lung (Lurain & Chou, 2010) and kidney (Myhre et al., 2011) transplants. The following section assesses the application of combination therapy in combating drug-resistant infections caused by HBV, influenza A virus, HSV, and cytomegalovirus. The antiviral efficacy of monotherapy vs that of combination therapy is discussed. The section also explores the advantage of combination therapy in effecting prolonged antiviral activity and survival in treated animals and individuals as well as preventing the development of drug-resistant viral strains.

9.3 The use of combination therapy to counter viral drug resistance

9.3.1 Combination therapy against HBV

Microbial genetic evolution, substandard clinical practices by medical personnel, and lack of or improper compliance with the prescribed treatment timeline by patients has massively contributed to the development of drug-resistant viral pathogens. Although application of monotherapy generally resulted in a marked inactivation of viral agents, it also enabled a few drug-resistant escape mutants to survive and flourish in the presence of the prescribed antimicrobial compound. Subsequent proliferation of these "drug-tolerating viruses" allowed transmission of drug-resistant genes from drug-resistant strains to susceptible microorganisms by a variety of strategies, mainly involving horizontal gene transfer. The presence of these drug-resistant pathogens led to development of viral diseases that do not respond to currently licensed antiviral agents. Ultimately, this promoted dissemination of drug-resistant microorganisms from infected patients to other patients, hospital personnel, and other individuals. Consequently, the scourge of these untreatable viral diseases caused by mutant strains has placed an enormous burden on both the public and private healthcare institutions.

Numerous methods have been surveyed with the aim of combating viral drug-resistant diseases. One such method is the development of novel antiviral molecules targeting new therapeutic targets on the viral genome. Although this practice may lead to synthesis of antimicrobial compounds capable of eradicating drug-resistant viruses, its major limitation is the extended period of time required for therapeutic production, in vitro and in vivo analysis, clinical trial setup, and data capturing as well as drug licensing. As a result, this method does not immediately benefit patients currently afflicted with untreatable viral infections who are in dire need of alternative therapy. Therefore combination therapy has been utilized as a speedy approach to treat and eradicate drug-resistant viral strains. This effective strategy involves simultaneous administration of licensed antiviral compounds targeting different regions of the pathogen that play a key role during viral entry, establishing infection, replication, pathogenicity, and dissemination.

A randomized trial consisting of 230 patients with chronic HBV infection was conducted to investigate the clinical benefits of using the combination of IFN-α and lamivudine relative to administration of either drug in isolation (Schalm et al., 2000). Patients were treated with a combination of IFN-α (10 million units) and 100 mg of lamivudine three times per week for up to 16 weeks following pretreatment with lamivudine for 8 weeks ($n=75$), IFN-α (10 million units) three times per week for up to 16 weeks ($n=69$), or 100 mg lamivudine for 52 weeks ($n=82$). Efficacy was assessed at 52 weeks following drug administration by measuring the rate of hepatitis B e antigen (HBeAg) seroconversion which is an indicator of active HBV replication (Schalm et al., 2000). Data emanating from this trial revealed that treatment with lamivudine alone reduced HBeAg seroconversion by 18%, while IFN-α monotherapy suppressed the HBeAg seroconversion rate by 19%. However, the combination of these two antiviral agents effected the highest suppression of HBV replication, by 29%. Importantly, side effects caused by combination therapy were comparable to those triggered by IFN-α, while lamivudine monotherapy resulted in minimal adverse effects (Schalm et al., 2000).

Combinatorial use of IFN-α and lamivudine has also been evaluated in children with chronic HBV infection (Dikici et al., 2001). Specifically, 30 children were treated with IFN-α 2b (10 MU/m^2 of body surface) and 100 mg lamivudine for 6 months (group 1), while 27 patients received the same dosage of IFN-α 2b and lamivudine for up to 12 months (group 2). Data analysis was performed at the end of each respective study and 6 months following completion of therapy. The combination of IFN-α 2b and lamivudine resulted in a considerable suppression of HBeAg secretion of about 33% and 59% at the end of therapy in group 1 and 2, respectively (Dikici et al., 2001). Impressively, simultaneous use of these antiviral agents also effected a marked decrease in HBeAg production, by 37% (group 1) and 56% (group 2) at 6 months following drug administration. Moreover, a remarkable decline in HBV DNA of about 97% and 100% was observed at the end of therapy in group 1 and 2, respectively. A marked decrease in viral DNA was also observed at 6 months following therapy where sustained levels of HBV DNA suppression of about 97% were reported. A significant increase in alanine aminotransferase, an indicator of cytotoxicity, was not observed, indicating safe application of combination therapy without inducing adverse side effects (Dikici et al., 2001).

In a closely related study, chronic HBV patients were provided a cocktail of PEGylated IFN-α 2b (100 g per week) and lamivudine at a daily dosage of 100 mg (Janssen et al., 2005). This drug combination was compared to anti-HBV activity of PEGylated IFN-α 2b monotherapy at a weekly dosage of 100 g for up to 52 weeks. About 36% of patients that received monotherapy and 35% that were given combination therapy had significantly reduced HBeAg, below detectable levels. Encouragingly, 44% of individuals assigned a combination of PEGylated IFN-α 2b and lamivudine, as opposed to 29% of patients offered PEGylated IFN-α 2b alone, efficiently cleared HBeAg at the end of the clinical trial (Janssen et al., 2005). Taken together, these results illustrated that, although treatment with a single antimicrobial agent is capable of reducing markers of HBV replication, combination therapy offers sustained antiviral activity capable of inducing efficient HBV DNA and HBeAg clearance, leading to symptom alleviation in patients with chronic drug-resistant HBV infection (Janssen et al., 2005). Moreover, this strategy results in positive clinical outcomes in patients with drug-resistant infections and may be conveniently used to prevent development of drug-resistant viral strains.

9.3.2 Combination therapy against influenza A virus

The utility of employing combination therapy instead of conventional monotherapy has been explored with the aim of treating and eradicating drug-resistant influenza A viral pathogens. To achieve this goal, a combination of different antiviral compounds targeting various regions of the influenza A virus has been applied to inhibit viral replication and prevent emergence and dissemination of drug-resistant escape mutants.

In a study conducted by Govorkova et al. (2004), a combination of neuraminidase inhibitors (oseltamivir carboxylate, zanamivir, or peramivir) and rimantadine was evaluated in cultured MDCK cells expressing H1N1 and H3N2 strains of influenza A (Govorkova et al., 2004). All three drug combinations, representing oseltamivir carboxylate-rimantadine, zanamivir-rimantadine, or peramivir-rimantadine, displayed synergistic and additive antiinfluenza activity in cultured cells expressing H1N1 and H3N2 subtypes of influenza A virus (Govorkova et al., 2004). A concentration of < 2.5 μM rimantadine coupled with a low dosage of either neuraminidase inhibitor caused a considerable decline in markers associated with active H1N1 replication. In addition to this, all drug combinations resulted in a marked decrease in the extracellular production of the H3N2 influenza strain (Govorkova et al., 2004).

In a separate investigation, 127 patients with pandemic influenza A infection placed in the intensive care unit were treated with oseltamivir monotherapy (103 patients) or a combination of oseltamivir, ribavirin, and amantadine (24 patients) (Kim et al., 2011). At 14 days post-drug administration, 35% mortality was observed in individuals that received oseltamivir alone, while a significantly low fatality of 17% was reported in patients assigned a triple drug combination (Kim et al., 2011). Furthermore, no toxicity emanating from drug application was reported in either cohort (Kim et al., 2011).

In a recent in vivo analysis, nude mice were infected with influenza A virus (Kiso et al., 2018). Subsequently, these animals were treated with a polymerase inhibitor (favipiravir) or neuraminidase inhibitors (oseltamivir and laninamivir) or a combination of polymerase and neuraminidase inhibitors (Kiso et al., 2018). Relative to mice that received monotherapy, a combination of polymerase and neuraminidase antiviral agents significantly prolonged the survival times of infected animals (Kiso et al., 2018).

In another current investigation, the antimicrobial activity of coadministration of zanamivir and oseltamivir as therapy against drug-resistant H1N1 influenza viruses was assessed using the hollow fiber infection model (Pires de Mello et al., 2018). Constant dosages of 600 mg zanamivir and 75 mg oseltamivir were applied as monotherapy or a double-drug combination. A significant decline in viral replication was induced following application of oseltamivir or zanamivir as monotherapy. Similarly, coadministration of these antiviral compounds as combination therapy effected a marked decrease in influenza replication at a level that was comparable to that achieved by monotherapy (Pires de Mello et al., 2018). In contrast to monotherapy, combination therapy prevented the development and dissemination of drug-resistant viral mutants. This encouraging observation indicated that simultaneous administration of antiviral agents is critical for triggering prolonged viral suppression and is also essential for preventing emergence of drug-resistant mutant strains that would result in drug-resistant viral infections. Therefore data pertaining to utilization of different drug combinations to inhibit influenza A virus demonstrate that, although in some cases mono- and combination therapy may exhibit the same level of antiviral activity, the latter is advantageous as it provides long-term efficacy and prevents the emergence and spread of drug-resistant viral strains (Pires de Mello et al., 2018).

9.3.3 Combination therapy against HSV

The beneficial clinical impact of applying a combination of antiviral agents as opposed to a single therapeutic molecule has been investigated for treatment of viral diseases that remain untreatable by conventional anti-HSV drugs. To this end, combinatorial administration of various antimicrobial agents has been employed to target various HSV sites to achieve significant viral suppression, eradication, and subsequent treatment of drug-resistant HSV diseases.

A non-randomized study involving 45 individuals with HSV encephalitis was conducted (Kamei et al., 2005). The patients were treated with acyclovir and the outcome was assessed and analyzed at 3 months following drug administration (Kamei et al., 2005). Treatment of infected patients with acyclovir led to a decrease in markers of HSV replication. However, codelivery of acyclovir and corticosteroid achieved the highest suppression of viral pathogenicity (Kamei et al., 2005).

Tan et al. (2015) evaluated the utility of combining bevacizumab, a monoclonal antibody targeting VEGFA and the HSV HF10 as therapy for breast cancer using the human breast carcinoma xenograft model (Tan et al., 2015). A significant increase in anticancer activity was observed following administration of both anti-HSV agents (Tan et al., 2015). Specifically, bevacizumab inhibited angiogenesis and tumor proliferation, while the oncolytic HSV HF10 significantly induced cell death in the cancerous cell population (Tan et al., 2015).

The beneficial clinical effects of the oncolytic HF10 was also evaluated in a more current study (Tanaka et al., 2017). In this investigation, HF10 and dacarbazine, a cytotoxic molecule employed as chemotherapy against advanced melanoma, were assessed for their anti-HSV activity in cultured mammalian cells and in immunocompetent mice. The antiviral activity of these molecules was determined based on subcutaneous tumor growth, immunological and histopathological analyses, and length of survival. In cultured cells, delivery of dacarbazine or HF10 as monotherapy led to increased cytotoxic effects in melanoma cells. However, a more pronounced cytotoxicity in cancerous cells was achieved following codelivery of these therapeutic agents (Tanaka et al., 2017). In infected mice, a considerable reduction in tumor growth and proliferation of noninoculated contralateral tumors were observed after intratumoral HF10 application. Furthermore, inflammatory cell infiltration and lysis of tumor cells were markedly increased post-HF10 application in infected animals (Tanaka et al., 2017). Encouragingly, coapplication of HF10 and dacarbazine triggered the strongest anticancer immunological response and prolonged the survival of treated mice. Collectively, anti-HSV combination therapy is more clinically beneficial relative to monotherapy with conventional antimicrobial agents, as it is capable of inhibiting viral replication and dissemination, thereby effecting the highest antiviral activity, mounting the strongest host immune response and prolonging the survival of treated animals.

9.3.4 Combination therapy against CMV

The advantage of employing combination therapy to suppress and eradicate CMV diseases that do not respond to monotherapy has been examined. To test the effectiveness of this approach, clinically relevant CMV drug-resistant isolates have been exposed to different anti-CMV agents with the primary aim of achieving marked viral inhibition and symptom alleviation while preventing the emergence of drug escape mutants.

To evaluate the clinical impact of combination therapy against CMV, six organ transplant patients with ganciclovir-resistant CMV infection were given a daily intravenous dose of foscarnet that was steadily increased to 125 mg/kg and a low dosage of ganciclovir (Mylonakis, Kallas, & Fishman, 2002). Following simultaneous application of these therapeutic compounds, a substantial decline in magnesium coupled with a positive clinical response was achieved in all treated individuals within 3–4 days (Mylonakis et al., 2002). This positive clinical outcome was strengthened by the lack of relapse, which was not observed for up to 30 months of follow-up and highlighted that coutilization of ganciclovir and foscarnet may be employed as an effective strategy to treat ganciclovir-resistant CMV infection in organ transplant recipients (Mylonakis et al., 2002).

Cai et al. (2014) evaluated the clinical impact of utilizing a combination of different antiviral compounds with the aim of inhibiting proliferation of drug-resistant ganciclovir strains in cultured cells (Cai et al., 2014). The tested compounds included: (a) compounds that inhibit viral DNA polymerase, (b) compounds targeting cellular proteins and viral DNA polymerase, and (c) compounds with unknown cellular targets. Simultaneous application of foscarnet and ganciclovir resulted in slight synergism; however, the highest synergism was observed after coadministration of ganciclovir with artemisinin-derived molecules or the mitogen-activated protein kinase (MEK) inhibitor U0126 (Cai et al., 2014). Similarly, the combination of cardiac glycosides (ouabain, digitoxin, and digoxin) with ganciclovir also resulted in improved antiviral activity (Cai et al., 2014).

Recently, the clinical outcome accompanying concurrent application of ganciclovir and foscarnet was examined in infants of ≤ 6 months (Vora et al., 2018). The patients were treated with foscarnet only, ganciclovir alone, or a combination of foscarnet and ganciclovir. Infants that received both therapeutic molecules were severely immunocompromised and were considered for hematopoietic cell transplantation. Compared to patients that received either foscarnet or ganciclovir monotherapy, a remarkable and sustained viral suppression was achieved in infants treated with the double-drug combination. Of prime importance was also the absence of clinically significant mutations associated with CMV drug resistance. Unfortunately, adverse side effects such as electrolyte abnormalities, neutropenia, and thrombocytopenia were common (Vora et al., 2018) and may be attributed to the underdeveloped immune response of the infants. Generally, anti-CMV combination therapy leads to superior antiviral activity relative to monotherapy and results in improved clinical outcomes irrespective of patient age and host immune and CMV disease status, and also prevents the emergence and transmission of drug-resistant viral mutants (Vora et al., 2018).

The utility of employing combination therapy has been surveyed due to shortcomings related to long-term monotherapy, such as exertion of high selection pressure and development of drug-resistant infections. The previously mentioned combination therapy studies relating to HBV, influenza A virus, HSV, and cytomegalovirus highlighted the clinical advantage of employing combination therapy over monotherapy. These studies covered a broad range of research and clinical settings, including in vitro, in vivo, and clinical trial studies. Although monotherapy is capable of treating some drug-resistant viral infections, it remains highly susceptible and incapable of preventing or repelling the emergence of drug-resistant viral mutants. Conversely, combination therapy relies on simultaneous administration of a cocktail of therapeutic agents targeting different viral regions. This results in low selection pressure being applied on the therapeutic agents and dramatically decreases development of mutations associated with drug resistance. Consequently, this strategy has been widely employed to treat a number of diseases caused by drug-resistant pathogens and to prevent emergence of drug-resistant viruses (Cai et al., 2014; Dikici et al., 2001; Govorkova et al., 2004; Janssen et al., 2005; Kamei et al., 2005; Kim et al., 2011; Kiso et al., 2018; Mylonakis et al., 2002; Pires de Mello et al., 2018; Schalm et al., 2000; Tan et al., 2015; Tanaka et al., 2017; Vora et al., 2018).

9.4 Conclusion

The development of drug-resistant viral pathogens has placed a huge burden on both public and private healthcare institutions. This has resulted in manifestation of viral diseases that remain untreatable even in the presence of licensed and powerful antiviral agents. The high mutation rates of some viral pathogens and horizontal gene transfer strategies have promoted the emergence and spread of drug-resistant viral strains. Moreover, incorrect diagnosis, improper prescriptions, and overuse of antimicrobial drugs as well as substandard clinical practices by medical staff have also massively allowed development and dissemination of drug-resistant pathogens.

Long-term use of a single antimicrobial molecule or monotherapy has enabled viruses to exert selection pressure on the drug, leading to drug tolerance, drug resistance, and subsequent development of viral diseases that are unresponsive to conventional antiviral therapy. Therefore adherence to the treatment plan by patients and implementation of accurate diagnostic methods and improved medical training and practices by medical staff may minimize the occurrence of drug-resistant viruses. Furthermore, medical interventions capable of eradicating drug-resistant diseases have to be urgently researched and implemented in both the private and public healthcare settings.

In the future, synthesis of novel and effective antiviral therapeutics may treat and prevent drug-resistant viral diseases. Unfortunately, this approach is generally time consuming, as it requires a number of testing phases and analysis from drug development, cell culture and animal screening, clinical trials, and drug approval by relevant authorities. Consequently, it does not immediately benefit patients that are currently afflicted with drug-resistant infections and are in desperate need of urgent medical intervention.

Therefore the combination of already-approved antiviral drugs has been employed as a strategy of alleviating disease symptoms caused by drug-resistant viruses and preventing the transmission of a wide array of drug-resistant viral strains. To date, combination therapy has been effectively applied to counter the replication and spread of clinically significant drug-resistant viral infections that were generally not responding to conventional treatment regimens (Cai et al., 2014; Dikici et al., 2001; Govorkova et al., 2004; Janssen et al., 2005; Kamei et al., 2005; Kim et al., 2011; Kiso et al., 2018; Mylonakis et al., 2002; Pires de Mello et al., 2018; Schalm et al., 2000; Tan et al., 2015; Tanaka et al., 2017; Vora et al., 2018). The reported positive clinical outcomes following application of combination therapy have been mainly attributed to its versatility in targeting and inhibiting different therapeutic drug targets concurrently. As a result, this multitargeting procedure has a considerable negative impact on the various viral signaling pathways needed to establish host infection, replication, and dissemination. Subsequently, this leads to a drastic decrease in the amount of selection pressure exerted on the antiviral compounds by viral pathogens. Ultimately, this has beneficial clinical implications, because it eradicates drug-resistant diseases, preventing the emergence and proliferation of drug-resistant strains. Generally, the use of combination therapy reduces the drug MIC values needed to inhibit microbial replication and also displays synergistic effects that improve antimicrobial activity.

An alternative therapeutic method such as RNA interference (RNAi) may also be employed to treat drug-resistant viral infections. The RNAi mechanism is naturally occurring and plays a critical role in regulating gene expression in various eukaryotic organisms. The double-stranded substrates such as primary microRNAs (pri-miRNAs), precursor miRNAs (pre-miRNAs), and microRNAs (miRNAs) are involved at various stages of the RNAi pathway. Complete base-pairing between the guide strand of these substrates and the target mRNA molecule results in mRNA degradation, while partial base-pairing causes translational suppression. Impressively, when synthetic molecules that resemble pri-miRNAs, pre-miRNAs, or miRNAs are introduced into a cell, the RNAi pathway is activated and reprogrammed, leading to target gene silencing (Benitez et al., 2015; Estrin et al., 2018; Ufaz et al., 2018). Therefore this technique may be utilized to target and suppress the expression of drug-resistant genes, thus blocking the replication and transmission of drug-resistant viruses.

Gene editing strategies such as the clustered regularly interspaced short palindromic repeat (CRISPR)/CRISPR-associated protein 9 (cas9) may be applied to counter the evolution of drug resistant strains viruses and diseases. The CRISPR-cas9 approach may be specifically employed to cleave the viral DNA (Lebbink et al., 2017; Wang et al., 2018; Yuan et al., 2015). Cleavage of target DNA by CRISPR-cas9 results in the formation of double-stranded breaks. Subsequent DNA binding and cleavage cycles are activated following DNA repair mechanisms. The homology directed repair (HDR) and nonhomologous end joining (NHEJ) are involved in repairing the cleaved target DNA molecule. The HDR process results in accurate repair of cleaved target DNA, while the NHEJ procedure is error-prone and leads to introduction of deletions and insertions in the cleaved DNA sequence. As a result, this mechanism may be effectively applied to decrease viral replication following repeated target DNA cleavage events. The resulting mutations in the target DNA coupled with the subsequent decline in viral proliferation may bide the host enough time to mount a powerful immune response, leading to eradication of drug-resistant infections.

The production of immune-stimulatory compounds may also be a feasible strategy to combat drug resistance. The use of these molecules may trigger the host immune signaling pathways to mount a competent immune response, leading to efficient clearance of infectious viral agents. Improved host immune competence achieved following application of the agents is also crucial for secretion of immune cells that play a key role in preventing and/or repelling future microbial infections. Furthermore, production of novel vaccines capable of conferring improved immune protection against a broad range of viral pathogens may prevent infection by drug-resistant strains. Specifically, stimulation of the host adaptive and humoral immune responses through vaccination may generate long-term immune protection against susceptible and drug-resistant microorganisms. Therefore production of immune-stimulatory compounds and novel vaccines are jointly essential for improving and maintaining a competent host immune response capable of:

(a) Preventing host infection,
(b) Effective clearance of infectious microbial agents from circulation, and
(c) Maintaining high numbers of memory T-cells and other immune cells that prevent and repel future host infections.

In the long run, synthesis of novel therapeutics targeting new viral targets that play a critical role during viral entry, infection, replication, and transmission such as virulence genes and proteins may also be an effective approach to treat and counter the emergence and spread of drug-resistant infections and diseases. Ideally, discovery of new drug targets required for inducing viral pathogenicity that are highly conserved is essential for development of potent antiviral drugs. This is because novel drug targets that have a low mutation rate generally have low selection pressure generated against them by microbial agents and may serve as important therapeutic targets for synthesis of highly effective antimicrobial agents capable of eradicating and preventing dissemination of drug-resistant viral infections.

Acknowledgments

We gratefully acknowledge financial assistance provided by the National Research Foundation of South Africa (NRF) Research Development Grant for Y-Rated Researchers (RDYR180418322304; Grant No: 116339), the Department of Science and Technology (DST)/NRF Innovation Postdoctoral Fellowship (Grant no: 112051), the University of the Witwatersrand Faculty of Health Sciences Research Committee (FRC, Grant no: 001254 8464101 5121105 000000 0000000000 5254) and the National Health Laboratory Service Research grant (NHLSRT-1/3/17-1/9/19).

Competing Interests

We have no competing interests to declare.

References

Allen, M. I., et al. (1998). Identification and characterization of mutations in hepatitis B virus resistant to lamivudine. Lamivudine Clinical Investigation Group. *Hepatology*, *27*(6), 1670–1677.

Angus, P., et al. (2003). Resistance to adefovir dipivoxil therapy associated with the selection of a novel mutation in the HBV polymerase. *Gastroenterology*, *125*(2), 292–297.

Bacon, T. H., et al. (2003). Herpes simplex virus resistance to acyclovir and penciclovir after two decades of antiviral therapy. *Clinical Microbiology Reviews*, *16*(1), 114–128.

Belo, F., et al. (2012). Cytomegalovirus encephalitis in an immunocompetent child: A sceptic diagnosis. *BMJ Case Reports*, *2012*, 1–5.

Benitez, A. A., et al. (2015). In vivo RNAi screening identifies MDA5 as a significant contributor to the cellular defense against influenza A virus. *Cell Reports*, *11*(11), 1714–1726.

Benvegnu, L., et al. (1994). Concurrent hepatitis B and C virus infection and risk of hepatocellular carcinoma in cirrhosis. A prospective study. *Cancer*, *74*(9), 2442–2448.

Bestman-Smith, J., & Boivin, G. (2003). Drug resistance patterns of recombinant herpes simplex virus DNA polymerase mutants generated with a set of overlapping cosmids and plasmids. *Journal of Virology*, *77*(14), 7820–7829.

Breton, G., et al. (1998). Acyclovir-resistant herpes zoster in human immunodeficiency virus-infected patients: Results of foscarnet therapy. *Clinical Infectious Diseases: An Official Publication of the Infectious Diseases Society of America*, *27*(6), 1525–1527.

Bright, R. A., et al. (2006). Adamantane resistance among influenza A viruses isolated early during the 2005–2006 influenza season in the United States. *JAMA*, *295*(8), 891–894.

Bryan, C. J., et al. (2008). Acyclovir-resistant chronic verrucous vaccine strain varicella in a patient with neuroblastoma. *The Pediatric Infectious Disease Journal*, *27*(10), 946–948.

Burrel, S., et al. (2013). Molecular characterization of herpes simplex virus 2 strains by analysis of microsatellite polymorphism. *Journal of Clinical Microbiology, 51*(11), 3616–3623.
Cady, S. D., et al. (2010). Structure of the amantadine binding site of influenza M2 proton channels in lipid bilayers. *Nature, 463*(7281), 689–692.
Cai, H., et al. (2014). In vitro combination of anti-cytomegalovirus compounds acting through different targets: Role of the slope parameter and insights into mechanisms of action. *Antimicrobial Agents and Chemotherapy, 58*(2), 986–994.
Chen, E., et al. (2013). Human infection with avian influenza A(H7N9) virus re-emerges in China in winter 2013. *Euro surveillance: bulletin Europeen sur les maladies transmissibles = European communicable disease bulletin, 18*(43), 1–10.
Chen, D. S., et al. (2016). Horizontal gene transfer events reshape the global landscape of arm race between viruses and *Homo sapiens*. *Scientific Reports, 6*, 26934.
Chou, S., Miner, R. C., & Drew, W. L. (2000). A deletion mutation in region V of the cytomegalovirus DNA polymerase sequence confers multidrug resistance. *The Journal of Infectious Diseases, 182*(6), 1765–1768.
Chou, S., et al. (1998). Mutation in region III of the DNA polymerase gene conferring foscarnet resistance in cytomegalovirus isolates from 3 subjects receiving prolonged antiviral therapy. *The Journal of Infectious Diseases, 178*(2), 526–530.
Chou, S., et al. (2002). Cytomegalovirus UL97 phosphotransferase mutations that affect susceptibility to ganciclovir. *The Journal of Infectious Diseases, 185*(2), 162–169.
Chou, S., et al. (2003). Viral DNA polymerase mutations associated with drug resistance in human cytomegalovirus. *The Journal of Infectious Diseases, 188*(1), 32–39.
Chou, S., et al. (2008). Contrasting drug resistance phenotypes resulting from cytomegalovirus DNA polymerase mutations at the same exonuclease locus. *Journal of Clinical Virology: The Official Publication of the Pan American Society for Clinical Virology, 43*(1), 107–109.
Cohen, P. (2000). The regulation of protein function by multisite phosphorylation—A 25 year update. *Trends in Biochemical Sciences, 25*(12), 596–601.
Crassard, N., et al. (2000). Acyclovir-resistant varicella infection with atypical lesions in a non-HIV leukemic infant. *Acta Paediatrica, 89*(12), 1497–1499.
Das, K., et al. (2001). Molecular modeling and biochemical characterization reveal the mechanism of hepatitis B virus polymerase resistance to lamivudine (3TC) and emtricitabine (FTC). *Journal of Virology, 75*(10), 4771–4779.
De Clercq, E. (1984). The antiviral spectrum of (E)-5-(2-bromovinyl)-2′-deoxyuridine. *The Journal of Antimicrobial Chemotherapy, 14*(Suppl. A), 85–95.
De Clercq, E. (2008). The discovery of antiviral agents: Ten different compounds, ten different stories. *Medicinal Research Reviews, 28*(6), 929–953.
De Clercq, E., & Holy, A. (2005). Acyclic nucleoside phosphonates: A key class of antiviral drugs. *Nature Reviews. Drug Discovery, 4*(11), 928–940.
Dikici, B., et al. (2001). Lamivudine and interferon-alpha combination treatment of childhood patients with chronic hepatitis B infection. *The Pediatric Infectious Disease Journal, 20*(10), 988–992.
Dong, G., et al. (2015). Adamantane-resistant influenza a viruses in the world (1902–2013): Frequency and distribution of M2 gene mutations. *PLoS One, 10*(3), e0119115.
Doong, S. L., et al. (1991). Inhibition of the replication of hepatitis B virus in vitro by 2′,3′-dideoxy-3′-thiacytidine and related analogues. *Proceedings of the National Academy of Sciences of the United States of America, 88*(19), 8495–8499.
Drake, A. L., et al. (2012). Valacyclovir suppressive therapy reduces plasma and breast milk HIV-1 RNA levels during pregnancy and postpartum: A randomized trial. *The Journal of Infectious Diseases, 205*(3), 366–375.
Elion, G. B. (1993). Acyclovir: discovery, mechanism of action, and selectivity. *Journal of Medical Virology*, (Suppl. 1), 2–6.
Elton, D., et al. (1999a). Identification of amino acid residues of influenza virus nucleoprotein essential for RNA binding. *Journal of Virology, 73*(9), 7357–7367.

Elton, D., et al. (1999b). Oligomerization of the influenza virus nucleoprotein: Identification of positive and negative sequence elements. *Virology, 260*(1), 190–200.

Estrin, M. A., et al. (2018). Host-directed combinatorial RNAi improves inhibition of diverse strains of influenza A virus in human respiratory epithelial cells. *PLoS One, 13*(5), e0197246.

Furuse, Y., Suzuki, A., & Oshitani, H. (2009). Large-scale sequence analysis of M gene of influenza A viruses from different species: Mechanisms for emergence and spread of amantadine resistance. *Antimicrobial Agents and Chemotherapy, 53*(10), 4457–4463.

Geisinger, E., & Isberg, R. R. (2017). Interplay between antibiotic resistance and virulence during disease promoted by multidrug-resistant bacteria. *The Journal of Infectious Diseases, 215*(Suppl. 1), S9–S17.

Gilbert, C., Bestman-Smith, J., & Boivin, G. (2002). Resistance of herpesviruses to antiviral drugs: Clinical impacts and molecular mechanisms. *Drug Resistance Updates: Reviews and Commentaries in Antimicrobial and Anticancer Chemotherapy, 5*(2), 88–114.

Gilbert, C., & Cordaux, R. (2017). Viruses as vectors of horizontal transfer of genetic material in eukaryotes. *Current Opinion in Virology, 25*, 16–22.

Gilbert, C., et al. (2011). Recombinant phenotyping of cytomegalovirus UL54 mutations that emerged during cell passages in the presence of either ganciclovir or foscarnet. *Antimicrobial Agents and Chemotherapy, 55*(9), 4019–4027.

Gitlin, N. (1997). Hepatitis B: Diagnosis, prevention, and treatment. *Clinical Chemistry, 43*(8 Pt. 2), 1500–1506.

Gokulgandhi, M. R., et al. (2012). Transporter-targeted lipid prodrugs of cyclic cidofovir: A potential approach for the treatment of cytomegalovirus retinitis. *Journal of Pharmaceutical Sciences, 101*(9), 3249–3263.

Gopal, M. G., et al. (2013). A comparative study to evaluate the efficacy and safety of acyclovir and famciclovir in the management of herpes zoster. *Journal of Clinical and Diagnostic Research: JCDR, 7*(12), 2904–2907.

Govorkova, E. A., et al. (2004). Neuraminidase inhibitor-rimantadine combinations exert additive and synergistic anti-influenza virus effects in MDCK cells. *Antimicrobial Agents and Chemotherapy, 48*(12), 4855–4863.

Griffiths, A., et al. (2003). Translational compensation of a frameshift mutation affecting herpes simplex virus thymidine kinase is sufficient to permit reactivation from latency. *Journal of Virology, 77*(8), 4703–4709.

Griffiths, A., et al. (2006). Low-level expression and reversion both contribute to reactivation of herpes simplex virus drug-resistant mutants with mutations on homopolymeric sequences in thymidine kinase. *Journal of Virology, 80*(13), 6568–6574.

Gulen, T. A., et al. (2015). Clinical importance and cost of bacteremia caused by nosocomial multi drug resistant acinetobacter baumannii. *International Journal of Infectious Diseases: IJID: Official Publication of the International Society for Infectious Diseases, 38*, 32–35.

Hadchouel, M., et al. (1985). Presence of HBV DNA in spermatozoa: A possible vertical transmission of HBV via the germ line. *Journal of Medical Virology, 16*(1), 61–66.

Hadziyannis, S. J., et al. (2006). Long-term therapy with adefovir dipivoxil for HBeAg-negative chronic hepatitis B for up to 5 years. *Gastroenterology, 131*(6), 1743–1751.

Horimoto, T., & Kawaoka, Y. (2005). Influenza: Lessons from past pandemics, warnings from current incidents. *Nature Reviews. Microbiology, 3*(8), 591–600.

Hupalo, D. N., et al. (2016). Population genomics studies identify signatures of global dispersal and drug resistance in *Plasmodium vivax*. *Nature Genetics, 48*(8), 953–958.

Jagger, B. W., et al. (2012). An overlapping protein-coding region in influenza A virus segment 3 modulates the host response. *Science, 337*(6091), 199–204.

Janssen, H. L., et al. (2005). Pegylated interferon alfa-2b alone or in combination with lamivudine for HBeAg-positive chronic hepatitis B: A randomised trial. *Lancet, 365*(9454), 123–129.

Jiang, Y. C., et al. (2016). New strategies against drug resistance to herpes simplex virus. *International Journal of Oral Science, 8*(1), 1–6.

Jones, M., et al. (2014). RNA-seq analysis of host and viral gene expression highlights interaction between varicella zoster virus and keratinocyte differentiation. *PLoS Pathogens, 10*(1), e1003896.

Kamei, S., et al. (2005). Evaluation of combination therapy using aciclovir and corticosteroid in adult patients with herpes simplex virus encephalitis. *Journal of Neurology, Neurosurgery, and Psychiatry, 76*(11), 1544–1549.

Kim, W. R., et al. (2008). Serum activity of alanine aminotransferase (ALT) as an indicator of health and disease. *Hepatology, 47*(4), 1363–1370.

Kim, W. Y., et al. (2011). Triple-combination antiviral drug for pandemic H1N1 influenza virus infection in critically ill patients on mechanical ventilation. *Antimicrobial Agents and Chemotherapy, 55*(12), 5703–5709.

Kiso, M., et al. (2018). Combination therapy with neuraminidase and polymerase inhibitors in nude mice infected with influenza virus. *The Journal of Infectious Diseases, 217*(6), 887–896.

Kobashi, H., et al. (2009). Two cases of development of entecavir resistance during entecavir treatment for nucleoside-naive chronic hepatitis B. *Hepatology International, 3*(2), 403–410.

Kotton, C. N., et al. (2010). International consensus guidelines on the management of cytomegalovirus in solid organ transplantation. *Transplantation, 89*(7), 779–795.

Krumbholz, A., et al. (2009). High prevalence of amantadine resistance among circulating European porcine influenza A viruses. *The Journal of General Virology, 90*(Pt. 4), 900–908.

Lamb, R. A., & Choppin, P. W. (1983). The gene structure and replication of influenza virus. *Annual Review of Biochemistry, 52*, 467–506.

Langley, D. R., et al. (2007). Inhibition of hepatitis B virus polymerase by entecavir. *Journal of Virology, 81*(8), 3992–4001.

Latallo, M. J., et al. (2017). Predicting allosteric mutants that increase activity of a major antibiotic resistance enzyme. *Chemical Science, 8*(9), 6484–6492.

Lebbink, R. J., et al. (2017). A combinational CRISPR/Cas9 gene-editing approach can halt HIV replication and prevent viral escape. *Scientific Reports, 7*, 41968.

Levin, M. J., et al. (2003). Development of resistance to acyclovir during chronic infection with the Oka vaccine strain of varicella-zoster virus, in an immunosuppressed child. *The Journal of Infectious Diseases, 188*(7), 954–959.

Li, Y., et al. (2018). Phosphorylation and dephosphorylation of threonine 188 in nucleoprotein is crucial for the replication of influenza A virus. *Virology, 520*, 30–38.

Liaw, Y. F., et al. (2009). 2-Year GLOBE trial results: Telbivudine is superior to lamivudine in patients with chronic hepatitis B. *Gastroenterology, 136*(2), 486–495.

Lischka, P., et al. (2010). In vitro and in vivo activities of the novel anticytomegalovirus compound AIC246. *Antimicrobial Agents and Chemotherapy, 54*(3), 1290–1297.

Liu, D., Shi, W., & Gao, G. F. (2014). Poultry carrying H9N2 act as incubators for novel human avian influenza viruses. *Lancet, 383*(9920), 869.

Liu, H., et al. (2010). Widespread horizontal gene transfer from double-stranded RNA viruses to eukaryotic nuclear genomes. *Journal of Virology, 84*(22), 11876–11887.

Ljungman, P., et al. (1990). Acyclovir-resistant herpes simplex virus causing pneumonia after marrow transplantation. *The Journal of Infectious Diseases, 162*(1), 244–248.

Lok, A. S. (2008). How to diagnose and treat hepatitis B virus antiviral drug resistance in the liver transplant setting. *Liver Transplantation: Official Publication of the American Association for the Study of Liver Diseases and the International Liver Transplantation Society, 14*(Suppl. 2), S8–S14.

Lurain, N. S., & Chou, S. (2010). Antiviral drug resistance of human cytomegalovirus. *Clinical Microbiology Reviews, 23*(4), 689–712.

Margolis, H. S. (1998). Hepatitis B virus infection. *Bulletin of the World Health Organization, 76*(Suppl. 2), 152–153.

Markus, A., et al. (2011). Varicella-zoster virus (VZV) infection of neurons derived from human embryonic stem cells: Direct demonstration of axonal infection, transport of VZV, and productive neuronal infection. *Journal of Virology*, *85*(13), 6220–6233.

Martinez, J. L., & Baquero, F. (2000). Mutation frequencies and antibiotic resistance. *Antimicrobial Agents and Chemotherapy*, *44*(7), 1771–1777.

Martinez, V., et al. (2006). Topical imiquimod for recurrent acyclovir-resistant HSV infection. *The American Journal of Medicine*, *119*(5), e9–11.

Melendez, D. P., & Razonable, R. R. (2015). Letermovir and inhibitors of the terminase complex: A promising new class of investigational antiviral drugs against human cytomegalovirus. *Infection and Drug Resistance*, *8*, 269–277.

Mesquita, P. M. M., et al. (2017). Role of interleukin 32 in human immunodeficiency virus reactivation and its link to human immunodeficiency virus-herpes simplex virus coinfection. *The Journal of Infectious Diseases*, *215*(4), 614–622.

Micallef, S., & Galea, R. (2018). CMV encephalitis in an immune-competent patient. *BMJ Case Reports*, *2018*. bcr-2018-224740.

Minces, L. R., et al. (2014). Ganciclovir-resistant cytomegalovirus infections among lung transplant recipients are associated with poor outcomes despite treatment with foscarnet-containing regimens. *Antimicrobial Agents and Chemotherapy*, *58*(1), 128–135.

Momper, J. D., et al. (2013). Pharmacokinetics of low-dose cidofovir in kidney transplant recipients with BK virus infection. *Transplant Infectious Disease: An Official Journal of the Transplantation Society*, *15*(1), 34–41.

Morfin, F., et al. (1999). Phenotypic and genetic characterization of thymidine kinase from clinical strains of varicella-zoster virus resistant to acyclovir. *Antimicrobial Agents and Chemotherapy*, *43*(10), 2412–2416.

Muramoto, Y., et al. (2013). Identification of novel influenza A virus proteins translated from PA mRNA. *Journal of Virology*, *87*(5), 2455–2462.

Myhre, H. A., et al. (2011). Incidence and outcomes of ganciclovir-resistant cytomegalovirus infections in 1244 kidney transplant recipients. *Transplantation*, *92*(2), 217–223.

Mylonakis, E., Kallas, W. M., & Fishman, J. A. (2002). Combination antiviral therapy for ganciclovir-resistant cytomegalovirus infection in solid-organ transplant recipients. *Clinical Infectious Diseases: An Official Publication of the Infectious Diseases Society of America*, *34*(10), 1337–1341.

Nelson, M. I., et al. (2009). The origin and global emergence of adamantane resistant A/H3N2 influenza viruses. *Virology*, *388*(2), 270–278.

Ormrod, D., Scott, L. J., & Perry, C. M. (2000). Valaciclovir: A review of its long term utility in the management of genital herpes simplex virus and cytomegalovirus infections. *Drugs*, *59*(4), 839–863.

Perti, T., et al. (2013). High-dose valacyclovir decreases plasma HIV-1 RNA more than standard-dose acyclovir in persons coinfected with HIV-1 and HSV-2: A randomized crossover trial. *Journal of Acquired Immune Deficiency Syndromes*, *63*(2), 201–208.

Perz, J. F., et al. (2006). The contributions of hepatitis B virus and hepatitis C virus infections to cirrhosis and primary liver cancer worldwide. *Journal of Hepatology*, *45*(4), 529–538.

Pielak, R. M., Schnell, J. R., & Chou, J. J. (2009). Mechanism of drug inhibition and drug resistance of influenza A M2 channel. *Proceedings of the National Academy of Sciences of the United States of America*, *106*(18), 7379–7384.

Pires de Mello, C. P., et al. (2018). Oseltamivir-zanamivir combination therapy suppresses drug-resistant H1N1 influenza A viruses in the hollow fiber infection model (HFIM) system. *European Journal of Pharmaceutical Sciences: Official Journal of the European Federation for Pharmaceutical Sciences*, *111*, 443–449.

Plosa, E. J., et al. (2012). Cytomegalovirus infection. *Pediatrics in Review*, *33*(4), 156–163. quiz 163.

Portela, A., & Digard, P. (2002). The influenza virus nucleoprotein: A multifunctional RNA-binding protein pivotal to virus replication. *The Journal of General Virology*, *83*(Pt. 4), 723–734.

Prichard, M. N. (2009). Function of human cytomegalovirus UL97 kinase in viral infection and its inhibition by maribavir. *Reviews in Medical Virology*, *19*(4), 215–229.

Prichard, M. N., et al. (1999). A recombinant human cytomegalovirus with a large deletion in UL97 has a severe replication deficiency. *Journal of Virology*, *73*(7), 5663–5670.

Safrin, S., et al. (1991). A controlled trial comparing foscarnet with vidarabine for acyclovir-resistant mucocutaneous herpes simplex in the acquired immunodeficiency syndrome. The AIDS Clinical Trials Group. *The New England Journal of Medicine*, *325*(8), 551–555.

Saint-Leger, E., et al. (2001). Clinical and virologic characterization of acyclovir-resistant varicella-zoster viruses isolated from 11 patients with acquired immunodeficiency syndrome. *Clinical Infectious Diseases: An Official Publication of the Infectious Diseases Society of America*, *33*(12), 2061–2067.

Saliba, W. R., et al. (2004). Cytomegalovirus encephalitis in an immunocompetent pregnant woman. *European Journal of Clinical Microbiology & Infectious Diseases: Official Publication of the European Society of Clinical Microbiology*, *23*(7), 563–566.

Schalm, S. W., et al. (2000). Lamivudine and alpha interferon combination treatment of patients with chronic hepatitis B infection: A randomised trial. *Gut*, *46*(4), 562–568.

Schmidtke, M., et al. (2006). Amantadine resistance among porcine H1N1, H1N2, and H3N2 influenza A viruses isolated in Germany between 1981 and 2001. *Intervirology*, *49*(5), 286–293.

Schnell, J. R., & Chou, J. J. (2008). Structure and mechanism of the M2 proton channel of influenza A virus. *Nature*, *451*(7178), 591–595.

Scott, R. M., et al. (1980). Experimental transmission of hepatitis B virus by semen and saliva. *The Journal of Infectious Diseases*, *142*(1), 67–71.

Scott, G. M., et al. (2007). Multidrug resistance conferred by novel DNA polymerase mutations in human cytomegalovirus isolates. *Antimicrobial Agents and Chemotherapy*, *51*(1), 89–94.

Sharon, A., & Chu, C. K. (2008). Understanding the molecular basis of HBV drug resistance by molecular modeling. *Antiviral Research*, *80*(3), 339–353.

Sims, C. R., et al. (2007). Oral topical cidofovir: Novel route of drug delivery in a severely immunosuppressed patient with refractory multidrug-resistant herpes simplex virus infection. *Transplant Infectious Disease: An Official Journal of the Transplantation Society*, *9*(3), 256–259.

Strasfeld, L., & Chou, S. (2010). Antiviral drug resistance: Mechanisms and clinical implications. *Infectious Disease Clinics of North America*, *24*(2), 413–437.

Suazo, P. A., et al. (2015). Evasion of early antiviral responses by herpes simplex viruses. *Mediators of Inflammation*, *2015*, 593757.

Tabor, E., et al. (1985). Horizontal transmission of hepatitis B virus among children and adults in five rural villages in Zambia. *Journal of Medical Virology*, *15*(2), 113–120.

Tan, G., et al. (2015). Combination therapy of oncolytic herpes simplex virus HF10 and bevacizumab against experimental model of human breast carcinoma xenograft. *International Journal of Cancer*, *136*(7), 1718–1730.

Tanaka, R., et al. (2017). The efficacy of combination therapy with oncolytic herpes simplex virus HF10 and dacarbazine in a mouse melanoma model. *American Journal of Cancer Research*, *7*(8), 1693–1703.

Tenney, D. J., et al. (2004). Clinical emergence of entecavir-resistant hepatitis B virus requires additional substitutions in virus already resistant to Lamivudine. *Antimicrobial Agents and Chemotherapy*, *48*(9), 3498–3507.

Tenney, D. J., et al. (2007). Two-year assessment of entecavir resistance in Lamivudine-refractory hepatitis B virus patients reveals different clinical outcomes depending on the resistance substitutions present. *Antimicrobial Agents and Chemotherapy*, *51*(3), 902–911.

Thomaston, J. L., & DeGrado, W. F. (2016). Crystal structure of the drug-resistant S31N influenza M2 proton channel. *Protein Science: A Publication of the Protein Society*, *25*(8), 1551–1554.

Tuttleman, J. S., Pourcel, C., & Summers, J. (1986). Formation of the pool of covalently closed circular viral DNA in hepadnavirus-infected cells. *Cell*, *47*(3), 451–460.

Ufaz, S., et al. (2018). Anti-viral RNAi nanoparticles protect shrimp against white spot disease. *Molecular Systems Design & Engineering*, *3*(1), 38–48.

Villet, S., et al. (2008). Impact of hepatitis B virus rtA181V/T mutants on hepatitis B treatment failure. *Journal of Hepatology*, *48*(5), 747–755.

Vora, S. B., et al. (2018). Antiviral combination therapy for cytomegalovirus infection in high-risk infants. *Antiviral Therapy*, *23*(6), 505–511.

Vossen, M. G., et al. (2014). Single-dose pharmacokinetics of cidofovir in continuous venovenous hemofiltration. *Antimicrobial Agents and Chemotherapy*, *58*(4), 1952–1955.

Wang, G., et al. (2018). CRISPR-Cas based antiviral strategies against HIV-1. *Virus Research*, *244*, 321–332.

Wise, H. M., et al. (2009). A complicated message: Identification of a novel PB1-related protein translated from influenza A virus segment 2 mRNA. *Journal of Virology*, *83*(16), 8021–8031.

Wolf, D. G., et al. (2001). Distinct and separate roles for herpesvirus-conserved UL97 kinase in cytomegalovirus DNA synthesis and encapsidation. *Proceedings of the National Academy of Sciences of the United States of America*, *98*(4), 1895–1900.

World Health Organization. (2017a). *Global priority list of antibiotic-resistance bacteria to guide research, discovery, and development of new antibiotics*. Geneva: World Health Organization. https://www.who.int/medicines/publications/global-priority-list-antibiotic-resistant-bacteria/en/. last accessed on 24 June 2019.

World Health Organization. (2017b). *Global hepatitis report 2017*. Geneva: World Health Organization. Licence: CC BY-NC-SA 3.0 IGO https://www.who.int/hepatitis/publications/global-hepatitis-report2017/en/. last accessed on 24 June 2019.

World Health Organization. (2017c). *WHO influenza mortality estimate based on respiratory diseases*. Geneva: World Health Organization. https://www.who.int/news-room/detail/14-12-2017-up-to-650-000-people-die-of-respiratory-diseases-linked-to-seasonal-flu-each-year. last accessed on 24 June 2019.

Wu, Y., et al. (2014). Bat-derived influenza-like viruses H17N10 and H18N11. *Trends in Microbiology*, *22*(4), 183–191.

Yang, H., et al. (2004). In vitro antiviral susceptibility of full-length clinical hepatitis B virus isolates cloned with a novel expression vector. *Antiviral Research*, *61*(1), 27–36.

Yang, H., et al. (2005). Cross-resistance testing of next-generation nucleoside and nucleotide analogues against lamivudine-resistant HBV. *Antiviral Therapy*, *10*(5), 625–633.

Yang, W. L., et al. (2015). Persistence of transmitted HIV-1 drug resistance mutations associated with fitness costs and viral genetic backgrounds. *PLoS Pathogens*, *11*(3), e1004722.

Yuan, M., et al. (2015). Efficiently editing the vaccinia virus genome by using the CRISPR-Cas9 system. *Journal of Virology*, *89*(9), 5176–5179.

Zhao, L., et al. (2018). Hepatitis B virus rtA181T/sW172non-stop mutation may increase resistance fold to adefovir- and entecavir-resistant mutants compared to rtA181T/sW172* mutation. *Antiviral Research*, *154*, 26–34.

Zuckerman, A. J. (1984). Perinatal transmission of hepatitis B. *Archives of Disease in Childhood*, *59*(11), 1007–1009.

CHAPTER

10

Combination therapy against human infections caused by viruses

Rifat Munir

Department of Clinical Microbiology and Infectious Diseases, University of Witwatersrand, Johannesburg, Gauteng, South Africa; Department of Biosystems, University of Manitoba, Winnipeg, MB, Canada; Department of Immunology, University of Witwatersrand, Johannesburg, Gauteng, South Africa

Abbreviations

ALT	alanine aminotransferase enzyme
ART	antiretroviral therapy
AZT	zidovudine
CDC	Centers for Disease Control and Prevention
CHB	chronic hepatitis B
ddC	zalcitabine
ddI	didanosine
DNA	deoxyribonucleic acid
HA	hemagglutinin
HAART	highly active antiretroviral therapy
HBeAg	hepatitis B envelope antigen
HBsAg	hepatitis B surface antigen
HBV	hepatitis B virus
HCC	hepatocellular carcinoma
HIV	human immune deficiency virus
HIVDR	HIV drug resistance
IFN	interferon
IV	influenza virus
MDCK	Madin-Darby canine kidney cells
NA	neuraminidase
NNRTI	nonnucleoside reverse transcriptase inhibitors
NRTIs	nucleoside/nucleotide reverse transcriptase inhibitors
RNA	ribonucleic acid
TCAD	triple combination antiviral drug

Combination Therapy Against Multidrug Resistance
https://doi.org/10.1016/B978-0-12-820576-1.00010-2

10.1 Human influenza virus

10.1.1 Introduction

The influenza virus (IV) is the causative agent of the highly infectious respiratory disease known as influenza (Martín-Benito & Ortín, 2013). The IV has been classified into four types of viruses, on the basis of their core proteins. However, types A, B, and C are the only classes known to affect humans (Martín-Benito & Ortín, 2013). IV types A and B are the only ones known to cause serious illness and outbreaks in humans in recent history (Hay, Gregory, Douglas, & Yi, 2001; WHO Influenza Factsheet, 2018) and will therefore form part of the main discussion that follows. The IV belongs to the Orthomyxovirus family. These viruses have a segmented single-stranded RNA genome in their core and four viral envelope proteins: neuraminidase (NA), hemagglutinin (HA), M1 matrix protein, and M2 ion channel protein (Horimoto & Kawaoka, 2005). Based on the distinct antigenic characteristics of their HA and NA glycoproteins, Type A influenza viruses have further been subdivided into 18 HA (H1–H18) and 11 NA (N1–N11) serotypes (Hamilton, Whittaker, & Daniel, 2012; Skehel, 2009; Tong et al., 2013). While IV B mainly infect humans, IV A are typically avian viruses that are known to occasionally cross infect a wide range of mammalian species, which includes humans (Hay et al., 2001).

Influenza is primarily transmitted through droplet aerosols produced as a result of coughing, sneezing, spitting, or talking by an infected individual; however, direct personal contact with contaminated surfaces is another means of spread of the virus (Treanor, 2014). Viral shedding (the ability of an infected individual to infect others) starts before the onset of symptoms, usually within 24 h of infection, and continues for approximately 5 days in otherwise healthy individuals. In immunocompromised and the elderly, viral shedding may last for over 2 weeks (Lee et al., 2009). Infected children usually shed viruses from several days before the onset of clinical symptoms until 2 weeks after onset (Suess et al., 2012).

Symptoms of influenza infection can be mild to severe. Classical symptoms usually set in within 24–48 h postinfection and include fever, chills, headache, sore throat, and malaise. Respiratory symptoms such as cough and nasal discharge usually appear after 48 h of infection and may last for over a week (Treanor, 2014). If antiviral therapy is not initiated within 48 h of infection, complications can arise, which include sinus infection and viral and bacterial pneumonia. Other rare complications, such as myositis, rhabdomyolysis, and encephalitis, have also been described. However, pulmonary complications are usually responsible for much of the associated mortalities (Tao et al., 2018; Treanor, 2014). While these complications are seen across all age groups, they are more common in the elderly and immunocompromised individuals as well as in children (Treanor, 2014).

10.1.2 Global public health and economic impact

Influenza is a major cause of severe respiratory infections, which lead to excessive hospitalizations and deaths worldwide. Seasonal outbreaks have a significant impact on public health and remain an important source of economic loss worldwide. The CDC estimates that approximately 50 million individuals are affected with influenza every year in the United States alone (CDC, 2018). According to the World Health Organization, between 3 million and 5 million people become ill every year with severe forms of influenza (WHO Influenza Factsheet, 2018). An estimate of 290,000–650,000 deaths (4–9 per 100,000 individuals) due to respiratory diseases associated with seasonal influenza occur annually worldwide. According to a study carried out by Iuliano et al. (2018), the highest mortality rates occur in Sub-Saharan Africa (2.8–16.5 per 100,000 individuals) and in Southeast Asia (3.5–9.2 per 100,000 individuals). Individuals aged 75 years or older (51.3–99.4 per 100,000 individuals), and children younger than 5 years have the highest mortality rates annually associated with influenza (Iuliano et al., 2018).

Influenza and its associated outbreaks have had a negative impact on the economy worldwide. This socioeconomic burden results from costs related to patient medical care both in and out of hospital, as well as indirect costs as a result of productivity losses. According to Molinari et al. (2007), the total economic expense due to influenza in the United States is approximately $87.1 billion per year. In another study, the annual economic impact of influenza varied across counties in the United States and was estimated to be between $14,000 to $960 million, with a median of approximately $3 million (Mao, Yang, Qiu, & Yang, 2012). In Europe, influenza is responsible for 10% of absenteeism from work and costs the continent approximately 300 million euros per year (Ryan, Zoellner, Gradl, Palache, & Medema, 2006). Clearly the impact of influenza on both public health and the economy worldwide is massive, thus the need to implement therapeutic and preventative measures to avert major outbreaks.

10.1.3 Treatment strategies against influenza virus

Influenza infection is caused by a variety of strains of influenza viruses (Li, Chan, & Lee, 2015). The rapid and continuous evolution of these viruses, mainly due to mutations, results in annual worldwide seasonal epidemics (IV Type A/H3N2, A/H1N1, and Type B), unpredictable pandemics (Type A/H1N1pdm09), and deadly zoonotic influenza viral infections (Type A/H5N1, Type AH7N9) (Lee & Ison, 2012). While administration of the annual influenza vaccination program can prevent and control seasonal epidemics, three classes of antiviral drugs are approved to treat IV infections:

(1) M2 ion channel inhibitors. These prevent viral replication by blocking the ion channel activity of the matrix M2 protein. Amantadine and rimantadine fall under this class of drugs and are known to be active only against IV type A;
(2) Neuraminidase inhibitors. These block viral neuraminidases and prevent viral reproduction by budding from the host cell. Antiviral agents in this category include zanamivir, oseltamivir, peramivir, and laninamivir. These are active against both IV type A and B;
(3) The inhibitors of influenza polymerase (baloxavir marboxil, favipiravir, and pimodivir active against both IV type A and B) (Hayden & Shindo, 2019; Hayden et al., 2018; Ng, 2019).

The emergence of antiviral drug resistance, which has been reported worldwide, is, however, of major concern when a single antiviral drug is used for treatment and provides a major challenge for global healthcare. Due to high rates of resistance, antivirals rimantadine and amantadine are currently not recommended for treatment of circulating influenza A (H3N2) and influenza A (H1N1) pdm09 viruses in the United States. Antiviral resistance and low susceptibility to the neuraminidase inhibitors and to baloxavir have been reported but at a much lower level and are therefore the treatment choice currently recommended for influenza (Ait-Aissa et al., 2018; CDC, 2019b; Deyde et al., 2007; Hayden & Shindo, 2019; Sheu et al., 2008). The combined use of different antiviral agents with different mechanisms of action (combination therapy) can optimize IV treatments and reduce the development of resistance (Monod et al., 2015). These agents are, however, not administered clinically as of yet.

Several studies have been carried out over the years that demonstrate the advantages of using combined therapy to treat influenza. Nguyen et al. (2009, 2010) evaluated the combined effects of three compounds of antiviral drugs, amantadine, ribavirin, and oseltamivir (TCAD), with different mechanisms of action against amantadine- and oseltamivir-resistant H1N1 IV both in vivo in mouse models and in vitro in Madin-Darby canine kidney (MDCK) cells. Their results suggested that the combination of three drugs was highly superior and acted synergistically against drug-resistant viruses. The synergy of the three drugs was significantly higher than the effects of any dual combinations tested and superior to single-drug regimens. Beigel et al. (2017) similarly used TCAD versus oseltamivir monotherapy in a trial study and concluded that, while combination therapy resulted in a significant decrease in viral shedding compared to monotherapy, the difference in the outcome was not associated with improved benefits clinically. Seo et al. (2013) investigated the safety and pharmacokinetics of TCAD therapy in immunocompromised individuals. Their results suggested that TCAD can safely be administered in immunocompromised patients and further showed a similarity in pharmacokinetics between combination therapy and monotherapy with ribavirin, amantadine, or oseltamivir. More recently, Popov, Shchelkanov, Dmitrenko, and Simakova (2018) assessed the efficacy of oseltamivir or umifenovir monotherapy on IV A H1N1 against dual therapy using antiviral drugs that differed in their mode of action (Kagocel with either umifenovir or oseltamivir) in patients confirmed to be infected with H1N1 virus. The study demonstrated that the use of a combination of drugs significantly increased the therapeutic efficacy by reducing the duration of clinical symptoms and decreased the incidence of clinical manifestations of indicators when compared with monotherapy using the corresponding drugs. Furthermore, a combination of rimantadine and zanamivir decreased H1N1 and H3N2 IV yields more efficiently in MDCK cells when compared with drugs used alone (Govorkova, Fang, Tan, & Webster, 2004). The combination of amantadine + oseltamivir carboxylate and amantadine + ribavirin similarly resulted in a synergistic outcome in vitro against the pathogenic avian IV strain H5N1 in MDCK cells (Smee, Hurst, Wong, Bailey, & Morrey, 2009). Additionally, Ilyushina, Hoffmann, Salomon, Webster, and Govorkova (2007) studied the effects of oseltamivir + ribavirin combination therapy against H5N1 IV infection in mice and observed that the combination therapy at optimal concentrations inhibited virus replication significantly, prevented the spread of the viruses, and abrogated the cytokine response. Therapy with a combination of oseltamivir and favipiravir (a nonspecific inhibitor of viral polymerases) similarly showed a reduction in viral loads and prevented the spread of the virus in lung tissues in mice. This resulted in reduced lung damage and decreased cytokine production, thus providing more time for treating infections caused by the highly pathogenic H5N1 IV, compared to animals receiving monotherapy (Marathe et al., 2016).

Even though combination therapy against IV using antiviral agents with different modes of action has mainly been tested in vitro and in preclinical models, the beneficial effects clearly outweigh the use of monotherapy in most cases. It is, however, crucial to determine their therapeutic benefit and clinical relevance by testing these combinations further in human patients. A number of antiviral drugs are in advanced phases of clinical development (Hayden & Shindo, 2019) and potentially offer promising clinical outcomes by reducing the emergence of resistance, improving antiviral potency, and extending the treatment window.

10.2 Human immune deficiency virus (HIV)

10.2.1 Introduction

Despite considerable prevention efforts and advances in antiretroviral therapy (ART) research, HIV continues to remain a worldwide public health problem. Over 35 million people have died globally from HIV-related diseases since its discovery in the early 1980s, and there were approximately 37 million people living with HIV in 2017, with approximately 2 million new infections annually (WHO HIV Factsheet, 2018).

Once an individual is infected, the virus attacks the immune system and, after an initial period of acute infection and clinical latency, the infection progresses and, if left untreated, leads to acquired immunodeficiency syndrome (AIDS), which results in multiple complications and comorbidities (Chu & Selwyn, 2011). As of yet, there is no cure for HIV infection; however, due to effective chemoprophylaxis and ART, new infections have declined by approximately 47% since the peak in 1996 and AIDS-related deaths have decreased by > 50% since 2004 (CDC, 2019a; Pirrone, Thakkar, Jacobson, Wigdahl, & Krebs, 2011; UNAIDS Factsheet, 2018). As more individuals become infected every year, over 20% of patients infected with HIV remain undiagnosed and pose an additional threat to new infections. Continuous efforts through education and lifestyle changes to reduce the spread of the virus have helped considerably to prevent new infections.

10.2.2 The purpose of implementing antiretroviral therapy

In the hope of finding a cure for HIV, intense research has focused on understanding the structure/morphology of the virus and its associated pathogenesis, leading to the discovery of various drugs that can inhibit some of the proteins/stages of the HIV replication cycle (Pirrone et al., 2011). Eradication of HIV, however, has still not been possible with the current available antiretroviral agents (Awi & Teow, 2018). While a cure, the ultimate goal of HIV treatment, may not be feasible now or in the near future, available medication provides considerable benefits in terms of delaying the progression of the disease to AIDS, maintaining health, and prolonging life by significantly reducing disease-associated mortality as well as reducing the transmission of HIV and consequently reducing new infections (Cohen et al., 2016; Gunthard et al., 2016). Once initiated, ART should be continued for life because treatment interruptions have been associated with drug-resistant mutations, rebound viremia, reduced immunity, and increased morbidity and mortality (El-Sadr, Lundgren, Neaton, Gordin, & Abrams, 2006).

10.2.3 Classification of antiretroviral treatment drugs

Efforts to interrupt the HIV replication cycle and stop its spread resulted in the development of over 30 antiretroviral agents that have been grouped into seven classes based on how the drugs act on the HIV cycle. These seven classes are:

i. Nucleoside/nucleotide reverse transcriptase inhibitors (NRTIs). These were one of the first drugs developed that focused on competitively inhibiting HIV-1 reverse transcriptase, which is an important enzyme for retrovirus replication. This enzyme converts viral RNA, which is single stranded, into double-stranded complementary proviral DNA (Cihlar & Ray, 2010; Prasad & Goff, 1990). Some drugs that belong to this class are zidovudine (AZT), abacavir, didanosine (ddI), zalcitabine (ddC), tenofovir, and lamivudine.

ii. Nonnucleoside reverse transcriptase inhibitors (NNRTIs). Drugs in this class prevent HIV-1 replication by noncompetitively inhibiting reverse transcriptase enzyme and include the first-generation NNRTIs (delavirdine, nevirapine, and efavirenz), second-generation NNRTIs (etravirine and rilpivirine) (Usach, Melis, & Peris, 2013), and the recently formulated next-generation doravirine (Hwang, Lai, & Hazuda, 2018)

iii. Protease inhibitors (PIs) act by inhibiting the HIV protease, an enzyme essential for the assembly of viral particles, viral maturation, and replication. The HIV protease inhibitors prevent the viruses from maturing, consequently resulting in the production of viruses that are immature and noninfectious (Scholar, 2011). Atazanavir, darunavir, fosamprenavir, indinavir, nelfinavir, and ritonavir are some members that fall into this class of antiretrovirals.

iv. Integrase strand transfer inhibitors (INSTIs). These are a class of antiretroviral agents that block the viral integrase enzyme. This hinders the integration of proviral DNA into the genetic material of healthy uninfected host cells and consequently prevents HIV replication. Antiretroviral drugs in this class include bictegravir, dolutegravir, elvitegravir, and raltegravir (Messiaen et al., 2013; Spagnuolo, Castagna, & Lazzarin, 2018).

v. Fusion inhibitors, such as enfuvirtide, are drugs that prevent binding of HIV to human cells and in turn prevent fusion and entry of the virus into the cell (Haqqani & Tilton, 2013).

vi. CCR5 antagonists are binding and entry inhibitors that block interactions between the proteins of the viral envelope and the CCR5 coreceptor. Maraviroc is the first CCR5-receptor antagonist approved for use in HIV infections. Similarly, drugs that inhibit interactions of HIV-1 gp120 with CD4 cell receptors have also been developed, which have been shown to have no cytotoxic effects (Meanwell et al., 2018; Tiberi et al., 2014).

vii. More recently, humanized monoclonal antibodies such as ibalizumab have been developed for adults resistant to currently available antiretroviral drugs. Ibalizumab binds to the CD4 HIV receptor, and inhibits HIV from entering cells (Beccari et al., 2019; Emu et al., 2018).

10.2.4 Current HIV treatment strategies

It is recommended that ART be initiated in all individuals with HIV infection regardless of their CD4 counts, so as to inhibit viral replication, reduce HIV-associated morbidity and mortality, and improve the overall quality of life. Delaying therapy until the CD4 counts have declined puts an individual at risk of AIDS-defining illnesses (Palella et al., 2014).

Antiretroviral monotherapy with AZT in earlier days resulted in the emergence of drug-resistant HIV isolates and most of the available drugs used had little effect on HIV-1. To curb the problem of drug resistance and the associated toxicity, many existing and new drugs were evaluated as combination partners. Research and clinical trials conducted in patients with a combination of NRTIs AZT and either ddI and or ddC, resulted in weight gain, an increase in CD4 counts, and decreased levels of p24. Disease progression and mortality rates were significantly decreased in patients who had CD4 counts ≥ 150 cells/mm^3 when initiating therapy but not in patients who had lower cell numbers at the start of therapy. Combination therapy resulted in a decrease in toxicity with high numbers of patients maintaining negative cultures for up to a year when compared to patients on monotherapy with AZT (Antonelli et al., 1994; Caliendo & Hirsch, 1994; Eron, Johnson, Merrill, Chou, & Hirsch, 1992; Gerstoft et al., 2003; Van Leeuwen et al., 2003). While the results of these studies suggested the beneficial effects of combination therapy using two or more drugs over monotherapy, AZT-resistant mutants, however, continued to emerge with rebounding viral loads. It was later determined that combination therapy using drugs from the same class with the same functional principle resulted in ART failure and that combination treatment should include drugs from varying classes that target the HIV at distinct stages of its replication cycle (Clay, 2004). Subsequent therapies incorporating drugs with three distinct mechanisms of action, termed triple therapy or highly active ART (HAART), resulted in a substantial decline in mortality rates (Hicks & Gulick, 2009; Palella et al., 2006; Summa et al., 2008). The development of new ARV drugs continues to improve management and clinical outcomes for individuals with HIV infection. Combining three different ARV drugs from seven distinct classes, the treatment regimen currently implemented, is therefore more effective at keeping the HIV virus under control and less likely to promote drug resistance than monotherapy.

10.2.4.1 *Antiretroviral combination therapy in treatment naïve adults*

The US Department of Health and Human Services (2018) guidelines recommend that treatment naïve patients need to be started off with an antiretroviral treatment strategy that includes two NRTIs in combination with a third active antiretroviral drug from one of three drug classes: an INSTI, an NNRTI, or a PI with a pharmacokinetic booster drug such as cobicistat and ritonavir. Table 10.1 summarizes the current recommended regimen. Some of these drugs occur as combinations in pill form and are available as fixed doses. In order to provide an effective, safe, tolerable, and easy-to-adhere-to regimen, a number of factors specific for both patients and regimen should be taken into consideration before initiating therapy. According to the Department of Health and Human Services (2018), some of these factors include CD4 cell counts, pretreatment viral load, genotyping results for drug resistance tests, HLA-B*5701 status, comorbidities, coinfections, and pregnancy status. Recent studies suggest that dolutegravir poses a health risk for infants and is therefore not recommended in expectant patients and young women of child-bearing age (Zash, Makhema, & Shapiro, 2018). The combination of eltagravir plus cobicistat is also not recommended for pregnant women in the second and third trimesters, because lower concentrations of the drug have been reported, which may result in lower exposure and increase the risk of virologic failure and may increase the risk of mother to child transmission (Momper et al., 2018). Use of raltegravir, however, seems to have lower fetal malformation rates and is therefore recommended for use in pregnancy.

10.2.4.2 *Management of treatment experienced patients*

Virologic failure in HIV-infected individuals

Virologic failure is a condition that usually occurs when ART is unable to repress and maintain the viral load of an HIV-infected individual to < 200 viral copies/mL (Ahmed, Merga, & Jarso, 2019). There are many reasons for ART to fail. Studies suggest that CD4 cell count of $< 200\,mm^3$, low body mass index at the start of therapy, drug intolerance/toxicity, suboptimal virologic potency, presence of acquired ART-resistant HIV, and poor compliance with ART contribute to virologic failure (Ahmed et al., 2019; Babo, Alemie, & Fentaye, 2017; Paredes et al., 2010). Expert advice to construct an effective regimen is required to manage patients that experience ART failure. It is, however, crucial to identify the potential cause of ART failure before modifying an existing regimen. A new antiretroviral treatment regimen should contain two to three fully active drugs that have been shown to have unwavering activity based on the treatment history and drug resistance results of the patient and should ideally be of a new mechanistic action (Cahn et al., 2013; Fätkenheuer et al., 2008; Hicks et al., 2006).

TABLE 10.1 Current recommended combination treatment regimen for HIV treatment naïve patients.

	Combination regimen component						
Regimen type	**NRTI**	**NRTI**	**NNRTI**	**Integrase inhibitor**	**Protease inhibitor**	**Protease inhibitor**	**Booster**
Strongly recommended	Tenofovir alafenamide	Emtricitabine		Bictegravir			
_[a]	Lamivudine	Abacavir		Dolutegravir			
	Tenofovir alafenamide	Emtricitabine		Dolutegravir			
_[b]	Lamivudine			Dolutegravir			
	Tenofovir alafenamide	Emtricitabine			Darunavir		Cobicistat
Moderately recommended	Tenofovir alafenamide	Emtricitabine		Raltegravir			
_[c]				Raltegravir	Darunavir	Ritonavir	
_[c]	Lamivudine				Darunavir	Ritonavir	
	Tenofovir alafenamide	Emtricitabine		Elvitegravir			Cobicistat
	Tenofovir alafenamide	Emtricitabine			Atazanavir		Cobicistat
	Lamivudine	Tenofovir disoproxil fumarate	Doravirine				
	Tenofovir alafenamide	Emtricitabine	Doravirine				

[a]*Strongly recommended for patients who are HLA*B5701 negative. HLA-B*5701 has been linked to abacavir hypersensitivity.*
[b]*Recommended therapy when abacavir, tenofovir alafenamide, or tenofovir disoproxil fumarate are not optimal.*
[c]*Recommended when plasma viral RNA copy number/mL is < 100,000 and CD4 cell count is > 200/mm^3.*
NNRTI, nonnucleoside reverse transcriptase inhibitors; NRTI, nucleoside/nucleotide reverse transcriptase inhibitors.

Antiretroviral drug resistance

Antiretroviral drug resistance (HIVDR) in patients is known to be as a result of mutations in the genome of HIV. According to the WHO HIV Drug Resistance Report (2017b), data from 14 countries showed that HIVDR is on the rise in every region studied, with Africa, Latin America, and Asia the most affected. HIVDR negatively impacts the antiretroviral's ability to halt the replication of HIV and there are concerns that current antiretroviral drugs are at risk of becoming nonfunctioning due to the emergence of resistant strains. HIVDR can either be acquired or transmitted. Acquired HIVDR occurs when HIV mutations randomly emerge in individuals receiving ARV drugs. It can also occur in individuals (both mother and child) exposed to antiretroviral drugs at birth centers around the time of delivery, and in individuals that have had an interruption in antiretroviral treatment (Clutter, Jordan, Bertagnolio, & Shafer, 2016; Kiertiburanakul et al., 2016; Kityo et al., 2016; Wang et al., 2011). Transmitted HIVDR, on the other hand, occurs in antiretroviral drug naïve individuals who have no previous record of exposure to ARV drugs. This happens when HIV with drug-resistance mutations is transmitted to previously uninfected individuals (Coetzee et al., 2017; Poppe et al., 2017). It is crucial to combat and/or minimize the emergence and transmission of HIVDR. According to the Global Action Plan on HIV Drug Resistance 2017–2021, monitoring levels of HIVDR by testing for drug resistance will help determine the HIV treatment regimen to be implemented (World Health Organization, 2017a). Furthermore, compliance and appropriate use of recommended antiretroviral drugs reduce the risk of drug resistance. The utmost virologic suppression may not be possible in patients with large-scale drug resistance; ART should, however, be continued with treatment strategy designed to preserve CD4 counts, reduce toxic effects, and delay disease progression (De Luca et al., 2013).

10.2.4.3 *Adverse effects of antiretroviral therapy*

Use of ARV drugs has been reported to have adverse effects on patients and this is the most common reason for interrupting treatment and for noncompliance (O'Brien, Clark, Besch, Myers, & Kissinger, 2003). Fortunately, over the years, antiretroviral drugs have improved and are associated with fewer serious and life-threatening side effects. Medication for comorbid conditions, such as tuberculosis and hepatitis, may interact with ART and increase toxicities and exacerbate adverse effects (Phung, Sogni, & Launay, 2014; Upton, Taiwo, & Robertson, 2013). Certain genetic factors have been identified that put individuals at risk of developing hypersensitivity reactions, neuropsychiatric toxicity, and hyperbilirubinemia (Gounden, van Niekerk, Snyman, & George, 2010; Rodríguez-Nóvoa et al., 2007). Each class of antiretroviral drugs is known to have side effects. The US Department of Health and Human Services (2018) summarizes some common and adverse effects associated with ART. NRTIs are associated with bone marrow suppression, psychiatric effects, cardiovascular diseases, lactic acidosis, lipodystrophy, and hyperlipidemia; NNRTIs are associated with hypersensitivity reactions, neuropsychiatric symptoms, rash, hepatic toxicity, and lipid abnormalities; protease inhibitors are associated with spontaneous bleeding, decrease in bone mineral density, cholelithiasis, cardiovascular complications, gastrointestinal intolerance, and glucose and lipid abnormalities. The CCR5 antagonist maraviroc has been associated with hypersensitivity reactions, hepatotoxicity, and bronchitis. The integrase inhibitor raltegravir has been implicated in Stevens-Johnson syndrome, myopathy, lipodystrophy, and hypersensitivity reactions. In addition, NNRTIs and protease inhibitors have been implicated in various drug-drug interactions. These adverse effects can be managed in the general population by monitoring individuals through complete blood count and comprehensive metabolic profiles every 3–6 months and annual analysis of lipid profile and urinalysis. Long-term morbid conditions such as cardiovascular and bone disease can be managed by making changes to treatment regimens and adjusting dosages.

10.3 Antiviral therapy against chronic hepatitis B

10.3.1 Introduction

Hepatitis B is potentially a life-threatening infection of the liver that is caused by the hepatitis B virus (HBV). Even though a safe and effective vaccine has been available since 1982, HBV infection still remains a major health problem worldwide, causing chronic liver infection leading to liver cancer, liver cirrhosis, and death (Seto, Lo, Pawlotsky, & Yuen, 2018). According to the World Health Organization Hepatitis B Factsheet (2018), an estimated 257 million people are currently living with HBV infection and over 800,000 die annually from hepatitis B-related complications. It is ranked at number 15 as a cause of death globally (Lozano et al., 2012; Ott, Ullrich, Mascarenhas, & Stevens, 2011).

Following an infection, HBV spreads throughout the liver, bypassing the host's innate immune response and thus preventing any inflammatory reaction from occurring. Around 95% of adults, however, recover from HBV due to the activation of an immune response after a period of 6 or more weeks. This immune response does not significantly damage the host and thus maintains overall functionality of the liver. The HBV immune response in infants and young children is less thorough and, as a result, > 90% of infants under 1 year will be chronically infected (Fattovich, Stroffolini, Zagni, & Donato, 2004; Saffar et al., 2014). Of the chronically infected individuals, approximately 14%–40% will acquire hepatocellular carcinoma (HCC) or liver cirrhosis and 15%–25% will experience death (Lozano et al., 2012).

In terms of geographical prevalence, chronic HBV infections range from < 2% in North America and Western Europe to 2%–7% in the Mediterranean region, Japan and Southeast Asia. Western Africa and South Sudan have the highest prevalence, where ≥ 8% of the adult population is infected (Schweitzer, Horn, Mikolajczyk, Krause, & Ott, 2015). Because of shared risk factors and modes of transmission, approximately 3 million people living with HBV are coinfected with HIV (WHO Hepatitis B Factsheet, 2018). HIV coinfection in individuals with HBV infection increases the rate of developing liver disease, which leads to increased rate of morbidity and mortality (Petty, Steinbeck, Pursell, & Jensen, 2014; Thio et al., 2002).

10.3.2 Preventing infection with hepatitis B virus

The prevention of hepatitis B is mainly dependent on the hepatitis B vaccine. Progress made toward the prevention of HCC caused by HBV has resulted in administering comprehensive vaccination programs against hepatitis B. This program involves universal infant vaccination, currently implemented in many countries around the world, where all infants are required to be vaccinated with the hepatitis B vaccine within 24h of birth. This has resulted in a significant decline in the rate of hepatitis B surface antigen (HBsAg) positivity among children and infants and a significant drop in the rate of childhood HBV-related HCC (Chang et al., 2005). Once vaccinated, these children will be protected as they grow into adults and consequently will be protected against the risk of developing complications from hepatitis B infection, including HCC. It is, however, recommended that all children not vaccinated and young adults under 19 years receive the vaccine. In addition, vaccinations are crucial for people who frequently require blood transfusions and organ transplant patients. Unvaccinated travelers, drug addicts, promiscuous individuals, and healthcare workers exposed to blood and blood products are at high risk for contracting HBV and require vaccination (Schillie et al., 2018; WHO Hepatitis B Factsheet, 2018).

According to the CDC viral hepatitis report (2019), two types of vaccines are currently authorized for use. These are usually administered intramuscularly in three doses. The second and third doses are dispensed at 1 and 6 months, respectively, after administering the first dose. The completed vaccine regimen induces at least 20 years to lifelong immunity in 95% of individuals. The vaccines include both monovalent and combination vaccines. The monovalent vaccines are specific for protection against HBV. Of the combination vaccines, the bivalent targets both hepatitis A and hepatitis B infections, while the polyvalent vaccine protects against hepatitis B, diphtheria, tetanus, acellular pertussis, and inactivated poliovirus.

10.3.3 Treatment of hepatitis B infection

10.3.3.1 Treatment goals in chronic HBV infection

HBV is classified into its own unique family, Hepadnaviridae, which are mostly associated with and restricted to hepatocytes within the host organism. They have a unique replication cycle where double-stranded DNA is replicated via an RNA intermediate (Datta, Chatterjee, Veer, & Chakravarty, 2012). The HBV genome either forms mini-chromosomes within the nucleus of infected hepatocytes or integrates into the cell genome. This results in suppression of the host immunity and enables the infection to persist, leading to disease progression (Op Den Brouw et al., 2009).

The main aim of anti-HBV therapy is to enhance the survival of chronically infected patients by eradicating infection and preventing the progression of disease to HCC, thereby preventing mortality. Additionally, preventing the transmission of HBV from mother to child and preventing reactivation of hepatitis B infections are important aspects to be considered when implementing therapy. This can be attained through long-term treatment to initially control infection followed by complete elimination of HBV. Eradication of HBV, however, is rarely obtained because the intrahepatic mini-chromosomes and the HBV genome that has integrated into the host DNA continue to exist, even after HBsAg seroclearance. This causes a risk for infection reactivation throughout an individual's life. HBsAg/HbeAg loss in serum samples and sustaining HBV DNA suppression without truly eradicating it is the optimal endpoint of long-term therapy (Lampertico et al., 2017; Scaglione & Lok, 2012).

10.3.3.2 Treatment strategies in chronic hepatitis B

According to the American Association for the Study of Liver Diseases (AASLD), six agents are recommended for the treatment of chronic hepatitis B (CHB) in adults and five in children with CHB. These include nucleot(s)ide analogs tenofovir, entecavir, lamivudine, adefovir, and telbivudine. Telbivudine has not been approved for children with CHB. These analogs are viral polymerases and reverse transcriptase inhibitors that suppress HBV replication and consequently result in efficient reduction of HBV DNA. Nucleot(s)ide therapy is generally well tolerated and is considered safe, but has a high rate of posttreatment virologic relapse, making its use a lifelong commitment. Interferon (IFN) therapy, on the other hand, prevents multiplication of HBV by intercepting the formation of replication-competent pregenomic RNA-containing HBV capsids. Therapy with INF includes pegylated interferon 2a for adults and interferon-α-2b for children (Baran, 2015; Terrault et al., 2016; Wieland, Eustaquio, Whitten-Bauer, Boyd, & Chisari, 2005; Wu & Ning, 2017). While these therapeutic agents are usually administered individually, the availability of these various agents with different targets allows combination therapy to be implemented. Combination therapy in HBV infection may provide an additive effect, which may result in an effective and sustained viral suppression as well as lower rates of drug resistance (Monto et al., 2010). Currently, however, combination therapy is not recommended for use in CHB infection (Terrault et al., 2016). Based on the levels of alanine aminotransferase enzyme (ALT), HBV DNA and a positive/negative HBeAg test, the AASLD has characterized the progression of CHB infection into four phases with treatment strategies implemented accordingly. These phases include an immune-tolerant phase, HBeAg-positive immune active phase, HBeAg-negative immune activation phase, and inactive CHB phase (Terrault et al., 2016).

The immune-tolerant phase

This is the early phase of infection that is highly replicative. It is usually characterized as a phase with a positive HBeAg test, normal ALT levels, and elevated HBV DNA levels. The AASLD guidelines recommend against treatment of infected individuals in this phase. The rationale behind this is that no studies exist that demonstrate the beneficial effects of antiviral treatment in this phase in terms of lowering the incidence of HCC, cirrhosis, and liver-related death. Additionally, the side effects of treatment agents and the risk of development of resistance outweigh its potential advantages. However, it is recommended to test the levels of ALT enzyme twice a year for individuals in this phase to monitor and exclude transition to the next phase, the immune-active or -inactive CHB.

HBeAg-positive immune active phase

This phase is usually presented by raised levels of both ALT and HBV DNA (above 20,000 IU/mL) and an HBeAg-positive test result. It is recommended to initiate treatment with any of the available drugs; the preferred monotherapies, however, are Peg-IFN-α, tenofovir, and entecavir. Studies to date suggest that combination therapies have no added benefits when compared to monotherapy (Paul & Han, 2011). Despite the lack of superiority, combination treatment as initial therapy is given primarily to combat resistance induced by monotherapy (Marcellin et al., 2010). Treatment, however, does not exclude the possibility of developing HCC, and monitoring individuals who are at risk is recommended (Terrault et al., 2016).

HBeAg-negative immune active phase

In this phase, the levels of both ALT and HBV DNA are elevated while the HBeAg is negative. Recommended treatment is usually as in the HBeAg-positive immune active phase with Peg-IFN-α, tenofovir, and entecavir as the current standard of therapy.

Inactive CHB phase

This phase is characterized by low levels of HBV DNA, normal levels of ALT enzyme, negative for HBeAg, and the presence of anti-HBe antibodies.

10.3.3.3 Prevention of perinatal transmission and therapy in expectant women

Perinatal transmission from mother to child usually occurs in mothers who are infected, have very high levels of blood HBV DNA, and are HBsAg/HBeAg seropositive. A child born to an infected mother has a 100% risk of acquiring HBV at birth. Over 90% of infants infected perinatally risk developing CHB if left unvaccinated (Brown et al., 2016; Burk, Hwang, Ho, Shafritz, & Beasley, 1994). It is therefore of utmost importance to administer the hepatitis B vaccine at birth so as to reduce the transmission of HBV to infants born to infected mothers and to prevent new infections in all children born to seronegative individuals (Brown et al., 2016). While administering the hepatitis B vaccine soon after birth considerably reduces the chances of developing chronic hepatitis, 5%–10% of newborns, however, do acquire HBV infection, which leads to CHB infection (Zhang et al., 2014).

It is recommended by the AASLD that antiviral therapy, in pregnant women who have high levels of HBV DNA (over 200,000 IU/mL), be administered in their third trimester. This has shown to reduce the incidence of HBV transmission to the child at birth. Lamivudine, telbivudine, and tenofovir are effective antivirals that are considered safe in pregnancy and thus recommended (Terrault et al., 2016).

References

Ahmed, M., Merga, H., & Jarso, H. (2019). Predictors of virological treatment failure among adult HIV patients on first-line antiretroviral therapy in Woldia and Dessie hospitals, Northeast Ethiopia: A case-control study. *BMC Infectious Diseases*, *19*, 305. https://doi.org/10.1186/s12879-019-3924-4.

Ait-Aissa, A., Derrar, F., Hannoun, D., Gradi, E. A., Scaravelli, D., & Bouslama, Z. (2018). Surveillance for antiviral resistance among influenza viruses circulating in Algeria during five consecutive influenza seasons (2009-2014). *Journal of Medical Virology*, *90*, 844–853. https://doi.org/10.1002/jmv.25029.

Antonelli, G., Dianzani, F., Bellarosa, D., Turriziani, O., Riva, E., & Gentile, A. (1994). Drug combination of AZT and ddI: Synergism of action and prevention of appearance of AZT-resistance. *Antiviral Chemistry and Chemotherapy, 5,* 51–55. https://doi.org/10.1177/095632029400500108.

Awi, N. J., & Teow, S.-Y. (2018). Antibody-mediated therapy against HIV/AIDS: Where are we standing now? *Journal of Pathogens, 2018,* 8724549. https://doi.org/10.1155/2018/8724549.

Babo, Y. D., Alemie, G. A., & Fentaye, F. W. (2017). Predictors of first-line antiretroviral therapy failure amongst HIV-infected adult clients at Woldia Hospital, Northeast Ethiopia. *PLoS One, 12,* e0187694. https://doi.org/10.1371/journal.pone.0187694.

Baran, B. (2015). Nucleos(t)ide analogs in the prevention of hepatitis B virus related hepatocellular carcinoma. *World Journal of Hepatology, 7,* 1742–1754. https://doi.org/10.4254/wjh.v7.i13.1742.

Beccari, M. V., Mogle, B. T., Sidman, E. F., Mastro, K. A., Asiago-Reddy, E., & Kufel, W. D. (2019). Ibalizumab: A novel monoclonal antibody for the management of multidrug resistant HIV-1 infection. *Antimicrobial Agents and Chemotherapy, 63.* https://doi.org/10.1128/aac.00110-19. e00110-19.

Beigel, J. H., Bao, Y., Beeler, J., Manosuthi, W., Slandzicki, A., Dar, S. M., … Alzogaray, M. F. (2017). Oseltamivir, amantadine, and ribavirin combination antiviral therapy versus oseltamivir monotherapy for the treatment of influenza: A multicentre, double-blind, randomised phase 2 trial. *The Lancet Infectious Diseases, 17*(12), 1255–1265. https://doi.org/10.1016/S1473-3099(17)30476–0.

Brown, R. S., Mcmahon, B. J., Lok, A. S. F., Wong, J. B., Ahmed, A. T., Mouchli, M. A., … Mohammed, K. (2016). Antiviral therapy in chronic hepatitis B viral infection during pregnancy: A systematic review and meta-analysis. *Hepatology, 63,* 319–333. https://doi.org/10.1002/hep.28302.

Burk, R. D., Hwang, L. Y., Ho, G. Y. F., Shafritz, D. A., & Beasley, R. P. (1994). Outcome of perinatal hepatitis B virus exposure is dependent on maternal virus load. *Journal of Infectious Diseases, 170,* 1418–1423. https://doi.org/10.1093/infdis/170.6.1418.

Cahn, P., Pozniak, A. L., Mingrone, H., Shuldyakov, A., Brites, C., Andrade-Villanueva, J. F., … Min, S. (2013). Dolutegravir versus raltegravir in antiretroviral-experienced, integrase-inhibitor-naive adults with HIV: Week 48 results from the randomised, double-blind, non-inferiority SAILING study. *The Lancet, 382,* 700–708. https://doi.org/10.1016/S0140-6736(13)61221-0.

Caliendo, A. M., & Hirsch, M. S. (1994). Combination therapy for infection due to human immunodeficiency virus type 1. *Clinical Infectious Diseases, 18,* 516–524. https://doi.org/10.1093/clinids/18.4.516.

CDC. (2018). *Estimated influenza illnesses, medical visits, hospitalizations, and deaths in the United States—2017–2018 influenza season* Available at https://www.cdc.gov/flu/about/burden/2017a-2018.htm (Accessed June 20, 2019).

CDC. (2019a). *HIV prevention progress report.* 1–92. Available at https://www.cdc.gov/hiv/pdf/policies/progressreports/cdc-hiv-preventionprogressreport.pdf (Accessed July 6, 2019).

CDC. (2019b). *Influenza antiviral drug resistance* Available at https://www.cdc.gov/flu/treatment/antiviralresistance.htm (Accessed May 29, 2019).

CDC Viral Hepatitis. (2019). Available at https://www.cdc.gov/hepatitis/hbv/hbvfaq.htm#vaccFAQ (Accessed July 8, 2019).

Chang, M. H., Chen, T. H. H., Hsu, H. M., Wu, T. C., Kong, M. S., Liang, D. C., … Chen, D. S. (2005). Prevention of hepatocellular carcinoma by universal vaccination against hepatitis B virus: The effect and problems. *Clinical Cancer Research, 11,* 7953–7957. https://doi.org/10.1158/1078-0432.CCR-05-1095.

Chu, C., & Selwyn, P. A. (2011). Complications of HIV infection: A systems-based approach. *American Family Physician, 83,* 395–406.

Cihlar, T., & Ray, A. S. (2010). Nucleoside and nucleotide HIV reverse transcriptase inhibitors: 25 years after zidovudine. *Antiviral Research, 85,* 39–58. https://doi.org/10.1016/j.antiviral.2009.09.014.

Clay, P. G. (2004). Single-class therapy for HIV is not optimal care. *Annals of Pharmacotherapy, 38*, 1307–1309. https://doi.org/10.1345/aph.1E030.

Clutter, D. S., Jordan, M. R., Bertagnolio, S., & Shafer, R. W. (2016). HIV-1 drug resistance and resistance testing. *Infection, Genetics and Evolution, 46*, 292–307. https://doi.org/10.1016/j.meegid.2016.08.031.

Coetzee, J., Hunt, G., Jaffer, M., Otwombe, K., Scott, L., Bongwe, A., ... Gray, G. E. (2017). HIV-1 viraemia and drug resistance amongst female sex workers in Soweto, South Africa: A cross sectional study. *PLoS One, 12*, e0188606. https://doi.org/10.1371/journal.pone.0188606.

Cohen, M. S., Chen, Y. Q., McCauley, M., Gamble, T., Hosseinipour, M. C., Kumarasamy, N., ... HPTN 052 Study Team. (2016). Antiretroviral therapy for the prevention of HIV-1 transmission. *The New England Journal of Medicine, 375*, 830–839. https://doi.org/10.1056/NEJMoa1600693.

Datta, S., Chatterjee, S., Veer, V., & Chakravarty, R. (2012). Molecular biology of the hepatitis B virus for clinicians. *Journal of Clinical and Experimental Hepatology, 2*, 353–365. https://doi.org/10.1016/j.jceh.2012.10.003.

De Luca, A., Dunn, D., Zazzi, M., Camacho, R., Torti, C., Fanti, I., ... Di Giambenedetto, S. (2013). Declining prevalence of HIV-1 drug resistance in antiretroviral treatment-exposed individuals in Western Europe. *Journal of Infectious Diseases, 207*, 1216–1220. https://doi.org/10.1093/infdis/jit017.

Department of Health and Human Services. (2018). *Guidelines for the use of antiretroviral agents in adults and adolescents living with HIV*. https://aidsinfo.nih.gov/guidelines/html/1/adult-and-adolescent-arv.

Deyde, V. M., Xu, X., Bright, R. A., Shaw, M., Smith, C. B., Zhang, Y., ... Klimov, A. I. (2007). Surveillance of resistance to adamantanes among influenza A (H3N2) and A(H1N1) viruses isolated worldwide. *The Journal of Infectious Diseases, 196*, 249–257. https://doi.org/10.1086/518936.

El-Sadr, W. M., Lundgren, J. D., Neaton, J. D., Gordin, F., & Abrams, D. (2006). CD4 + count-guided interruption of antiretroviral therapy. *The New England Journal of Medicine, 355*, 2283–2296.

Emu, B., Fessel, J., Schrader, S., Kumar, P., Richmond, G., Win, S., ... Lewis, S. (2018). Phase 3 study of ibalizumab for multidrug-resistant HIV-1. *The New England Journal of Medicine, 379*, 645–654. https://doi.org/10.1056/NEJMoa1711460.

Eron, J. J., Johnson, V. A., Merrill, D. P., Chou, T. C., & Hirsch, M. S. (1992). Synergistic inhibition of replication of human immunodeficiency virus type 1, including that of a zidovudine-resistant isolate, by zidovudine and 2′,3′-dideoxycytidine in vitro. *Antimicrobial Agents and Chemotherapy, 36*, 1559–1562. https://doi.org/10.1128/AAC.36.7.1559.

Fätkenheuer, G., Nelson, M., Lazzarin, A., Konourina, I., Hoepelman, A. I. M., Lampiris, H., ... van der Ryst, E. (2008). Subgroup analyses of maraviroc in previously treated R5 HIV-1 infection. *The New England Journal of Medicine, 359*, 1442–1455. https://doi.org/10.1056/nejmoa0803154.

Fattovich, G., Stroffolini, T., Zagni, I., & Donato, F. (2004). Hepatocellular carcinoma in cirrhosis: Incidence and risk factors. *Gastroenterology, 127*, S35–S50. https://doi.org/10.1053/j.gastro.2004.09.014.

Gerstoft, J., Kirk, O., Obel, N., Pedersen, C., Mathiesen, L., Nielsen, H., ... Lundgren, J. D. (2003). Low efficacy and high frequency of adverse events in a randomized trial of the triple nucleoside regimen abacavir, stavudine and didanosine. *AIDS, 17*, 2045–2052. https://doi.org/10.1097/00002030-200309260-00005.

Gounden, V., van Niekerk, C., Snyman, T., & George, J. A. (2010). Presence of the CYP2B6 516G > T polymorphism, increased plasma efavirenz concentrations and early neuropsychiatric side effects in South African HIV-infected patients. *AIDS Research and Therapy, 7*, 32. https://doi.org/10.1186/1742-6405-7-32.

Govorkova, E. A., Fang, H. B., Tan, M., & Webster, R. G. (2004). Neuraminidase inhibitor-rimantadine combinations exert additive and synergistic anti-influenza virus effects in MDCK cells. *Antimicrobial Agents and Chemotherapy, 48*, 4855–4863. https://doi.org/10.1128/AAC.48.12.4855-4863.2004.

Gunthard, H. F., Saag, M. S., Benson, C. A., del Rio, C., Eron, J. J., Gallant, J. E., … Volberding, P. A. (2016). Antiretroviral drugs for treatment and prevention of HIV infection in adults. *Journal of the American Medical Association, 316*, 191–210. https://doi.org/10.1001/jama.2016.8900.

Hamilton, B. S., Whittaker, G. R., & Daniel, S. (2012). Influenza virus-mediated membrane fusion: Determinants of hemagglutinin fusogenic activity and experimental approaches for assessing virus fusion. *Viruses, 4*, 1144–1168. https://doi.org/10.3390/v4071144.

Haqqani, A. A., & Tilton, J. C. (2013). Entry inhibitors and their use in the treatment of HIV-1 infection. *Antiviral Research, 98*, 158–170. https://doi.org/10.1016/j.antiviral.2013.03.017.

Hay, A. J., Gregory, V., Douglas, A. R., & Yi, P. L. (2001). The evolution of human influenza viruses. *Philosophical Transactions of the Royal Society of London Series B, Biological Sciences, 356*, 1861–1870. https://doi.org/10.1098/rstb.2001.0999.

Hayden, F. G., & Shindo, N. (2019). Influenza virus polymerase inhibitors in clinical development. *Current Opinion in Infectious Diseases, 32*, 176–186. https://doi.org/10.1097/qco.0000000000000532.

Hayden, F. G., Yamada, K., Portsmouth, S., Kawaguchi, K., Shishido, T., Arai, M., … Uehara, T. (2018). Baloxavir marboxil for uncomplicated influenza in adults and adolescents. *The New England Journal of Medicine, 379*, 913–923. https://doi.org/10.1056/NEJMoa1716197.

Hicks, C. B., Cahn, P., Cooper, D. A., Walmsley, S. L., Katlama, C., Clotet, B., … Valdez, H. (2006). Durable efficacy of tipranavir-ritonavir in combination with an optimized background regimen of anti-retroviral drugs for treatment-experienced HIV-1-infected patients at 48 weeks in the Randomized Evaluation of Strategic Intervention in multi-drug resistant patients with Tipranavir (RESIST) studies: An analysis of combined data from two randomised open-label trials. *The Lancet, 368*, 466–475. https://doi.org/10.1016/S0140-6736(06)69154-X.

Hicks, C., & Gulick, R. M. (2009). Raltegravir: The first HIV type 1 integrase inhibitor. *Clinical Infectious Diseases, 48*, 931–939. https://doi.org/10.1086/597290.

Horimoto, T., & Kawaoka, Y. (2005). Influenza: Lessons from past pandemics, warnings from current incidents. *Nature Reviews Microbiology, 3*, 591–600. https://doi.org/10.1038/nrmicro1208.

Hwang, C., Lai, M.-T., & Hazuda, D. (2018). 545. Rational design of doravirine (DOR): A review of development from bench to patients. *Open Forum Infectious Diseases, 5*, S203. https://doi.org/10.1093/ofid/ofy210.553.

Ilyushina, N. A., Hoffmann, E., Salomon, R., Webster, R. G., & Govorkova, E. A. (2007). Amantadine-oseltamivir combination therapy for H5N1 influenza virus infection in mice. *Antiviral Therapy, 12*, 363–370.

Iuliano, A. D., Roguski, K. M., Chang, H. H., Muscatello, D. J., Palekar, R., Tempia, S., … Mustaquim, D. (2018). Estimates of global seasonal influenza-associated respiratory mortality: A modelling study. *The Lancet, 391*(10127), 1285–1300. https://doi.org/10.1016/S0140-6736(17)33293-2.

Kiertiburanakul, S., Pinsai, S., Chantratita, W., Pasomsub, E., Leechawengwongs, M., Thipmontree, W., … Sungkanuparph, S. (2016). Prevalence of primary HIV drug resistance in Thailand detected by short reverse transcriptase genotypic resistance assay. *PLoS One, 11*, e0147945. https://doi.org/10.1371/journal.pone.0147945.

Kityo, C., Sigaloff, K. C. E., Boender, T. S., Kaudha, E., Kayiwa, J., Musiime, V., … Mugyenyi, P. N. (2016). HIV drug resistance among children initiating first-line antiretroviral treatment in Uganda. *AIDS Research and Human Retroviruses, 32*, 628–635. https://doi.org/10.1089/aid.2015.0215.

Lampertico, P., Agarwal, K., Berg, T., Buti, M., Janssen, H. L. A., Papatheodoridis, G., & Tacke, F. (2017). EASL 2017 Clinical Practice Guidelines on the management of hepatitis B virus infection. *Journal of Hepatology, 67*, 370–398. https://doi.org/10.1016/j.jhep.2017.03.021.

Lee, N., Chan, P. K. S., Hui, D. S. C., Rainer, T. H., Wong, E., Choi, K., ... Sung, J. J. Y. (2009). Viral loads and duration of viral shedding in adult patients hospitalized with influenza. *The Journal of Infectious Diseases, 200*, 492–500. https://doi.org/10.1086/600383.

Lee, N., & Ison, M. G. (2012). Diagnosis, management and outcomes of adults hospitalized with influenza. *Antiviral Therapy, 17*, 143–157. https://doi.org/10.3851/IMP2059.

Li, T. C. M., Chan, M. C. W., & Lee, N. (2015). Clinical implications of antiviral resistance in influenza. *Viruses, 7*, 4929–4944. https://doi.org/10.3390/v7092850.

Lozano, R., Naghavi, M., Foreman, K., Lim, S., Shibuya, K., Aboyans, V., ... Murray, C. J. L. (2012). Global and regional mortality from 235 causes of death for 20 age groups in 1990 and 2010: A systematic analysis for the Global Burden of Disease Study 2010. *The Lancet, 380*, 2095–2128. https://doi.org/10.1016/S0140-6736(12)61728-0.

Mao, L., Yang, Y., Qiu, Y., & Yang, Y. (2012). Annual economic impacts of seasonal influenza on US counties: Spatial heterogeneity and patterns. *International Journal of Health Geographics, 11*, 16. https://doi.org/10.1186/1476-072X-11-16.

Marathe, B. M., Wong, S. S., Vogel, P., Garcia-Alcalde, F., Webster, R. G., Webby, R. J., ... Govorkova, E. A. (2016). Combinations of oseltamivir and T-705 extend the treatment window for highly pathogenic influenza a (H5N1) virus infection in mice. *Scientific Reports, 6*, 26742. https://doi.org/10.1038/srep26742.

Marcellin, P., Cadranel, J. F., Fontanges, T., Poynard, T., Pol, S., Trepo, C., ... Zarski, J. P. (2010). High rate of adefovir-lamivudine combination therapy in nucleoside-naïve patients with chronic hepatitis B in France: Results of a national survey in 1730 patients. *European Journal of Gastroenterology and Hepatology, 22*, 1290–1296. https://doi.org/10.1097/MEG.0b013e32832fba4f.

Martín-Benito, J., & Ortín, J. (2013). Influenza virus transcription and replication. *Advances in Virus Research, 87*, 113–137. https://doi.org/10.1016/B978-0-12-407698-3.00004-1.

Meanwell, N. A., Krystal, M. R., Nowicka-Sans, B., Langley, D. R., Conlon, D. A., Eastgate, M. D., ... Kadow, J. F. (2018). Inhibitors of HIV-1 attachment: The discovery and development of temsavir and its prodrug fostemsavir. *Journal of Medicinal Chemistry, 61*, 62–80. https://doi.org/10.1021/acs.jmedchem.7b01337.

Messiaen, P., Wensing, A. M. J., Fun, A., Nijhuis, M., Brusselaers, N., & Vandekerckhove, L. (2013). Clinical use of HIV integrase inhibitors: A systematic review and meta-analysis. *PLoS One, 8*, e52562. https://doi.org/10.1371/journal.pone.0052562.

Molinari, N. A. M., Ortega-Sanchez, I. R., Messonnier, M. L., Thompson, W. W., Wortley, P. M., Weintraub, E., & Bridges, C. B. (2007). The annual impact of seasonal influenza in the US: Measuring disease burden and costs. *Vaccine, 25*, 5086–5096. https://doi.org/10.1016/j.vaccine.2007.03.046.

Momper, J. D., Best, B. M., Wang, J., Capparelli, E. V., Stek, A., Barr, E., et al.Mirochnick, M. (2018). Elvitegravir/cobicistat pharmacokinetics in pregnant and postpartum women with HIV. *AIDS, 32*, 2507–2516. https://doi.org/10.1097/QAD.0000000000001992.

Monod, A., Swale, C., Tarus, B., Tissot, A., Delmas, B., Ruigrok, R. W., ... Slama-Schwok, A. (2015). Learning from structure-based drug design and new antivirals targeting the ribonucleoprotein complex for the treatment of influenza. *Expert Opinion on Drug Discovery, 10*, 345–371. https://doi.org/10.1517/17460441.2015.1019859.

Monto, A., Schooley, R. T., Lai, J. C., Sulkowski, M. S., Chung, R. T., Pawlotsky, J. M., ... Jacobson, I. M. (2010). Lessons from HIV therapy applied to viral hepatitis therapy: Summary of a workshop. *American Journal of Gastroenterology, 105*, 989–1004. https://doi.org/10.1038/ajg.2009.726.

Ng, K. E. (2019). Xofluza (baloxavir marboxil) for the treatment of acute uncomplicated influenza. *P &T: A Peer-Reviewed Journal for Formulary Management, 44*(1), 9–11.

Nguyen, J. T., Hoopes, J. D., Le, M. H., Smee, D. F., Patick, A. K., Faix, D. J., ... Went, G. T. (2010). Triple combination of amantadine, ribavirin, and oseltamivir is highly active and synergistic against drug resistant influenza virus strains in vitro. *PLoS One, 5*, e9332. https://doi.org/10.1371/journal.pone.0009332.

Nguyen, J. T., Hoopes, J. D., Smee, D. F., Prichard, M. N., Driebe, E. M., Engelthaler, D. M., … Went, G. T. (2009). Triple combination of oseltamivir, amantadine, and ribavirin displays synergistic activity against multiple influenza virus strains in vitro. *Antimicrobial Agents and Chemotherapy*, *53*, 4115–4126. https://doi.org/10.1128/AAC.00476-09.

O'Brien, M. E., Clark, R. A., Besch, C. L., Myers, L., & Kissinger, P. (2003). Patterns and correlates of discontinuation of the initial HAART regimen in an urban outpatient cohort. *Journal of Acquired Immune Deficiency Syndromes*, *34*, 407–414. https://doi.org/10.1097/00126334-200312010-00008.

Op Den Brouw, M. L., Binda, R. S., Van Roosmalen, M. H., Protzer, U., Janssen, H. L. A., Van Der Molen, R. G., & Woltman, A. M. (2009). Hepatitis B virus surface antigen impairs myeloid dendritic cell function: A possible immune escape mechanism of hepatitis B virus. *Immunology*, *126*, 280–289. https://doi.org/10.1111/j.1365-2567.2008.02896.x.

Ott, J. J., Ullrich, A., Mascarenhas, M., & Stevens, G. A. (2011). Global cancer incidence and mortality caused by behavior and infection. *Journal of Public Health*, *33*, 223–233. https://doi.org/10.1093/pubmed/fdq076.

Palella, F., Armon, C., Chmiel, J. S., Brooks, J. T., Debes, R., Novak, R. M., … Buchacz, K. (2014). Higher Cd4 at art initiation predicts greater long term likelihood of Cd4 normalization. *Value in Health*, *17*, A269. https://doi.org/10.1016/j.jval.2014.03.1567.

Palella, F. J., Baker, R. K., Moorman, A. C., Chmiel, J. S., Wood, K. C., Brooks, J. T., & Holmberg, S. D. (2006). Mortality in the highly active antiretroviral therapy era: Changing causes of death and disease in the HIV outpatient study. *Journal of Acquired Immune Deficiency Syndromes*, *43*, 27–34. https://doi.org/10.1097/01.qai.0000233310.90484.16.

Paredes, R., Lalama, C. M., Ribaudo, H. J., Schackman, B. R., Shikuma, C., Giguel, F., … Kuritzkes, D. R. (2010). Pre-existing minority drug-resistant HIV-1 variants, adherence, and risk of antiretroviral treatment failure. *The Journal of Infectious Diseases*, *201*, 662–671. https://doi.org/10.1086/650543.

Paul, N., & Han, S. H. (2011). Combination therapy for chronic hepatitis B: Current indications. *Current Hepatitis Reports*, *10*, 98–105. https://doi.org/10.1007/s11901-011-0095-1.

Petty, L. A., Steinbeck, J. L., Pursell, K., & Jensen, D. M. (2014). Human immunodeficiency virus and coinfection with hepatitis B and C. *Infectious Disease Clinics of North America*, *28*, 477–499. https://doi.org/10.1016/j.idc.2014.05.005.

Phung, B. C., Sogni, P., & Launay, O. (2014). Hepatitis B and human immunodeficiency virus co-infection. *World Journal of Gastroenterology*, *49*, S138–S145. https://doi.org/10.3748/wjg.v20.i46.17360.

Pirrone, V., Thakkar, N., Jacobson, J. M., Wigdahl, B., & Krebs, F. C. (2011). Combinatorial approaches to the prevention and treatment of HIV-1 infection. *Antimicrobial Agents and Chemotherapy*, *55*, 1831–1842. https://doi.org/10.1128/aac.00976-10.

Popov, A. F., Shchelkanov, M. Y., Dmitrenko, K. A., & Simakova, A. I. (2018). Combined therapy of influenza with antiviral drugs with a different mechanism of action in comparison with monotherapy. *Journal of Pharmaceutical Sciences and Research*, *10*, 357–360.

Poppe, L. K., Chunda-Liyoka, C., Kwon, E. H., Gondwe, C., West, J. T., Kankasa, C., … Wood, C. (2017). HIV drug resistance in infants increases with changing prevention of mother-to-child transmission regimens. *AIDS*, *31*, 1885–1889. https://doi.org/10.1097/QAD.0000000000001569.

Prasad, V. R., & Goff, S. P. (1990). Structure-function studies of HIV reverse transcriptase. *Annals of the New York Academy of Sciences*, *616*, 11–21. https://doi.org/10.1111/j.1749-6632.1990.tb17823.x.

Rodríguez-Nóvoa, S., Martín-Carbonero, L., Barreiro, P., González-Pardo, G., Jiménez-Nácher, I., González-Lahoz, J., & Soriano, V. (2007). Genetic factors influencing atazanavir plasma concentrations and the risk of severe hyperbilirubinemia. *AIDS*, *21*, 41–46. https://doi.org/10.1097/QAD.0b013e328011d7c1.

Ryan, J., Zoellner, Y., Gradl, B., Palache, B., & Medema, J. (2006). Establishing the health and economic impact of influenza vaccination within the European Union 25 countries. *Vaccine*, *24*, 6812–6822. https://doi.org/10.1016/j.vaccine.2006.07.042.

Saffar, H., Ajami, A., Saffar, M. J., Shojaei, J., Sotudeh-Anvari, M., Shams-Esfandabad, K., & Khalilian, A. R. (2014). Prevalence of hepatitis B virus seromarkers in young adults vaccinated at birth; impact on the epidemiology of hepatitis B infection in Iran. *Hepatitis Monthly*, *14*, e17263. https://doi.org/10.5812/hepatmon.17263.

Scaglione, S. J., & Lok, A. S. F. (2012). Effectiveness of hepatitis B treatment in clinical practice. *Gastroenterology*, *142*. https://doi.org/10.1053/j.gastro.2012.01.044. 1360–1368.e1.

Schillie, S., Vellozzi, C., Reingold, A., Harris, A., Haber, P., Ward, J. W., & Nelson, N. P. (2018). Prevention of hepatitis B virus infection in the United States: Recommendations of the Advisory Committee on Immunization Practices. *MMWR. Recommendations and Reports*, *67*, 1–31. https://doi.org/10.15585/mmwr.rr6701a1.

Scholar, E. (2011). HIV protease inhibitors. In *xPharm: The comprehensive pharmacology reference*. https://doi.org/10.1016/B978-008055232-3.61025-6.

Schweitzer, A., Horn, J., Mikolajczyk, R. T., Krause, G., & Ott, J. J. (2015). Estimations of worldwide prevalence of chronic hepatitis B virus infection: A systematic review of data published between 1965 and 2013. *The Lancet*, *386*, 1546–1555. https://doi.org/10.1016/S0140-6736(15)61412-X.

Seo, S., Englund, J. A., Nguyen, J. T., Pukrittayakamee, S., Lindegardh, N., Tarning, J., & Boeckh, M. J. (2013). Combination therapy with amantadine, oseltamivir and ribavirin for influenza A infection: Safety and pharmacokinetics. *Antiviral Therapy*, *18*, 377–386. https://doi.org/10.3851/IMP2475.

Seto, W. K., Lo, Y. R., Pawlotsky, J. M., & Yuen, M. F. (2018). Chronic hepatitis B virus infection. *The Lancet*, *392*, P2313–P2324. https://doi.org/10.1016/S0140-6736(18)31865-8.

Sheu, T. G., Deyde, V. M., Okomo-Adhiambo, M., Garten, R. J., Xu, X., Bright, R. A., … Gubareva, L. V. (2008). Surveillance for neuraminidase inhibitor resistance among human influenza A and B viruses circulating worldwide from 2004 to 2008. *Antimicrobial Agents and Chemotherapy*, *52*, 3284–3292. https://doi.org/10.1128/AAC.00555-08.

Skehel, J. (2009). An overview of influenza haemagglutinin and neuraminidase. *Biologicals*, *37*, 177–178. https://doi.org/10.1016/j.biologicals.2009.02.012.

Smee, D. F., Hurst, B. L., Wong, M. H., Bailey, K. W., & Morrey, J. D. (2009). Effects of double combinations of amantadine, oseltamivir, and ribavirin on influenza A (H5N1) virus infections in cell culture and in mice. *Antimicrobial Agents and Chemotherapy*, *53*, 2120–2128. https://doi.org/10.1128/AAC.01012-08.

Spagnuolo, V., Castagna, A., & Lazzarin, A. (2018). Bictegravir. *Current Opinion in HIV and AIDS*, *13*, 326–333. https://doi.org/10.1097/COH.0000000000000468.

Suess, T., Remschmidt, C., Schink, S. B., Schweiger, B., Heider, A., Milde, J., … Buchholz, U. (2012). Comparison of shedding characteristics of seasonal influenza virus (sub)types and influenza A(H1N1)pdm09; Germany, 2007-2011. *PLoS One*, *7*, e51653. https://doi.org/10.1371/journal.pone.0051653.

Summa, V., Petrocchi, A., Bonelli, F., Crescenzi, B., Donghi, M., Ferrara, M., … Rowley, M. (2008). Discovery of raltegravir, a potent, selective orally bioavailable HIV-integrase inhibitor for the treatment of HIV-AIDS infection. *Journal of Medicinal Chemistry*, *25*, 5843–5855. https://doi.org/10.1021/jm800245z.

Tao, R. J., Luo, X. L., Xu, W., Mao, B., Dai, R. X., Li, C. W., … Xu, J. F. (2018). Viral infection in community acquired pneumonia patients with fever: A prospective observational study. *Journal of Thoracic Disease*, *10*, 4387–4395. https://doi.org/10.21037/jtd.2018.06.33.

Terrault, N. A., Bzowej, N. H., Chang, K.-M., Hwang, J. P., Jonas, M. M., & Murad, M. H. (2016). AASLD guidelines for treatment of chronic hepatitis B. *Hepatology*, *63*, 261–283. https://doi.org/10.1002/hep.28156.

Thio, C. L., Seaberg, E. C., Skolasky, R., Phair, J., Visscher, B., Muñoz, A., & Thomas, D. L. (2002). HIV-1, hepatitis B virus, and risk of liver-related mortality in the Multicenter Cohort Study (MACS). *The Lancet*, *360*, 1921–1926. https://doi.org/10.1016/S0140-6736(02)11913-1.

Tiberi, M., Tintori, C., Ceresola, E. R., Fazi, R., Zamperini, C., Calandro, P., … Botta, M. (2014). 2-Aminothiazolones as anti-HIV agents that act as gp120-CD4 inhibitors. *Antimicrobial Agents and Chemotherapy*, *58*, 3043–3052. https://doi.org/10.1128/aac.02739-13.

Tong, S., Zhu, X., Li, Y., Shi, M., Zhang, J., Bourgeois, M., ... Donis, R. O. (2013). New world bats harbor diverse influenza A viruses. *PLoS Pathogens, 9*, e1003657. https://doi.org/10.1371/journal.ppat.1003657.

Treanor, J. J. (2014). Influenza (including avian influenza and swine influenza). In *Mandell, Douglas, and Bennett's principles and practice of infectious diseases*. https://doi.org/10.1016/B978-1-4557-4801-3.00167-3.

UNAIDS Factsheet. (2018). *Global HIV & AIDS statistics*. https://www.unaids.org/en/resources/fact-sheet (Accessed July 1, 2019).

Upton, C. T., Taiwo, B., & Robertson, K. R. (2013). Neurotoxicity of antiretroviral therapy. *Future Virology, 8*, 469–475. https://doi.org/10.2217/fvl.13.30.

Usach, I., Melis, V., & Peris, J. E. (2013). Non-nucleoside reverse transcriptase inhibitors: A review on pharmacokinetics, pharmacodynamics, safety and tolerability. *Journal of the International AIDS Society, 16*, 1–14. https://doi.org/10.7448/IAS.16.1.18567.

Van Leeuwen, R., Katlama, C., Murphy, R. L., Squires, K., Gatell, J., Horban, A., ... Lange, J. M. A. (2003). A randomized trial to study first-line combination therapy with or without a protease inhibitor in HIV-1-infected patients. *AIDS, 17*, 987–999. https://doi.org/10.1097/00002030-200305020-00007.

Wang, D., Hicks, C. B., Goswami, N. D., Tafoya, E., Ribeiro, R. M., Cai, F., ... Gao, F. (2011). Evolution of drug-resistant viral populations during interruption of antiretroviral therapy. *Journal of Virology, 85*, 6403–6415. https://doi.org/10.1128/jvi.02389-10.

WHO HIV Factsheet. (2018). https://www.who.int/news-room/fact-sheets/detail/hiv-aids (Accessed July 1, 2019).

WHO Influenza Factsheet, 2018. Available at: https://www.who.int/news-room/fact-sheets/detail/influenza-(seasonal).

Wieland, S. F., Eustaquio, A., Whitten-Bauer, C., Boyd, B., & Chisari, F. V. (2005). Interferon prevents formation of replication-competent hepatitis B virus RNA-containing nucleocapsids. *Proceedings of the National Academy of Sciences of the United States of America, 102*, 9913–9917. https://doi.org/10.1073/pnas.0504273102.

World Health Organization. (2017a). *Global action plan on HIV drug resistance 2017-2021*. Available at https://apps.who.int/iris/bitstream/handle/10665/255883/9789241512848-eng.pdf?sequence=1 (Accessed July 6, 2019).

World Health Organization. (2017b). *HIV drug resistance report 2017*. Available at https://apps.who.int/iris/bitstream/handle/10665/255896/9789241512831-eng.pdf (Accessed July 6, 2019).

World Health Organization Hepatitis B Factsheet. (2018). Available at https://www.who.int/news-room/fact-sheets/detail/hepatitis-b (Accessed July 13, 2019).

Wu, D., & Ning, Q. (2017). Toward a cure for hepatitis B virus infection: Combination therapy involving viral suppression and immune modulation and long-term outcome. *The Journal of Infectious Diseases, 216*, S771–S777. https://doi.org/10.1093/infdis/jix355.

Zash, R., Makhema, J., & Shapiro, R. L. (2018). Neural-tube defects with dolutegravir treatment from the time of conception. *The New England Journal of Medicine, 379*, 979–981. https://doi.org/10.1056/NEJMc1807653.

Zhang, L., Gui, X. E., Teter, C., Zhong, H., Pang, Z., Ding, L., ... Zhang, L. (2014). Effects of hepatitis B immunization on prevention of mother-to-infant transmission of hepatitis B virus and on the immune response of infants towards hepatitis B vaccine. *Vaccine, 32*, 6091–6097. https://doi.org/10.1016/j.vaccine.2014.08.078.

Further reading

WHO n.d. https://www.who.int/biologicals/vaccines/influenza/en/.

CHAPTER

11

Phenotype screenings of drugs for combination therapy against multidrug resistance

Arif Mohammed, Othman A. Alghamdi

Department of Biological Sciences, Faculty of Science, University of Jeddah, Jeddah, Kingdom of Saudi Arabia

Abbreviations

ABC ATP-binding cassette
CSC cancer stem cell
MDR multidrug resistance
MMOAs molecular mechanisms of action
PDD phenotypic drug discovery
P-gp P-glycoprotein
TDD target-based drug discovery
TMDs transmembrane domains
WHO World Health Organization

11.1 Introduction

During the last few decades, the incidence of drug resistance has increased dramatically. Continuous deployment of antimicrobial drugs in treating infections has led to the emergence of resistance among the various strains of microorganisms. A recent report on the global spread of antimicrobial resistance estimated that 10 million people will die due to infections with resistant microbes by the year 2050 (O'Neill, 2014). A list of high-priority pathogens, including *Acinetobacter baumannii*, *Pseudomonas aeruginosa*, and *Enterobacteriaceae*, has been published by the World Health Organization (WHO) to prompt researchers to discover new antibiotics to fight infections with these bacteria (Shahzad, Willcox, & Shahzad, 2020; World Health Organization, 2017).

https://doi.org/10.1016/B978-0-12-820576-1.00011-4

As of 2017, the emergence and spread of antimicrobial resistance continued unabated around the world, leaving devastating health and economic outcomes in its wake. Those consequences will multiply if collaborative global action is not taken to address the spread of resistance. An influential report released in 2016 estimated that each year at least 700,000 people across the world die from infections that are resistant to current antibiotics, and by 2050, drug-resistant infections will take an estimated 10 million lives per year. The same report further predicted that the economic cost of lost global production caused by antimicrobial resistance will amount to approximately $100 trillion between now and 2050 if antimicrobial resistance is not tackled.

Despite this crisis, there are very few prospects for new antibiotics in the development pipeline that are likely to receive regulatory approval; new drugs will be critical as existing antibiotics lose their effectiveness against infections at an escalating rate. Additionally, improved antimicrobial stewardship practices that promote appropriate use of antimicrobials may be needed to be adopted widely to reduce drug resistance and achieve better patient outcomes. As the burden of resistance to antibiotics grows, even routine medical conditions and procedures will become life threatening because of the risk of untreatable infection. International multilateral organizations have recognized the threat of antimicrobial resistance and have developed plans for action, but there is an urgent need for the immediate implementation of collective actions ("Combating Antimicrobial Resistance," 2017). Major drivers of antimicrobial resistance in humans have been accelerated by inappropriate antimicrobial prescribing in healthcare practices, the inappropriate use of antimicrobials in livestock, and the promulgation of antibiotic resistance genes in the environment ("Combating Antimicrobial Resistance," 2017). There have been major gains in scientific knowledge, concepts, and professional guidance related to antimicrobial resistance. However, the expectation that science will provide an endless supply of new antibiotic drugs remains prevalent. The reality is that the pharmaceutical industry has been losing interest in delivering new medicines because new antibiotic drugs are not financially viable, which is juxtaposed with the continued increase in antimicrobial resistance trends worldwide. The first global survey of antimicrobial resistance conducted by WHO in 2014 examined a group of hospital and community infections and resistance patterns to study the magnitude of antimicrobial resistance across all WHO regions and countries. The report found high levels of antimicrobial resistance in all regions, as well as significant gaps in surveillance data and underreporting of key concerns such as multidrug-resistant tuberculosis ("Combating Antimicrobial Resistance," 2017).

Multidrug resistance (MDR) is defined as insensitivity or resistance of a microorganism to the administered antimicrobial medicines (which are structurally unrelated and have different molecular targets) despite earlier sensitivity to it (Bolan, Sparling, & Wasserheit, 2012; Menzies, Cohen, Lin, Murray, & Salomon, 2012). According to WHO, these resistant microorganisms (like bacteria, fungi, viruses, and parasites) can fight off attacks by antimicrobial drugs, which leads to ineffective treatment, resulting in persistence and spreading of infections. Although the development of MDR is a natural phenomenon, an extensive rise in the number of immunocompromised conditions, like HIV-infected patients, diabetic patients, individuals who have undergone organ transplantation, and severe burn patients, makes the body an easy target for hospital-acquired infectious diseases, thereby contributing to further spread of MDR. Studies from WHO report very high rates of resistance in bacteria causing common infections, such as *Escherichia coli* against antibiotics such as cephalosporin and fluoroquinolones, *Klebsiella pneumoniae* against cephalosporin and carbapenems, *Staphylococcus aureus* against methicillin, *Streptococcus pneumoniae* against penicillin, nontyphoidal Salmonella against fluoroquinolones, Shigella species against fluoroquinolones, *Neisseria gonorrhoeae* against cephalosporin, and *Mycobacterium tuberculosis* against rifampicin, isoniazid, and fluoroquinolone (Coates, Halls, & Hu, 2011; Pichichero & Casey, 2007). A limited number of antifungal drugs are available for the treatment of chronic fungal infections (Chang et al., 2015).

This chapter emphasizes the significance of MDR, various mechanisms contributing to its development, and problems associated with MDR, and it also provides details about drug-screening methods, with emphasis on the phenotypic drug-screening method.

11.2 MDR in bacterial pathogens

The discovery of penicillin in 1928 and the use of sulfonamides in 1937 marked the start of the era of several scientific discoveries related to microbial infections, antimicrobial resistance, and drug development. The next three decades saw the "golden era" of antibiotic development, discovery, and production that gave rise to the discovery of today's antibiotics classes. In the 1950s the first transferable resistance was identified, which was followed by WHO-led scientific group research on resistance in 1959 ("Combating Antimicrobial Resistance," 2017).

As already mentioned, MDR bacteria represent an increasing threat to public health, compounded by the dearth of drug candidates in the pipeline against such pathogens. Because MDR infections are often significantly harder and more expensive to treat, they represent a growing public health threat (Bolan et al., 2012; Menzies et al., 2012; Pichichero & Casey, 2007). The problem is exacerbated by the fact that still fewer antimicrobial agents using novel mechanisms of action are in clinical development

(Coates et al., 2011). However, for different pathogens, different underlying mechanisms are traditionally used to explain these observations, and it is unclear whether each bacterial taxon has its own mechanism(s) for MDR or whether there are common mechanisms between distantly related pathogens. Better understanding of the cause(s) of the excess of MDR is the first step to rational design of more effective interventions to prevent the origin and/or proliferation of MDR (Chang et al., 2015). Research shows that there is often a positive correlation within a bacterial population between resistance to one drug and resistance to one or more other drugs. The high frequency of MDR isolates among resistant strains represents an unsolved scientific problem (Chang et al., 2015).

The simplest explanation for observing an excess of MDR is that a single biochemical mechanism confers resistance to more than one drug: for example, the presence of bacterial efflux pumps (Alekshun & Levy, 2007; Poole, 2005), which extrude antibiotics out of cells in such a way that the intracellular antibiotic concentration decreases and resistance to the antibiotics occurs. Some efflux systems are antibiotic specific, but others confer resistance to multiple drug classes (Poole, 2005). Typically, efflux pumps provide low-level drug resistance (Chang et al., 2007; Pumbwe, Chang, Smith, & Wexler, 2006). Another example is cell-wall thickening in *S. aureus*, resulting in resistance to vancomycin and daptomycin, antibiotics with relatively large molecular sizes (Cui, Tominaga, Neoh, & Hiramatsu, 2006). Other causes may be due to genetic linkage, where MDR strains may arise because determinants of resistance to multiple drug classes are genetically linked, because they are either physically close on bacterial chromosomes or on the same horizontally transmitted element, such as a plasmid or conjugative transposon. In such cases, when a strain acquires one resistance phenotype, it acquires many (Chang et al., 2015). Another possibility is the presence of highly mutable bacterial lineages in which the rate of mutation is high, which is mainly due to variation in genes involved in DNA proofreading. Because mutation and recombination are the sources of drug resistance genes and alleles, more highly mutable lineages have higher frequencies of MDR determinants (Chang et al., 2015). Interestingly, the use of multidrug therapy can also contribute to MDR strains. Multidrug therapy has long been recommended for treating infections in which resistance is acquired by mutation, to ensure that mutants that are resistant to one drug are killed by other drugs in the mixture. The emergence of multiple resistance is much more likely from a singly resistant precursor. The first resistance mutation may facilitate the emergence and selection of further resistance mutations. In this mechanism, under multidrug therapy, the singly resistant state is unstable, and those strains will be replaced by strains that have acquired mutations for additional drug resistance, leading to an excess of MDR (Hingley-Wilson et al., 2013; van Rie et al., 2005).

11.3 Drug resistance in cancer

Anticancer drug resistance is a complex process that arises from an alteration in the drug targets. Advances in modern technologies like DNA microarray, proteomics technology, and the development of targeted therapies provide new strategies to overcome drug resistance. Despite several advances in cancer treatment and therapy, the most promising option for cancer treatment is chemotherapy. Currently, 90% of failures in chemotherapy occur during the invasion and metastasis of cancers related to drug resistance. In chemotherapy, by following the administration of a certain drug, a large number of patients' tumor cells become resistant to the drug. So, drug resistance appears as a serious problem in the field of cancer treatment (Mansoori, Mohammadi, Davudian, Shirjang, & Baradaran, 2017). Although designs for new chemotherapy agents are developing quickly, an effective chemotherapy agent has not been discovered against the advanced stage of cancer (such as metastasis and invasion). There are several mechanisms causing resistance to chemotherapy, including inactivation of the drug; MDR; inhibiting cell death (apoptosis suppression); changes in drug metabolism, epigenetic and drug targets; enhanced DNA repair; and gene amplification (Mansoori et al., 2017; Moffat, Rudolph, & Bailey, 2014).

11.3.1 MDR in cancer

MDR in cancer chemotherapy has been pointed out as the ability of cancer cells to survive against a wide range of anticancer drugs (Zahreddine & Borden, 2013). The MDR mechanism may develop from increased release of the drug outside the cells, so the drug absorption is reduced in these cells (Sampath et al., 2006).

This mechanism of MDR in cancer cells, namely increasing the release of drugs outside the cell, is due to a family of ATP-dependent transporters that are involved in the transporting of the nutrients and other molecules across the membrane. The ABC transporters are composed of two cytoplasmic domains that bind to ATP, known as the ATP-binding cassette (ABC), and two transmembrane domains (TMDs) (Sampath et al., 2006). The ABC family has three members: (1) P-glycoprotein (P-gp); (2) multidrug resistance-associated protein 1 (MRP1); and (3) breast cancer resistance protein (BCRP/ABCG2) (Townsend & Tew, 2003). P-gp, which is a multidrug membrane transporter that normally is known as a pump for moving chloride out of the cells, can bind to a variety of chemotherapy agents, including doxorubicin, vinblastine and taxol, and following binding ATP is hydrolyzed, and then the structure of P-gp is altered. As a result, the agent releases to the extracellular space. Following the second ATP hydrolysis, the transporter returns to its basic structure and is able to release the drug outside of the cell (Cumming et al., 2001; Manolitsas, Englefield, Eccles, & Campbell, 1997). The second mechanism of MDR in cancer cells occurs by reducing the absorption of the drug. The absorption of the anticancer agent into the tumor cells can occur by passive transfer (e.g., doxorubicin and vinblastine), facilitated diffusion, and active transport (for example, nucleoside analogs). The cytotoxic agents can enter the cells along the concentration gradient by the three ABC transporter molecules, as mentioned earlier. But the absorption of the drug into the cells via the direction of a high concentration gradient occurs only through active transport (Mansoori et al., 2017). Most of the membrane's transporters belong to solute carrier SLC transporters (transport minerals, vitamins, etc.). Reduction in the absorption of the drugs can occur in two ways: (1) reducing the tendency to drug binding; and/or (2) reducing the number of transporters. Some of the agents use specific transporters to enter the cells. Mutations in these transporters inhibit them and reduce the absorption of the drugs. For example, the resistance to methotrexate usually occurs from human folate carrier's gene mutation in patients with acute lymphoblastic leukemia. The mutation of the G point at nucleotide 133 and the substitution for lysine by glutamic acid in the first TMD of hRFC protein reduces the tendency of the drugs to bind the transporter (Mansoori et al., 2017).

11.4 Drug screening

In the past three decades, target-based drug discovery (TDD)—in which the starting point is a defined molecular target that is hypothesized to have an important role in disease—has been the dominant approach to drug discovery in the pharmaceutical industry, driven by advances in molecular biology and genomics. However, in recent years, there has been a revival in interest in phenotypic drug discovery (PDD) approaches, which do not rely on knowledge of the identity of a specific drug target or a hypothesis about its role in disease (Moffat, Vincent, Lee, Eder, & Prunotto, 2017). A landmark study by Swinney and Anthony (2011) found that among the 183 small-molecule drugs across all therapeutic areas approved between 1999 and 2008, 58 (32%) were discovered using phenotype-based approaches. Importantly, 28 (56%) of the 50 small-molecule first-in-class new molecular entities identified in their study resulted from phenotypic screening approaches (PDD), whereas 17 (34%) resulted from target-based approaches. Focusing on TDD in the absence of validated molecular mechanisms of action (MMOAs) was suggested to be a major technical factor contributing to the high attrition rates of small-molecule first-in-class medicines observed in the clinic (Moffat et al., 2014). The power of PDD as a tool to address the complexity of diseases that are poorly understood by the scientific community is also clear. Furthermore, there have been recent rapid advances in various technologies for cell-based phenotypic screening, including the development of induced pluripotent stem (iPS) cell technologies (Shi, Inoue, Wu, & Yamanaka, 2017), gene-editing tools such as CRISPR-Cas (Fellmann, Gowen, Lin, Doudna, & Corn, 2017), organoids, and imaging assay technologies. Such tools have enabled the development of novel cell-based disease models that promise to more realistically recapitulate human disease biology (Moffat et al., 2017).

11.4.1 Phenotypic vs. target-based drug screening

The screening of bioactive compounds involves a large number of analysis assays for assessing the potential of biological extracts or molecules. The assays can be performed at the whole animal, cell-based, or molecular levels. For drug discovery, two different complementary approaches can be applied: classical pharmacology, also known as PDD, which is the historical basis of drug discovery, and reverse pharmacology, or TDD (Lee, Uhlik, Moxham, Tomandl, & Sall, 2012). TDD is based on the formulation and testing of specific molecular hypotheses (Lee et al., 2012), while in PDD, extracts or compounds are assessed against, for instance, cell lines in a quantitative measurement of one or more cellular parameters. The approaches differ in their first focus. In the phenotypic approach, the function is primarily considered, and compounds are screened to find those that alter the phenotype, while the disease state and the mechanism will be identified later. In the target-based approach, the gene that codes for a protein target involved in the disease is first identified and compounds are screened to find high-affinity binding. After this, the active compounds need to be checked in tissues and whole organisms. These two strategies have advantages and disadvantages and promote very different screening assays. Complex and advanced high-content phenotypic drug screenings have been developed in which 384- or 1536-well high-content screening plates are used to screen disease-relevant cell types assembled in a biomimetic fashion (Lage et al., 2018). Organisms like fruit fly, zebrafish, or mouse are commonly used for phenotypic screenings at the whole animal level. Target-based discovery benefits from the advances and evolution of science in the fields of biochemistry, structural biology, chemistry, genomics, and technology (Lage et al., 2018).

11.4.2 Phenotype screening

Phenotypic assays are compound screening systems that focus on the modulation of a disease-linked phenotype in a target agnostic manner, in contrast to those that center on a specific protein (as is the case in TDD). The success of phenotypic approaches, rather unexpected in an era focused on TDD, has been ascribed to the unbiased nature of such screens

(Lee & Berg, 2013; Swinney & Anthony, 2011). Generally, phenotypic assays are more physiologically relevant than target-based ones, because they are minimally cell-based, if not tissue- or whole-organism-based. Furthermore, they offer the possibility of identifying compounds acting through either unknown targets or unprecedented MMOA for known targets. For example, a phenotypic screen using live bacteria led to linezolid, a member of a novel antibiotic class preventing the initiation of protein translation (Vincent et al., 2015; Wilson et al., 2008). Similarly, the antiepileptic lacosamide was identified through testing in an in vivo rat model of epilepsy; after its US Food and Drug Administration approval, it was documented to promote slow inactivation of voltage-gated sodium channels rather than directly blocking them (Choi, Stables, & Kohn, 1996; Errington, Stöhr, Heers, & Lees, 2008; Vincent et al., 2015).

By definition, TDD postulates a direct link between the modulation of a target through a given MMOA and the resolution or mitigation of a disease state. However, the high rate of failure of target-based projects due to lack of efficacy indicates that this underlying hypothesis is often incorrect (Allison, 2012). The target selection process usually uses a reductionist approach in which a hypothesized link from the disease to the target is built by using progressively simpler systems, with the assumption of translation to the clinic at each step. These assumptions—some known and others unknown to the researchers—are validated during the drug discovery process, as compounds are tested in progressively more complex systems up to, ultimately, patients. Unrecognized assumptions made during target selection may then play a substantial role in the documented high rate of project failure (Vincent et al., 2015).

By design, phenotypic approaches have the potential to mitigate our incomplete understanding of physiology and disease biology. To fully deliver on that promise, it follows that those phenotypic assays most faithfully replicating the disease state will incorporate fewer incorrect assumptions and will consequently be more likely to deliver clinically relevant leads. In other words, the value of a phenotypic assay lies entirely in its ability to predict the successful translation of a given compound and/or MMOA to patients. Practically, this statement can be used as a guiding philosophy to evaluate the potential value of proposed assays in a resource-constrained environment. From a technical standpoint, three criteria stand out as helpful in this evaluation: assay system, stimulus, and readout (Vincent et al., 2015).

11.4.2.1 Phenotype screening against malaria

Malaria is a mosquito-borne infectious disease affecting humans and other animals caused by parasitic protozoans belonging to the *Plasmodium* genus. Malaria is widespread in tropical and subtropical regions, including America, Asia, and Africa. According to the "World Malaria Report 2016" of the WHO, there were 212 million new cases of malaria worldwide in 2015. Most deaths from malaria are caused by *Plasmodium falciparum*, one of the five species of human infectious malaria parasites. The increasing resistance of *P. falciparum* to the available drugs and new efforts to eradicate malaria drive the need to develop new and effective antimalarial drugs (Ridley, 2002). Traditionally, plants have been excellent sources of antimalarial compounds. In fact, artemisinin, a natural product isolated from the *Artemisia annua*, used in traditional Chinese medicine, is the drug recommended by the WHO to be used in combination therapies as a first-line treatment of malaria. The combination therapy decreases drug resistance because the combination of antimalarial drugs with independent modes of action can impede the development of resistance to each individual component of the combination (Kokwaro, 2009), increasing the probability of success (Lage et al., 2018). Several drugs with antimalarial activity are under development. The Medicines for Malaria Venture Malaria Box is a collection of over 400 compounds representing families of structures identified in phenotypic screens of pharmaceutical and academic libraries (Spangenberg et al., 2013). This library was originated for screening against the asexual-stage of *P. falciparum* (Lage et al., 2018).

The phenotypic screening assays are generally developed in asexual erythrocytic-stage parasites with different readouts for the inhibition of parasite growth, such as lactate dehydrogenase (LDH) release, one of the most commonly used due to its robustness and specificity (Makler & Hinrichs, 1993), MitoTracker or SYBR Green dye incorporation, hypoxanthine incorporation or the 4′,6-diamidino-2-phenylindole (DAPI) imaging assay, or also using a transgenic *P. falciparum* line that stably expresses cytoplasmic firefly luciferase that was used to develop a cell-based luminescent method for testing antimalarial drugs that can be adapted to HTS format (Che, Cui, Kutsch, Cui, & Li, 2012; Lage et al., 2018).

11.4.2.2 Phenotype screening against cancer

Cancer is a broad term for several diseases affecting any part of the body, including lungs, liver, stomach, and breast. Cancer is defined as a serious disease caused when cells in the body grow in a way that is uncontrolled and not normal, killing normal cells and often causing death. The modification from normal cells into tumor cells is the initial multistage process that leads to a continual unregulated proliferation of the cells. The tumor cells are able to grow abnormally and without boundaries, spreading easily into neighboring tissues and organs (Lage et al., 2018). Cancer was responsible for 8.8 million deaths in 2015 worldwide, the leading cause of death in the world (Lage et al., 2018). Consequently, new anticancer agents with more effectiveness and less toxicity are urgently demanded. Natural microbial products have been explored as an answer to these pharmaceutical needs with promising results, in addition to the synthetic compounds libraries used for cancer drug discovery. Drug discovery for cancer is carried out using both target-based and phenotypic approaches. Target-based approaches to drug discovery are extensively used in the pharmaceutical industry, but there are few fully validated drug targets in some types of cancers that are dependent on the tumor microenvironment, such as in pancreatic cancer (Coleman et al., 2014). In others, such as breast cancer, the estrogen receptor is one of the targets screened or, in most types, the blocking of microtubule proteins such as tubulin is a pursued target for interfering with the mitosis of the cells (Sun et al., 2017). Due to the lack of avowed targets in some cancers and the advances in cell culture techniques, the most widely used approaches in cancer research are phenotypic.

In the phenotypic approach, the first steps of the research for anticancer leads rely on the in vitro assessment of anticancer activity. Established cell lines of each type of cancer are commonly used to carry out the research by HTS assays, while normal cell lines are used as control. In many types of cancer, such as pancreatic cancer, the subpopulation of cancer stem cells (CSCs) is highly enriched. CSCs are resistant to current chemotherapeutic drugs and therefore promote tumor recurrence (Kreso & Dick, 2014). CSCs, also called tumor-initiating cells, share many characteristics with normal stem cells, such as asymmetric cell division, in which each CSC generates one daughter cell with self-renewal capacity and another cell destined to differentiate. The self-renewal capacity helps to maintain the number of CSCs within the tumor. CSCs also exhibit unique features, such as their metastasis ability and the ability to remain in a quiescent state, which protect them from the chemotherapeutic drugs developed to target actively dividing cells. Nowadays, the use of CSCs in anticancer screening is becoming the most relevant phenotypic approach (Lage et al., 2018).

11.5 Conclusion and future direction

Rapid increase of severe systemic infections and the spread of resistant microorganisms are indisputable facts. The inadequacy of available antimicrobial drugs compels continuous development of newer drugs. These facts make new disease-relevant cell-based phenotypic assay methodologies necessary to enhance our knowledge of the dynamics of host-pathogen interactions in their natural environment and the development of new therapies. In addition, even though there have been many notable drug discovery and development achievements in recent years, disease areas such as aggressive cancers remain largely intractable. This failure can be attributed, in part, to our limited understanding of the targets of diseases and to a lack of robust disease-relevant screening methods that mimic the key pathophysiological features of the human disease. Therefore current efforts should be directed toward the development of new models, new assay formats, and innovative screening technologies that better summarize in vivo physiology (Horvath et al., 2016; Lage et al., 2018).

While all the early drugs were discovered by phenotypic screening, the past three decades have given rise to new technologies for making large chemical libraries (combinatorial chemistry) and high-throughput screening (robotics) that have since dominated the pharmaceutical industry. Future trends in drug discovery and development include systematic computational analyses mimicking ecosystems of the pathogens and patient-derived cell cultures, induced pluripotent stem cell (iPSC) technology, three-dimensional coculture, organotypic systems, advances in cell imaging, microfluidics, nanotechnologies, and gene-editing technologies (Lage et al., 2018).

Among the computational systems used for drug discovery, dynamic combinatorial chemistry is a powerful tool for hit identification and optimization for the discovery of binders to DNA, RNA, and protein targets, including biological macromolecular targets as receptors (Lage et al., 2018). New computational methods and concepts allow the handling of phenotypic models such that the added complexity arising from considering biological relevance can actually help to improve productivity of early drug discovery. However, there are challenges ahead. Among the technical challenges, image-based morphological profiling requires a certain minimum information in terms of the resolution of the optical system used to collect the data and the dimensions of the target. For microorganisms, cell-morphological information may be obscured, or even entirely lost, due to insufficient optical resolution. Emerging superresolution microscopy methods are only just beginning to show potential for high-throughput phenotypic screening. The application of cross-disciplinary phenotypic discovery to probe emerging biology is a powerful and exciting approach to discover new biological targets and molecules. The advances in screening technology and increased biological relevance along with effective practices will increase the clinical success of new candidate drugs and ultimately deliver transformative medicines to impact the lives of patients (Lage et al., 2018; Moffat et al., 2014, 2017).

A successful phenotypic screening campaign requires significant investments from different disciplines, including disease biology, screening capabilities, bioinformatics, chemical biology, and computational and medicinal chemistry. This also highlights the diverse skills required and the necessity for cross-disciplinary collaboration to make phenotypic discovery projects successful.

References

Alekshun, M. N., & Levy, S. B. (2007). Molecular mechanisms of antibacterial multidrug resistance. *Cell, 128*(6), 1037–1050.

Allison, M. (2012). Reinventing clinical trials. *Nature Biotechnology, 30*(1), 41–49.

Bolan, G. A., Sparling, P. F., & Wasserheit, J. N. (2012). The emerging threat of untreatable gonococcal infection. *The New England Journal of Medicine, 366*(6), 485–487.

Chang, H. H., Cohen, T., Grad, Y. H., Hanage, W. P., O'Brien, T. F., & Lipsitch, M. (2015). Origin and proliferation of multiple-drug resistance in bacterial pathogens. *Microbiology and Molecular Biology Reviews, 79*(1), 101–116.

Chang, T. M., Lu, P. L., Li, H. H., Chang, C. Y., Chen, T. C., & Chang, L. L. (2007). Characterization of fluoroquinolone resistance mechanisms and their correlation with the degree of resistance to clinically used fluoroquinolones among Escherichia coli isolates. *Journal of Chemotherapy, 19*(5), 488–494.

Che, P., Cui, L., Kutsch, O., Cui, L., & Li, Q. (2012). Validating a firefly luciferase-based high-throughput screening assay for antimalarial drug discovery. *Assay and Drug Development Technologies, 10*(1), 61–68.

Choi, D., Stables, J. P., & Kohn, H. (1996). Synthesis and anticonvulsant activities of N-benzyl-2-acetamidopropionamide derivatives. *Journal of Medicinal Chemistry, 39*(9), 1907–1916.

Coates, A. R., Halls, G., & Hu, Y. (2011). Novel classes of antibiotics or more of the same. *British Journal of Pharmacology, 163*(1), 184–194.

Coleman, S. J., Watt, J., Arumugam, P., Solaini, L., Carapuca, E., Ghallab, M., et al. (2014). Pancreatic cancer organotypics: High throughput, preclinical models for pharmacological agent evaluation. *World Journal of Gastroenterology, 20*(26), 8471–8481.

Combating Antimicrobial Resistance: A One Health Approach to a Global Threat: Proceedings of a Workshop. (2017). National Academies of Sciences, Engineering, and Medicine; Health and Medicine Division; Board on Global Health; Forum on Microbial Threats. Washington, DC: National Academies Press (US); 2017 PMID: 29227604.

Cui, L., Tominaga, E., Neoh, H. M., & Hiramatsu, K. (2006). Correlation between reduced daptomycin susceptibility and vancomycin resistance in vancomycin-intermediate Staphylococcus aureus. *Antimicrobial Agents and Chemotherapy, 50*(3), 1079–1082.

Cumming, R. C., Lightfoot, J., Beard, K., Youssoufian, H., O'Brien, P. J., & Buchwald, M. (2001). Fanconi anemia group C protein prevents apoptosis in hematopoietic cells through redox regulation of GSTP1. *Nature Medicine, 7*(7), 814–820.

Errington, A. C., Stöhr, T., Heers, C., & Lees, G. (2008). The investigational anticonvulsant lacosamide selectively enhances slow inactivation of voltage-gated sodium channels. *Molecular Pharmacology, 73*(1), 157–169.

Fellmann, C., Gowen, B. G., Lin, P. C., Doudna, J. A., & Corn, J. E. (2017). Cornerstones of CRISPR-Cas in drug discovery and therapy. *Nature Reviews Drug Discovery, 16*(2), 89–100.

Hingley-Wilson, S. M., Casey, R., Connell, D., Bremang, S., Evans, J. T., Hawkey, P. M., et al. (2013). Undetected multidrug-resistant tuberculosis amplified by first-line therapy in mixed infection. *Emerging Infectious Diseases, 19*(7), 1138–1141.

Horvath, P., Aulner, N., Bickle, M., Davies, A. M., Nery, E. D., Ebner, D., et al. (2016). Screening out irrelevant cell-based models of disease. *Nature Reviews Drug Discovery, 15*(11), 751–769.

Kokwaro, G. (2009). Ongoing challenges in the management of malaria. *Malaria Journal, 8*(Suppl. 1), S2.

Kreso, A., & Dick, J. E. (2014). Evolution of the cancer stem cell model. *Cell Stem Cell, 14*(3), 275–291.

Lage, O. M., Ramos, M. C., Calisto, R., Almeida, E., Vasconcelos, V., & Vicente, F. (2018). Current screening methodologies in drug discovery for selected human diseases. *Marine Drugs, 16*(8), E279.

Lee, J. A., & Berg, E. L. (2013). Neoclassic drug discovery: The case for lead generation using phenotypic and functional approaches. *Journal of Biomolecular Screening, 18*(10), 1143–1155.

Lee, J. A., Uhlik, M. T., Moxham, C. M., Tomandl, D., & Sall, D. J. (2012). Modern phenotypic drug discovery is a viable, neoclassic pharma strategy. *Journal of Medicinal Chemistry, 55*(10), 4527–4538.

Makler, M. T., & Hinrichs, D. J. (1993). Measurement of the lactate dehydrogenase activity of Plasmodium falciparum as an assessment of parasitemia. *The American Journal of Tropical Medicine and Hygiene, 48*(2), 205–210.

Manolitsas, T. P., Englefield, P., Eccles, D. M., & Campbell, I. G. (1997). No association of a 306-bp insertion polymorphism in the progesterone receptor gene with ovarian and breast cancer. *British Journal of Cancer, 75*(9), 1398–1399.

Mansoori, B., Mohammadi, A., Davudian, S., Shirjang, S., & Baradaran, B. (2017). The different mechanisms of cancer drug resistance: A brief review. *Advanced Pharmaceutical Bulletin, 7*(3), 339–348.

Menzies, N. A., Cohen, T., Lin, H. H., Murray, M., & Salomon, J. A. (2012). Population health impact and cost-effectiveness of tuberculosis diagnosis with Xpert MTB/RIF: A dynamic simulation and economic evaluation. *PLoS Medicine, 9*(11), e1001347.

Moffat, J. G., Rudolph, J., & Bailey, D. (2014). Phenotypic screening in cancer drug discovery—Past, present and future. *Nature Reviews Drug Discovery, 13*(8), 588–602.

Moffat, J. G., Vincent, F., Lee, J. A., Eder, J., & Prunotto, M. (2017). Opportunities and challenges in phenotypic drug discovery: An industry perspective. *Nature Reviews Drug Discovery, 16*(8), 531–543.

O'Neill, J. (2014). Antimicrobial resistance: Tackling a crisis for the health and wealth of nations. *The Review on Antimicrobial Resistance, 20*, 1–16.

Pichichero, M. E., & Casey, J. R. (2007). Emergence of a multiresistant serotype 19A pneumococcal strain not included in the 7-valent conjugate vaccine as an otopathogen in children. *Journal of the American Medical Association, 298*(15), 1772–1778.

Poole, K. (2005). Efflux-mediated antimicrobial resistance. *The Journal of Antimicrobial Chemotherapy, 56*(1), 20–51.

Pumbwe, L., Chang, A., Smith, R. L., & Wexler, H. M. (2006). Clinical significance of overexpression of multiple RND-family efflux pumps in Bacteroides fragilis isolates. *The Journal of Antimicrobial Chemotherapy, 58*(3), 543–548.

Ridley, R. G. (2002). Medical need, scientific opportunity and the drive for antimalarial drugs. *Nature, 415*(6872), 686–693.

Sampath, D., Cortes, J., Estrov, Z., Du, M., Shi, Z., Andreeff, M., et al. (2006). Pharmacodynamics of cytarabine alone and in combination with 7-hydroxystaurosporine (UCN-01) in AML blasts in vitro and during a clinical trial. *Blood, 107*(6), 2517–2524.

Shahzad, S., Willcox, M., & Shahzad, A. (2020). Identification of novel in vitro antibacterial action of cloprostenol and evaluation of other non-antibiotics against multi-drug resistant A. baumannii. *The Journal of Antibiotics, 73*, 72–75. https://doi.org/10.1038/s41429-019-0244-2.

Shi, Y., Inoue, H., Wu, J. C., & Yamanaka, S. (2017). Induced pluripotent stem cell technology: A decade of progress. *Nature Reviews Drug Discovery, 16*(2), 115–130.

Spangenberg, T., Burrows, J. N., Kowalczyk, P., McDonald, S., Wells, T. N., & Willis, P. (2013). The open access malaria box: A drug discovery catalyst for neglected diseases. *PLoS One, 8*(6), e62906.

Sun, J., Wei, Q., Zhou, Y., Wang, J., Liu, Q., & Xu, H. (2017). A systematic analysis of FDA-approved anticancer drugs. *BMC Systems Biology, 11*(Suppl. 5), 87.

Swinney, D. C., & Anthony, J. (2011). How were new medicines discovered. *Nature Reviews Drug Discovery, 10*(7), 507–519.

Townsend, D. M., & Tew, K. D. (2003). The role of glutathione-S-transferase in anti-cancer drug resistance. *Oncogene, 22*(47), 7369–7375.

van Rie, A., Victor, T. C., Richardson, M., Johnson, R., van der Spuy, G. D., Murray, E. J., et al. (2005). Reinfection and mixed infection cause changing Mycobacterium tuberculosis drug-resistance patterns. *American Journal of Respiratory and Critical Care Medicine, 172*(5), 636–642.

Vincent, F., Loria, P., Pregel, M., Stanton, R., Kitching, L., Nocka, K., et al. (2015). Developing predictive assays: The phenotypic screening "rule of 3". *Science Translational Medicine, 7*(293), 293ps15.

Wilson, D. N., Schluenzen, F., Harms, J. M., Starosta, A. L., Connell, S. R., & Fucini, P. (2008). The oxazolidinone antibiotics perturb the ribosomal peptidyl-transferase center and effect tRNA positioning. *Proceedings of the National Academy of Sciences of the United States of America, 105*(36), 13339–13344.

World Health Organization. (2017). *WHO publishes list of bacteria for which new antibiotics are urgently needed*. Geneva: WHO. http://www.who.int/mediacentre/news/releases/2017/bacteria-antibiotics-needed/en/.

Zahreddine, H., & Borden, K. L. (2013). Mechanisms and insights into drug resistance in cancer. *Frontiers in Pharmacology, 4*, 28.

CHAPTER

12

New approaches for targeting drug resistance through drug combination

Shailesh Kumar Singh[a], *Arif Mohammed*[b], *Othman A. Alghamdi*[b], *Syed Masood Husain*[a]

[a]Molecular Synthesis and Drug Discovery Unit, Centre of Biomedical Research, Lucknow, India, [b]Department of Biological Sciences, Faculty of Science, University of Jeddah, Jeddah, Kingdom of Saudi Arabia

12.1 Background

12.1.1 Antibiotics and antibiotic resistance

Antibiotics are drugs used to treat bacterial infections in humans and animals (Phillips, 2003). Antibiotics are safe for human use because they function by interfering with essential processes or structures in the bacterial cell, ultimately killing it or slowing down its growth. Major antibiotic targets in bacteria include the bacterial cell wall and the machinery that synthesizes the nucleic acids DNA, RNA, or proteins (Kohanski, Dwyer, & Collins, 2010). Antibiotics have been wonder drugs that have boosted general human health and protected us from life-threatening infections (Banin, Hughes, & Kuipers, 2017). However, inappropriate prescriptions and misuse of antibiotics against common infections have eroded their efficacy and fueled the emergence of bacteria that are able to thrive even in the presence of therapeutic levels of antibiotics, giving rise to antibiotic resistance (Fair & Tor, 2014; Fleming-Dutra et al., 2016). Data from epidemiological studies have suggested a direct link between antibiotic usage and antibiotic resistance (The antibiotic alarm, 2013). Bacteria can develop resistance to antibacterial drugs as a result of genetic mutation or exchange of resistance-conferring genes via conjugative plasmids or

Combination Therapy Against Multidrug Resistance
https://doi.org/10.1016/B978-0-12-820576-1.00012-6

transposons among different species (Neu, 1992). Resistant bacteria are the result of evolution under selective pressure of antibiotic treatments; these bacteria change in ways that reduce or eliminate the effectiveness of the antibiotics (Read & Woods, 2014). Various factors, including overuse, ill-advised prescriptions, extensive agricultural use of antibiotics, limited availability of new antibiotics, and regulatory barriers in discovery and commercialization of new drugs, have given rise to the crisis of antibiotic resistance (Ventola, 2015). The World Health Organization (WHO) and the US Centers for Disease Control and Prevention (CDC) have already recognized antibiotic resistance as a serious threat. The US CDC estimates that resistant bacteria are responsible for 25,000 deaths in Europe, 23,000 + deaths in the United States, and 38,000 + deaths in Thailand per year. In India alone, 58,000 babies died in 1 year due to infection with resistant bacteria from their mothers (CDC, 2017). A study on antimicrobial resistance commissioned by the government of the United Kingdom forecasts that by 2050 antibiotic-resistant infections could cost the world $100 trillion and kill 10 million people per year (one death every 3 s), many more than the combined total from all cancers and diabetes (O'Neill, 2014).

12.1.2 Spread and emergence of multidrug resistance (MDR)

Millions of tons of antibiotics have been used since their discovery to treat diseases caused by pathogenic bacteria. This usage has imposed high selection pressure on all infectious bacteria to evolve into antibiotic-resistant strains. The most evident example of the rapid evolution of bacterial resistance can be seen in the case of β-lactam antibiotics, where resistant bacteria possess a gene encoding β-lactamase that inactivates these antibiotics. Thousands of β-lactamases have been identified across the globe, indicating the rapid emergence of the resistance. Resistance genes such as Enterobacteriaceae-producing ESBL (extended-spectrum β-lactamase), NDM (New Delhi metallo)-β-lactamase 1, and KPC (*Klebsiella pneumoniae* carbapenemase) are found worldwide, indicating the spread of resistance (Laxminarayan et al., 2013).

Resistance to one specific antibiotic may confer resistance to a whole class of related antibiotics. The spread of resistance can unpredictably occur due to easy circulation of resistant bacteria in the human population through food, water, environment, migration, and travel. Poor hygiene, lack of sanitation, and poor infection control are triggering factors that contribute to the spread of resistant bacteria. However, the spread is not limited to a locality or province but rather international travel, trade, and globalization have helped the circulation of antibiotic-resistant bacteria across the globe, underlining the fact that it is a global issue now (World Health Organization, 2015).

There are several mechanisms through which a bacterium can become resistant to antibiotics:

- Bacteria can produce enzymes that inactivate antibiotics. These enzymes may be located on the cell surface or in cytosol. The genes encoding such enzymes are known as antibiotic resistance genes. Examples includes the genes that code for β-lactamases (e.g., penicillinase), which inactivate penicillin-like drugs (β-lactams) (Castanheira, Farrell, Krause, Jones, & Sader, 2014).
- Bacteria can completely replace or bypass the target site of the antibiotic for survival. Methicillin-resistant *Staphylococcus aureus* strains were reported in the early 1970s, where the resistance mechanism was distinct from that of penicillinase-mediated resistance. Further studies revealed that a specific gene (*mecA*), encoding the low-affinity penicillin-binding protein, is responsible for methicillin resistance. The methicillin resistance is broad in its spectrum of activity, conferring resistance to the entire class of β-lactam antibiotics (Chambers & DeLeo, 2009).
- Bacteria can utilize numerous efflux pump systems for active transportation of the antibiotic out of the bacteria, to remove it from the cell (Fernandez & Hancock, 2012). ABC (ATP-binding cassette) pumps like DrrAB, OtrC, TlrC, and MlbYZ are antibiotic efflux pumps that use energy from ATP hydrolysis to remove antibiotics from the cell (Peterson & Kaur, 2018).
- Bacteria can alter the target site of the antibiotic via structural changes, so that the drug cannot affect the normal functioning of the cell. For example, methicillin, a lab-synthesized relative of penicillin, is an antibiotic that binds cell wall proteins to enter into cells, but methicillin-resistant *S. aureus* (MRSA) can synthesize altered cell wall proteins with reduced affinity for methicillin, thereby minimizing its uptake (Berglund, Molling, Sjoberg, & Soderquist, 2005). Another example includes mutations in bacterial gyrases and topoisomerases, which are target sites of fluoroquinolones (Webber et al., 2013).
- Bacteria can alter the cell membrane properties in such a way that permeability for antibiotics decreases, resulting in reduced or no uptake of the antibiotics. The classic example includes the production of aberrant OprD (a porin protein) by *Pseudomonas aeruginosa* isolates during treatment with imipenem (a β-lactam antibiotic). The mutated OprD diminishes the uptake of imipenem and hence confers resistance against imipenem (Quinn, Dudek, DiVincenzo, Lucks, & Lerner, 1986).

Thus bacteria can directly inactivate antibiotics or alternatively change processes inside the cell to bypass the antibiotic action. They can also modify the cell membranes to reduce antibiotic uptake or remove the permeated antibiotics from the cell.

This story is not limited to bacteria resistant to a single antibiotic, as some are resistant to many different antibiotics. These are called multidrug-resistant (MDR) bacteria, because a single bacterium is resistant to multiple antibiotics at the same time. There are two distinct ways that a bacterium can become multidrug resistant. One possibility is that the bacterium could acquire several resistance genes, each conferring resistance to one particular antibiotic (Bennett, 2009). This can be facilitated by the sharing of antibiotic resistance genes via horizontal gene transfer among bacteria (Allen et al., 2010). Another possibility is that a single resistance mechanism employed by the bacterium is applicable to multiple antibiotics (Nikaido, 2009). For instance, the efflux pumps that recognize a particular antibiotic can also recognize many other antibiotics and hence can remove them all from the cell using active transportation. CmeABC, a multidrug efflux system present in different *Campylobacter* spp., confers resistance to multiple antibiotics such as rifampin, cephalothin, and cefoperazone (Guo, Lin, Reynolds, & Zhang, 2010). The term antimicrobial resistance (AMR) is also commonly used, which expands the scope of drug resistance beyond bacteria. AMR occurs when microorganisms such as bacteria, fungi, viruses, and parasites are excessively exposed to antimicrobial drugs and in response the microorganism undergoes mutations or develops different resistance mechanisms to combat the drug for its survival. The emergence of AMR is a natural phenomenon, but the selection pressure applied to microbes due to the misuse and overuse of antimicrobials is accelerating the process.

MDR infections are much more problematic, as they are hard to treat due to very limited or no treatment options. In some cases, drugs that are more toxic for the patient are prescribed, which may destroy the healthy microbiota and invite more infections. Healthcare facilities and large production animal farms are the hot spots for emergence of MDR bacteria, because of the excessive exposure to antibiotics. In hospitals, sick people are in close vicinity to each other, which increases the risk of spread of resistant bacteria enhanced by poor hygiene practices (Mody et al., 2019).

Still, bacteria are a normal part of our habitats, even with good hygiene practices. Food has become the common route for spread of MDR bacteria (Doyle, 2015). Antibiotics are used on animal farms to treat infections and to promote growth, but many farm animals colonize MDR bacteria, which can be transmitted to food products due to improper processing. MDR bacteria have been isolated from raw meat (Saud et al., 2019), fresh vegetables (Karumathil, Yin, Kollanoor-Johny, & Venkitanarayanan, 2016), and drinking water sources (Odonkor & Addo, 2018). The CDC has identified the drug-resistant foodborne bacteria *Campylobacter*, *Salmonella*, and *Shigella* as serious threats, with Salmonellae carrying multiple resistant genes being the most frequent causes of foodborne MDR outbreaks (CDC, 2013).

International travel is another factor that helps MDR bacteria spread globally. Trends show that a substantial proportion of visitors to international locations are colonized by MDR bacteria (Frost, Van Boeckel, Pires, Craig, & Laxminarayan, 2019). MDR bacteria such as extended-spectrum β-lactamase (ESBL)-producing *Escherichia coli* (*E. coli*) and *K. pneumoniae* are prevailing in different parts of the world because they are hard to treat. Various studies have shown that a large proportion of community populations are colonized with ESBL-producing Enterobacteriaceae as a result of travel (Woerther, Andremont, & Kantele, 2017).

The WHO Global Antimicrobial Resistance Surveillance System (GLASS)'s surveillance data on antibiotic resistance reveal high levels of resistance to a number of serious bacterial infections worldwide. The situation is very alarming due to widespread occurrence of antibiotic resistance among 500,000 people with suspected bacterial infections across 22 countries. In addition to *E. coli*, *K. pneumoniae*, *S. aureus*, *Streptococcus pneumoniae*, and *Salmonella* spp. are the most commonly reported resistant bacteria (WHO, 2017b). The WHO data on *Mycobacterium tuberculosis* estimated about 1.6 million deaths, and 9.0–11.1 million people (5.8 million men, 3.2 million women, and 1.0 million children) developed tuberculosis (TB) in 2017. TB is one of the top 10 causes of death and the leading cause from a single infectious agent (WHO, 2018).

12.1.3 Antibiotic discovery versus emergence of antibiotic resistance

The development of new antibiotics has been difficult because of the prevailing rate of emergence of MDR. The problem of antibiotic resistance is not new; in fact, resistant bacteria were identified shortly after the first antibiotic, penicillin, was introduced to market (CDC, 2013). Usage of penicillin started in the 1940s, and resistance became a worldwide problem by the 1950s. Tetracycline, a new class of antibiotic introduced to the market in 1950, lost its effectiveness even before the 1950s ended. Bacteria continued to adapt by creating subtle changes in their genetic material to counter the effects of the medicines being used against them. Erythromycin was discovered in 1952, and resistant Streptococcus was identified in 1968. Methicillin was developed in 1960 specifically to counter penicillin resistance, yet MRSA was identified in 1962. Further, ESBL bacteria outperformed not only penicillins but other β-lactam antibiotics, including cephalosporins (McKenna, 2017). Linezolid- and ceftaroline-resistant *Staphylococci* were reported within a year of their market introductions. Although in the case of vancomycin-resistant *Enterococcus* was first reported in 1988, 16 years after its market introduction in 1972, it is clear that any drug will eventually become ineffective due to the emergence of resistant strains (CDC, 2013).

Initially, scientific progress in antibiotic development provided us with new antibiotics to replace the obsolete ones, but now we have reached a point where bacterial resistance is spreading faster than antibiotic development. All the current antibiotics available for prescription are modifications and adaptations to existing antibiotics and no new class of antibiotics has been discovered since 1987 (World Economic Forum, 2013). Also, a single new antibiotic will not be the solution, because resistance mechanisms vary with different pathogenic bacterium, and resistance development makes constant innovation essential (Silver, 2011).

The timeline in Fig. 12.1 shows the relentless race of antibiotic discovery against the emergence of antibiotic resistance. Drug discovery is an expensive and time-consuming process. Finding a chemical that kills bacteria is easy, but to discover and develop substances that are not toxic to humans is a much more difficult challenge. If the candidate chemical gives promising results as an effective antibacterial and is found nontoxic to humans, only then can the years of clinical trials begin. It takes 10–20 years from finding a potent antibacterial to its clinical approval as a medicine. As a result, we are running out of antibiotics to treat even common infections. The WHO has warned that a "postantibiotic era" is approaching in which common infections, minor injuries, and even ordinary health problems can be deadly (Reardon, 2014). The promise to eradicate disease-causing bacteria is being replaced by warnings of more and more virulent ones, created by the same drugs that were supposed to save us (Angus, 2019). Nonetheless, scientific advancements in recent years are offering hope in combination therapies and nontraditional antibacterial therapeutics to counter multidrug resistance (Ghosh, Sarkar, Issa, & Haldar, 2019; Tehrani & Martin, 2018; Theuretzbacher & Piddock, 2019; Tyers & Wright, 2019).

12.2 Combination therapy to combat antibiotic resistance

Using more than one drug in concert, i.e., combination therapy, offers the simplest successful approach to tackling the problem of antibiotic discovery and antibiotic resistance. Due to the increasing failure rates of monotherapies, combination therapy is being used widely in medicine and offers a pivotal treatment of resistant bacterial infections. For example, MDR strains of *M. tuberculosis* are treated using a combination of four antibiotic drugs: isoniazid, rifampicin, pyrazinamide, and ethambutol administered simultaneously (WHO, 1999). Combination therapy is becoming a standard strategy to treat common infections and complex diseases. The US Food and Drug Administration (FDA) has already approved multiple combinations to treat AIDS, hepatitis, and cancer in addition to bacterial infections (Andrei, Valeanu, Chirvasuta, & Stefan, 2018; De Clercq & Li, 2016; Sun et al., 2017). Combination therapy

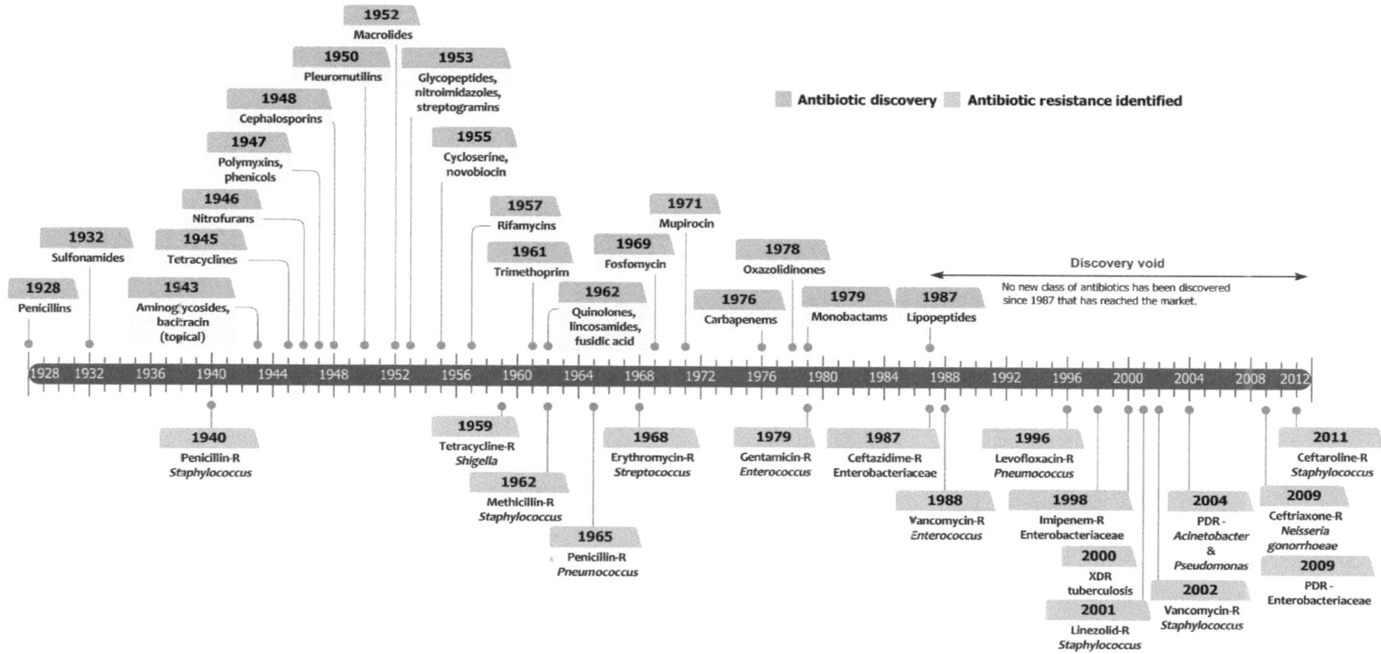

FIG. 12.1 The timeline of antibiotic discovery and identification of antibiotic resistance. The discovery void represents the time where no new class of antibiotics has been developed that has reached the market (CDC, 2013; Guo, Mandalapu, Ji, Gao, & Zhang, 2018; Silver, 2011).

aims to minimize the minimum inhibitory concentration (MIC) of individual drugs and to resensitize the resistant pathogens. Its ability to target the widest range of resistant organisms, along with the significant differences between the effectiveness of synergistic drug combinations and individual monotherapies, have driven scientists and clinicians to identify fruitful drug combinations (Singh & Yeh, 2017). While synergistic drug combinations have enhanced effectiveness in killing resistant pathogens, antagonistic drug combinations have shown the potential to reverse the evolution of antibiotic resistance through preferred selection of sensitive bacteria over resistant ones (Chait, Craney, & Kishony, 2007). Sometimes, finding potent combinations of clinically approved drugs has an added advantage, since these drugs have already been under clinical trials and proven safe for human use as monotherapy. Therefore, they are less likely to fail under clinical trials as combinations. Recently, advancements in robotics and high-throughput screening have paced the discovery of novel drug combinations, which otherwise was slow and laborious. Although combination therapy is in clinical use, the exact mechanisms of the many combinations remain unclear. Mechanistic insights into combination therapy are also necessary for rational discovery of potent drug combinations.

12.2.1 Benefits of combination therapy

Combination therapy has become common because of several benefits it offers over monotherapy. Combination therapy broadens the spectrum of susceptible pathogens and can help in management of polymicrobial infections when two or more different classes of antimicrobial agents are required to cover the pathogens. Another potential benefit of combination therapy is that it may delay or even prevent the emergence of resistance among pathogens, because the chances of emergence of resistance against two drugs are lower as compared with a single drug. If two drugs are given in combination, the first drug kills the strain resistant to the second drug and vice versa, hence preventing resistance to both of the drugs (Tamma, Cosgrove, & Maragakis, 2012). For instance, combination therapy for treating *M. tuberculosis* infections has prevented the emergence of drug resistance. Two antituberculosis drugs, isoniazid and rifampicin, were given in combination to treat tuberculosis in 183 patients; emergence of resistance against the former drug was significantly prevented, as only 0.5% of the treatments failed (Mitchison, 2012). The combination therapy can increase patient compliance and reduce the risk of emergence of resistance by shortening the treatment regimen. For instance, in tuberculosis, combining rifampicin with pyrazinamide reduces the treatment duration from 18 to 6 months (Boeree et al., 2017). The key motivation for using drug combinations is their potential to minimize or slow the evolution of resistance (Bollenbach, 2015). Further, combination therapy may help in choosing lower but still efficacious doses of drugs, minimizing toxicity and increasing safety for clinical use, along with slowing the development of antibiotic resistance. This could potentially revive obsolete antibiotics due to synergistic effects of the combining agent.

12.2.2 Combining antibiotics with nonantibiotic agents

Sometimes combination therapy involves the combination of an antibiotic with nonantibiotic drugs or agents with adjuvant activity. In such a case, the phenomenon of potentiation may be observed, resulting in improved effectiveness of the antibiotic partner. These adjuvant compounds can enhance effectiveness against resistant and antibiotic susceptible pathogens (Kalan & Wright, 2011). The combination of antibiotics with agents that target the bacterial resistance mechanism for respective antibiotics is the most commonly exercised strategy to combat MDR bacteria. For instance, Augmentin, launched in 1981, utilizes a combination of amoxicillin, a broad spectrum β-lactam antibiotic, with the β-lactamase inhibitor clavulanic acid, which renders β-lactamase-producing microorganisms susceptible to the action of the penicillin-derived antibiotic (Geddes, Klugman, & Rolinson, 2007). Since the launch of Augmentin, combinations of various inhibitors of β-lactamases with β-lactam antibiotics have been in clinical use to treat MDR infections (Drawz & Bonomo, 2010; Tehrani & Martin, 2018). Similarly, the combination of clavulanic acid with meropenem has also offered a novel strategy for treating extensively drug resistant (XDR) *M. tuberculosis* (Hugonnet, Tremblay, Boshoff, Barry, & Blanchard, 2009; Payen et al., 2018). Apart from β-lactamase combinations and β-lactam antibiotics, similar efforts at targeting resistance mechanisms have also been applied to block aminoglycoside-inactivating enzymes (Shakya et al., 2011) and inhibit antibiotic efflux pumps (Lomovskaya & Bostian, 2006; Sharma, Gupta, & Pathania, 2019) in vitro. However, clinical studies are still being awaited to prove their effectiveness for human use. Adjuvant compounds are not only limited to directly targeting the resistance mechanisms; in addition, some have been shown to potentiate antibiotic activity by enabling their uptake, by acting sequentially with them to weaken the pathogens, or by unknown mechanisms (Kalan & Wright, 2011). For instance, research findings have shown that silver can be used in combination with antibiotics specific to Gram-positive-specific antibiotics, such as vancomycin, to target resistant Gram-negative bacterial strains. Silver disrupts cellular processes in bacteria, resulting in high generation of reactive oxygen species, which alters the permeability of the cell membrane to increase antibiotic uptake (Morones-Ramirez, Winkler, Spina, & Collins, 2013). In 2011, a study reported the screening of the combination of an outdated antibiotic, minocycline, with more than 1000 clinically approved nonantibiotic drugs. Minocycline prevents protein synthesis and was widely used in the 1950s and 1960s, until resistance by bacteria developed. The combination of minocycline with loperamide, a nonantibiotic antidiarrheal drug, inhibited > 99% of *P. aeruginosa* that was clinically resistant to multiple antibiotics, including minocycline. Loperamide can dissipate the bacterial membrane potential and, in turn, potentiates minocycline by allowing its uptake by cells. The same study identified the potential of other nonantibiotic marketed drugs, such as benserazide, chloroxine, tegaserod, ascorbic acid, and mitomycin C, to potentiate growth inhibition by minocycline (Ejim et al., 2011). This ingenious strategy opened an entirely new field of mixing antibiotics with nonantibiotics, which has breathed fresh life into antibiotics weakened by bacterial resistance. Another study on MRSA reported that an antiplatelet drug, ticlopidine, can potentiate β-lactam cephalosporin cefuroxime. Ticlopidine acts by inhibiting the TarO enzyme, which is essential for biogenesis of early wall teichoic acid, and in turn destabilizes the cooperative functions of penicillin-binding proteins (PBPs) that contribute to resistance against β-lactam antibiotics (Farha et al., 2013). In a recent screening, scientists recorded the effect of 3000 drug combinations on six strains of Gram-negative MDR pathogens, *P. aeruginosa, E. coli,* and *Salmonella enterica,* including antibiotics, food additives, and human-targeted drugs. Over 500 combinations improved the antibiotic activity, out of which the most effective combination depicting promising potentiation was the pair consisting of the food additive vanillin and the antibiotic spectinomycin. Vanillin is the compound that gives vanilla its distinctive smell and taste. It revived the efficacy of spectinomycin, which had lost its effectiveness due to development of resistance against it (Brochado et al., 2018). This study clearly demonstrates that the efficacy of antibiotics can be improved by combining them with nonantibiotic drugs or even with food additives, and this can be used to defeat multidrug resistance. Some relevant combinations of antibiotics with nonantibiotic agents are listed in Table 12.1.

12.2.3 Synergy through antibiotic combinations

Individual antibiotics in combination can boost each other's effectiveness as a result of synergy, which is absent in monotherapy (Pletz, Hagel, & Forstner, 2017). A resistant bacteria can be made susceptible to weak antibiotics when used in combination (Whitfield, 2007). Further, this synergy can significantly reduce the individual dosage, which could minimize adverse effects of toxicity, making the combination therapy ideal for clinical approval (Shah & Chen, 2017). Combinations of various antibiotic drugs have been tested over decades in hopes of improved clinical results and combating bacterial resistance (Table 12.2).

TABLE 12.1 Key synergistic combinations of antibiotics with nonantibiotic agents and their mechanisms of action.

Antibiotic combinations	Mechanism of action	Effectiveness against pathogens	References
Amoxicillin and clavulanic acid	Clavulanic acid inhibits serine-β-lactamase rendering resistant bacteria susceptible to amoxicillin	*E. coli* and *Klebsiella aerogenes*	Ball, Geddes, Davey, Farrell, and Brookes (1980)
Meropenem and aspergillomarasmine A	Aspergillomarasmine A potentiates meropenem by inhibiting metallo-β-lactamase	*Pseudomonas* spp., *Acinetobacter* spp., *Enterobacteriaceae*, and *K. pneumoniae*	King et al. (2014)
Tobramycin and 7-hydroxytropolone	7-Hydroxytropolone inhibits aminoglycoside adenyltransferase and facilitates tobramycin uptake	*E. coli* and *K. pneumoniae*	Allen, Alborn, Hobbs, and Kirst (1982)
Berberine and 5′-methoxyhydnocarpin	5′-Methoxyhydnocarpin inhibits the multidrug efflux pump and potentiates the activity of berberine	*S. aureus*	Stermitz, Lorenz, Tawara, Zenewicz, and Lewis (2000)
Minocycline and loperamide	Loperamide perturbates the proton gradient of the bacterial cell membrane and increases uptake of minocycline	*P. aeruginosa*	Ejim et al. (2011)
β-Lactam antibiotics and 2-aminobenzimidazoles	Not reported	*Mycobacterium smegmatis* and *M. tuberculosis*	Nguyen et al. (2017)
Novobiocin and *S*-(3,4-dichlorobenzyl)-isothiourea (A22)	A22 inhibits MreB (an actin homolog) and potentiates novobiocin	*E. coli*	Taylor, Rossi, De Pascale, and Wright (2012)
Cefuroxime and ticlopidine	Ticlopidine inhibits wall teichoic acid biosynthesis and synergizes β-lactam antibiotic cefuroxime	MRSA	Farha et al. (2013)

Continued

TABLE 12.1 Key synergistic combinations of antibiotics with nonantibiotic agents and their mechanisms of action—cont'd

Antibiotic combinations	Mechanism of action	Effectiveness against pathogens	References
Imipenem and tricyclic indole acid L275	L275 inhibits wall teichoic acid biosynthesis and potentiates imipenem	MRSA	Wang et al. (2013)
Polymyxin B and closantel	Not reported	*Acinetobacter baumannii*	Tran et al. (2016)
Doxycycline and carprofen	Not reported	MRSA	Brochmann, Helmfrid, Jana, Magnowska, and Guardabassi (2016)
Rifampicin/novobiocin/erythromycin and pentamidine	Antiprotozoal drug pentamidine interacts with lipopolysaccharide and synergizes Gram-positive-specific antibiotics against Gram-negative bacteria	*A. baumanii, K. pneumonia, E. coli*	Stokes et al. (2017)
Spectinomycin and vanillin	Not reported	*E. coli*	Brochado et al. (2018)
Vancomycin and silver	Silver disrupts cellular processes and potentiates vancomycin by increasing its uptake	*E. coli*	Morones-Ramirez et al. (2013)
Colistin with ARB-002 (Helperby Therapeutics)	Not reported	*A. baumannii, P. aeruginosa,* and *Enterobacteriaceae*	Coates (2019) and WHO (2017a)

TABLE 12.2 Key synergistic combinations of antibiotics and their mechanisms of action.

Antibiotic combinations	Mechanism of action	Effectiveness against pathogens	References
Streptomycin and penicillin	Penicillin facilitates the entry of streptomycin by increasing the permeability of the cell membrane that targets 30S subunit of ribosome, disrupting protein synthesis	*Streptococcus viridans*	Plotz and Davis (1962)
Vancomycin and gentamicin	Vancomycin facilitates the entry of streptomycin by increasing the permeability of the cell membrane that targets 30S subunit of ribosome, disrupting protein synthesis	Penicillin-resistant pneumococci	Cottagnoud, Cottagnoud, and Tauber (2003)
Amoxicillin and erythromycin	Not reported	*Proteus vulgaris*, *Salmonella* Typhi, and *S. pyogenes*	Olajuyigbe and Animashaun (2012)
Phenothiazines and erythromycin	Accumulation of erythromycin in cells that may be due to the interference of proton gradient at the inner membrane by phenothiazines	*Burkholderia pseudomallei*	Chan, Ong, and Chua (2007)
Amoxicillin and gentamicin	Not reported	*Streptococcus pneumoniae*	Darras-Joly et al. (1996)
Tetracycline and amoxicillin	Not reported	*B. cereus*, *S. aureus*, *S.* Typhi, *A. calcaceticus*, *K. pneumonia*, and *E. faecalis*	Olajuyigbe (2012)
Trimethoprim and sulfamethoxazole	Both drugs target sequential steps in tetrahydrofolate (THF) biosynthesis	*E. coli* and MRSA	Minato et al. (2018)
Ciprofloxacin and imipenem	Not reported	*P. aeruginosa*	Giamarellou and Petrikkos (1987)

Continued

TABLE 12.2 Key synergistic combinations of antibiotics and their mechanisms of action—cont'd

Antibiotic combinations	Mechanism of action	Effectiveness against pathogens	References
Tigecycline and aminoglycoside	Not reported	*A. baumannii*	Guner, Hasanoglu, Keske, Kalem, and Tasyaran (2011)
Ampicillin and dicloxacillin	Dicloxacillin inhibits β-lactamase; in turn, ampicillin can penetrate through β-lactamase barriers into its target sites on the cytoplasmic membrane	*Citrobacter freundii*	Mizoguchi, Suginaka, and Kotani (1979)
Quinupristin and dalfopristin	Both drugs inhibit protein synthesis by targeting 50S ribosome subunits resulting in synergistic bactericidal activity	*Enterococcus faecium*	Jones, Ballow, Biedenbach, Deinhart, and Schentag (1998) and Winston et al. (2000)
Isoniazid, rifampicin, pyrazinamide, and ethambutol/streptomycin	Not reported	*M. tuberculosis*	WHO (2009)
Erythromycin and doxycycline	Both inhibit protein synthesis resulting in a synergistic bacteriostatic effect	*E. coli*	Chait et al. (2007) and Pena-Miller et al. (2013)

12.2.4 FDA-approved drug combinations for treatment of bacterial infections

Table 12.3 lists the FDA-approved drug combinations for treating bacterial infections (https://www.centerwatch.com/drug-information/fda-approved-drugs/).

12.2.5 Challenges of combination therapy

Failure to prevent emergence of resistance: Combination therapy cannot be successful all the time in preventing the emergence of resistance. For instance, co-trimoxazole (the combination of sulfonamide and trimethoprim) given to counter neonatal sepsis has failed to prevent resistance, as over 95% Gram-negative bacteria were found to be resistant (Viswanathan et al., 2012). A metaanalysis of randomized controlled trials also reported the failure of the combination of aminoglycoside/β-lactam as compared to β-lactam monotherapy in preventing the emergence of resistance (Bliziotis, Samonis, Vardakas, Chrysanthopoulou, & Falagas, 2005). The chances of superinfections due to emergence of MDR or XDR strains cannot be ignored when designing combination therapies.

Translation of in vitro results into clinical benefits: Evidence from clinical trials suggests that not all the in vitro synergy of drug combinations translates into clinical benefits (Tamma et al., 2012). This is due to the additional contribution of the complex immune system of the host, which changes the very nature of drug interactions in vivo. The availability of pharmacokinetic complementarity between various drugs could be another factor contributing to the differences between in vitro and clinical data.

Toxicity: Combinations involving compounds that have not been in clinical trials previously have a high chance of failing during clinical trials because of toxicity and various side effects.

Poor understanding of mechanism of synergy: The mechanism of synergy is often overlooked because we are more interested in getting the results. The interaction of drugs is a very complex process, and if more than two drugs are combined, it becomes much more complicated.

12.3 Future and perspectives

The golden era of antibiotic development has long gone, since the world is not able to produce new antibiotics fast enough to replace the ones that have stopped working. Now the focus is shifting to combination therapy to rejuvenate outdated antibiotics. Combination therapy

TABLE 12.3 FDA-approved drug combinations for the treatment of bacterial infections.

Drug combination	Treatment	Approval	Company
Recarbrio (imipenem, cilastatin, and relebactam)	Complicated urinary tract and intraabdominal infections	July 2019	Merck
Vabomere (meropenem and vaborbactam)	Complicated urinary tract infections	August 2017	The Medicines Company
Avycaz (ceftazidime-avibactam)	Complicated urinary tract and intraabdominal infections	February 2015	Actavis
Zerbaxa (tazobactam + ceftolozane)	Bacterial pneumonia and complicated urinary tract and intraabdominal infections	December 2014	Merck
Cefazolin and dextrose USP	Various bacterial infections, septicemia, endocarditis, and perioperative prophylaxis	July 2000	B Braun Medical
Synercid I.V. (quinupristin and dalfopristin)	For serious or life-threatening infections associated with vancomycin-resistant *E. faecium* (VREF) bacteremia	September 1999	Rhone Poulenc Rorer
Augmentin (amoxicillin/clavulanate)	Sinusitis, bronchitis, pneumonia	February 1996	SmithKline Beecham

has provided a promising treatment strategy, targeting a broad range of MDR pathogens, combating the emergence of resistance, and increasing patient compliance with shortened regimens. The combination of drugs has helped in reducing the dosage and toxicity of antibiotics, making them better candidates for clinical approval. The application of robotics and high-throughput screening is accelerating the process of identifying novel drug combinations. In a recent example, a high-throughput screen (HTS) has identified small adjunctive molecules that potentiate carbapenems and/or directly inhibit growth of carbapenem-resistant *Enterobacteriaceae* (CRE). This HTS strategy enables paced discovery of novel antimicrobials as well as potentiators to known antibiotics (Smith & Kirby, 2016).

The combination of already available antibiotics and drugs has reduced the burden on drug discovery, which is a time-consuming and quite expensive process. Successful drug combinations, from Augmentin (1981) to the recently FDA-approved Recarbrio (2019), have shown the underlying potential of combination therapy. Another promising class of compounds, antibiotic resistance breakers (ARBs), can rejuvenate old antibiotics (Laws, Shaaban, & Rahman, 2019). Like the rejuvenation of β-lactams with a β-lactamase inhibitor, other existing antibiotics can be made effective once again by targeting the resistance mechanism with new molecules. ARBs are novel, effective, and transferable, because one ARB can rejuvenate several antibiotics. This technique could rejuvenate all the existing 200 or so antibiotics just by combining with 20 or so ARBs. This may not reproduce the golden era of antibiotic discovery, but it could provide a solution to the antibiotic crisis (Coates & Hu, 2014).

Researchers at the University of Colorado Boulder have developed photoexcited quantum dots for killing multidrug-resistant bacteria through redox potentials of the photogenerated charge carriers (Courtney et al., 2016). They further demonstrated that superoxide-generating photoexcited quantum dots can potentiate antibiotics to kill drug-resistant bacteria. The effective antibiotic concentration of the clinical isolates was reduced up to 1000-fold below their respective sensitive/resistant breakpoint as a result of this potentiation (Levy et al., 2017). This study provided proof-of-concept that superoxide-generating quantum dots can potentiate older antibiotics to kill resistant pathogens.

In another proof-of-concept study, scientists from the Institut Pasteur, the CNRS, and the Universidad Politécnica de Madrid successfully programmed an advanced bacterial genetic structure to make it capable of specifically killing multiple antibiotic-resistant bacteria in mixed populations. In addition, unlike other approaches under development, this novel tool is associated with a minimal rate of emergence of new resistance (Koskiniemi & Virtanen, 2019). They used an engineered toxin-intein antimicrobial, which is only active against resistant bacteria, while sensitive bacteria would be killed by the regular antibiotic used in combination. This genetic weapon has been tested against antibiotic-resistant *Vibrio cholerae* bacteria and validated in their natural habitats such as water, zebrafish, and artemia (López-Igual, Bernal-Bayard, Rodríguez-Patón, Ghigo, & Mazel, 2019).

In addition to improving and advancing combination therapy, scientists have been developing alternative ways to kill resistant bacteria, such as the use of bacteriophages (Kortright, Chan, Koff, & Turner, 2019), antimicrobial peptides (Pfalzgraff, Brandenburg, & Weindl, 2018), genetic manipulations (Bikard et al., 2014), antibiotic hybrids (Domalaon, Idowu, Zhanel, & Schweizer, 2018), and other kinds of engineered nanoparticles (Baptista et al., 2018). New antibiotic classes and new antimicrobials are also being developed to replace antibiotics. A wide range of new antimicrobial compounds such as endolysins from various bacteriophages, liamocin from *Aureobasidium pullulans*, and laparaxin from *Lactobacillus paracasei* have been identified as possible suitable alternatives to replace traditional antibiotics (Seal et al., 2018). Another potential alternative could be bacteriocins, which are ribosomally synthesized peptides that function through different mechanisms frequently distinct from those used by antibiotics (Cotter, Ross, & Hill, 2013). Researchers have recently applied bioinformatics to convert structures of known antimicrobial peptides into a graphical representational database. Then an algorithm scoured the database, optimized to characteristics and repeat patterns that gave maximum antimicrobial potency, and proposed fresh models. The algorithm's proposed top five peptide structures were constructed and tested in the laboratory. The most effective peptide found during the screening was Omega 76 (Ω76). Ω76 adopts an α-helical structure in membranes, causing rapid membrane disruption, leakage, and bacterial death. Furthermore, the likelihood of drug resistance is much smaller because Ω76 causes actual physical damage to the bacterium (Nagarajan et al., 2019).

Teixobactin is a new class of antibiotics recently discovered that has shown promise in vivo and is in preclinical development. It binds lipid II and lipid III, which are involved in bacterial cell wall formation, thereby harming bacterial growth. Tests in mice show that teixobactin is effective against the drug-resistant bacterium methicillin-resistant *S. aureus*, as well as *Streptococcus pneumoniae*, which can cause pneumonia and meningitis. In culture, it has defeated other disease-causing bacteria, including *M. tuberculosis* and *Clostridium difficile* (Ling et al., 2015; Parmar et al., 2018). The important thing is that no bacteria has been able to develop resistance against teixobactin.

Another potential novel antibiotic is gepotidacin, which is currently in clinical trials for treating *Neisseria gonorrhoeae* infection (gonorrhea), including multidrug-resistant strains (Taylor et al., 2018). In another study, two novel antibiotics of a new family of cyclic peptidomimetics were found to be effective against Gram-positive and -negative multiresistant bacteria without triggering resistance. The researchers synthesized 20 molecules inspired by a natural peptide toxin produced by *S. aureus*, whose role is to facilitate infection and which is also capable of killing other bacteria present in the body. Out of 20 peptidomimetics synthesized, two (Pep16 and Pep19) were effective against resistant *S. aureus* and *P. aeruginosa* in mouse models of severe sepsis or skin infection. These compounds were well tolerated at their active doses without sign of renal toxicity. No toxicity to animals or human cells was observed even at doses 10–50 times higher than the effective dose (Nicolas, Bordeau, Bondon, Baudy-Floc'h, & Felden, 2019).

It is true that the pipeline for novel antibiotics is drying up, but extensive research is in progress to tackle the problem of multidrug resistance, with novel strategies being applied to produce antimicrobial solutions immune from the emergence of resistance. Hopefully, most of these innovative approaches will also show promise in clinical development.

12.4 Conclusion

Over several decades, the emergence of antibiotic resistance has demanded new antibiotics to replace the obsolete ones or the development of novel strategies to combat MDR pathogens and their emergence. In this race against the emergence of antibiotic resistance, the discovery of new antibiotics lags behind. All the major antibiotics have lost effectiveness due to the evolution of resistant bacteria. These outdated antibiotics are no longer able to cover even common bacterial infections. Global solutions to the problem of antibiotic resistance are urgently needed; otherwise the postantibiotic era is not far off. However, combination therapy, a simple yet powerful approach, offers a solution to this antibiotics crisis. Combining an antibiotic with another antibiotic, or nonantibiotic agents, has rejuvenated old antibiotics to become effective once again. Scientists are screening for fruitful drug combinations to target antibiotic resistance and treat deadly diseases. This has led to the identification of numerous efficacious drug combinations. Many are in clinical trials while some have already been approved for human use and are available for prescription. Despite the success of combination therapy, it still faces the challenges of overcoming the emergence of resistance, difficulties in obtaining clinical approval, and the lack of pharmacokinetic data on drug interactions.

References

Allen, N. E., Alborn, W. E., Hobbs, J. N., & Kirst, H. A. (1982). 7-Hydroxytropolone: An inhibitor of aminoglycoside-2″-O-adenylyltransferase. *Antimicrobial Agents and Chemotherapy*, *22*(5), 824–831. https://doi.org/10.1128/AAC.22.5.824.

Allen, H. K., Donato, J., Wang, H. H., Cloud-Hansen, K. A., Davies, J., & Handelsman, J. (2010). Call of the wild: Antibiotic resistance genes in natural environments. *Nature Reviews Microbiology*, *8*(4), 251–259. https://doi.org/10.1038/nrmicro2312.

Andrei, S., Valeanu, L., Chirvasuta, R., & Stefan, M.-G. (2018). New FDA approved antibacterial drugs: 2015-2017. *Discover*, *6*(1), e81, https://doi.org/10.15190/d.2018.1.

Angus, I. (2019). Superbugs in the Anthropocene. *Monthly Review*, *71*(2), 1–28. https://doi.org/10.14452/MR-071-02-2019-06_1.

Ball, A. P., Geddes, A. M., Davey, P. G., Farrell, I. D., & Brookes, G. R. (1980). Clavulanic acid and amoxycillin: A clinical, bacteriological, and pharmacological study. *Lancet (London, England)*, *1*(8169), 620–623. https://doi.org/10.1016/s0140-6736(80)91118-6.

Banin, E., Hughes, D., & Kuipers, O. P. (2017). Editorial: Bacterial pathogens, antibiotics and antibiotic resistance. *FEMS Microbiology Reviews*, *41*(3), 450–452. https://doi.org/10.1093/femsre/fux016.

Baptista, P. V., McCusker, M. P., Carvalho, A., Ferreira, D. A., Mohan, N. M., Martins, M., & Fernandes, A. R. (2018). Nano-strategies to fight multidrug resistant bacteria-"A Battle of the Titans". *Frontiers in Microbiology*, *9*, 1–26. https://doi.org/10.3389/fmicb.2018.01441.

Bennett, P. M. (2009). Plasmid encoded antibiotic resistance: Acquisition and transfer of antibiotic resistance genes in bacteria. *British Journal of Pharmacology*, *153*(S1), S347–S357. https://doi.org/10.1038/sj.bjp.0707607.

Berglund, C., Molling, P., Sjoberg, L., & Soderquist, B. (2005). Multilocus sequence typing of methicillin-resistant *Staphylococcus aureus* from an area of low endemicity by real-time PCR. *Journal of Clinical Microbiology*, *43*(9), 4448–4454. https://doi.org/10.1128/JCM.43.9.4448-4454.2005.

Bikard, D., Euler, C. W., Jiang, W., Nussenzweig, P. M., Goldberg, G. W., Duportet, X., … Marraffini, L. A. (2014). Exploiting CRISPR-Cas nucleases to produce sequence-specific antimicrobials. *Nature Biotechnology*, *32*(11), 1146–1150. https://doi.org/10.1038/nbt.3043.

Bliziotis, I. A., Samonis, G., Vardakas, K. Z., Chrysanthopoulou, S., & Falagas, M. E. (2005). Effect of aminoglycoside and β-lactam combination therapy versus β-lactam monotherapy on the emergence of antimicrobial resistance: A meta-analysis of randomized, controlled trials. *Clinical Infectious Diseases*, *41*(2), 149–158. https://doi.org/10.1086/430912.

Boeree, M. J., Heinrich, N., Aarnoutse, R., Diacon, A. H., Dawson, R., Rehal, S., … Hoelscher, M. (2017). High-dose rifampicin, moxifloxacin, and SQ109 for treating tuberculosis: A multi-arm, multi-stage randomised controlled trial. *The Lancet Infectious Diseases*, *17*(1), 39–49. https://doi.org/10.1016/S1473-3099(16)30274-2.

Bollenbach, T. (2015). Antimicrobial interactions: Mechanisms and implications for drug discovery and resistance evolution. *Current Opinion in Microbiology*, *27*, 1–9. https://doi.org/10.1016/j.mib.2015.05.008.

Brochado, A. R., Telzerow, A., Bobonis, J., Banzhaf, M., Mateus, A., Selkrig, J., … Typas, A. (2018). Species-specific activity of antibacterial drug combinations. *Nature*, *559*(7713), 259–263. https://doi.org/10.1038/s41586-018-0278-9.

Brochmann, R. P., Helmfrid, A., Jana, B., Magnowska, Z., & Guardabassi, L. (2016). Antimicrobial synergy between carprofen and doxycycline against methicillin-resistant *Staphylococcus pseudintermedius* ST71. *BMC Veterinary Research*, *12*(1), 126. https://doi.org/10.1186/s12917-016-0751-3.

Castanheira, M., Farrell, S. E., Krause, K. M., Jones, R. N., & Sader, H. S. (2014). Contemporary diversity of β-lactamases among Enterobacteriaceae in the nine U.S. census regions and ceftazidime-avibactam activity tested against isolates producing the most prevalent β-lactamase groups. *Antimicrobial Agents and Chemotherapy, 58*(2), 833–838. https://doi.org/10.1128/AAC.01896-13.

CDC. (2013). *Antibiotic resistance threats in the United States*. Retrieved from: https://www.cdc.gov/drugresistance/biggest_threats.html.

CDC. (2017). *CDC Global Health—Infographics—Antibiotic resistance the global threat*. Retrieved from: https://www.cdc.gov/globalhealth/infographics/antibiotic-resistance/antibiotic_resistance_global_threat.htm. [(Accessed 9 August, 2019)].

Chait, R., Craney, A., & Kishony, R. (2007). Antibiotic interactions that select against resistance. *Nature, 446*(7136), 668–671. https://doi.org/10.1038/nature05685.

Chambers, H. F., & DeLeo, F. R. (2009). Waves of resistance: *Staphylococcus aureus* in the antibiotic era. *Nature Reviews Microbiology, 7*(9), 629–641. https://doi.org/10.1038/nrmicro2200.

Chan, Y. Y., Ong, Y. M., & Chua, K. L. (2007). Synergistic interaction between phenothiazines and antimicrobial agents against *Burkholderia pseudomallei*. *Antimicrobial Agents and Chemotherapy, 51*(2), 623–630. https://doi.org/10.1128/AAC.01033-06.

Coates, A. (2019). The future of antibiotics lies in combination treatments. *Future Drug Discovery, 1*(1), FDD5. https://doi.org/10.4155/fdd-2019-0012.

Coates, A., & Hu, Y. (2014). Conventional antibiotics—Revitalized by new agents. In *Novel antimicrobial agents and strategies* (pp. 17–30). Weinheim, Germany: Wiley-VCH Verlag GmbH & Co. KGaA. https://doi.org/10.1002/9783527676132.ch2.

Cottagnoud, P., Cottagnoud, M., & Tauber, M. G. (2003). Vancomycin acts synergistically with gentamicin against penicillin-resistant pneumococci by increasing the intracellular penetration of gentamicin. *Antimicrobial Agents and Chemotherapy, 47*(1), 144–147. https://doi.org/10.1128/AAC.47.1.144-147.2003.

Cotter, P. D., Ross, R. P., & Hill, C. (2013). Bacteriocins—A viable alternative to antibiotics? *Nature Reviews Microbiology, 11*(2), 95–105. https://doi.org/10.1038/nrmicro2937.

Courtney, C. M., Goodman, S. M., McDaniel, J. A., Madinger, N. E., Chatterjee, A., & Nagpal, P. (2016). Photoexcited quantum dots for killing multidrug-resistant bacteria. *Nature Materials, 15*(5), 529–534. https://doi.org/10.1038/nmat4542.

Darras-Joly, C., Bédos, J. P., Sauve, C., Moine, P., Vallée, E., Carbon, C., & Azoulay-Dupuis, E. (1996). Synergy between amoxicillin and gentamicin in combination against a highly penicillin-resistant and -tolerant strain of *Streptococcus pneumoniae* in a mouse pneumonia model. *Antimicrobial Agents and Chemotherapy, 40*(9), 2147–2151. https://doi.org/10.1128/AAC.40.9.2147.

De Clercq, E., & Li, G. (2016). Approved antiviral drugs over the past 50 years. *Clinical Microbiology Reviews, 29*(3), 695–747. https://doi.org/10.1128/CMR.00102-15.

Domalaon, R., Idowu, T., Zhanel, G. G., & Schweizer, F. (2018). Antibiotic hybrids: The next generation of agents and adjuvants against Gram-negative pathogens? *Clinical Microbiology Reviews, 31*(2), https://doi.org/10.1128/CMR.00077-17.

Doyle, M. E. (2015). Multidrug-resistant pathogens in the food supply. *Foodborne Pathogens and Disease, 12*(4), 261–279. https://doi.org/10.1089/fpd.2014.1865.

Drawz, S. M., & Bonomo, R. A. (2010). Three decades of β-lactamase inhibitors. *Clinical Microbiology Reviews*, https://doi.org/10.1128/CMR.00037-09.

Ejim, L., Farha, M. A., Falconer, S. B., Wildenhain, J., Coombes, B. K., Tyers, M., … Wright, G. D. (2011). Combinations of antibiotics and nonantibiotic drugs enhance antimicrobial efficacy. *Nature Chemical Biology*, https://doi.org/10.1038/nchembio.559.

Fair, R. J., & Tor, Y. (2014). Antibiotics and bacterial resistance in the 21st century. *Perspectives in Medicinal Chemistry, 6*, PMC.S14459. https://doi.org/10.4137/PMC.S14459.

Farha, M. A., Leung, A., Sewell, E. W., D'Elia, M. A., Allison, S. E., Ejim, L., … Brown, E. D. (2013). Inhibition of WTA synthesis blocks the cooperative action of PBPs and sensitizes MRSA to β-lactams. *ACS Chemical Biology, 8*(1), 226–233. https://doi.org/10.1021/cb300413m.

Fernandez, L., & Hancock, R. E. W. (2012). Adaptive and mutational resistance: Role of porins and efflux pumps in drug resistance. *Clinical Microbiology Reviews*, *25*(4), 661–681. https://doi.org/10.1128/CMR.00043-12.

Fleming-Dutra, K. E., Hersh, A. L., Shapiro, D. J., Bartoces, M., Enns, E. A., File, T. M., ... Hicks, L. A. (2016). Prevalence of inappropriate antibiotic prescriptions among US ambulatory care visits, 2010-2011. *JAMA*, *315*(17), 1864. https://doi.org/10.1001/jama.2016.4151.

Frost, I., Van Boeckel, T. P., Pires, J., Craig, J., & Laxminarayan, R. (2019). Global geographic trends in antimicrobial resistance: The role of international travel. *Journal of Travel Medicine*, https://doi.org/10.1093/jtm/taz036.

Geddes, A. M., Klugman, K. P., & Rolinson, G. N. (2007). Introduction: Historical perspective and development of amoxicillin/clavulanate. *International Journal of Antimicrobial Agents*, *30*, 109–112. https://doi.org/10.1016/j.ijantimicag.2007.07.015.

Ghosh, C., Sarkar, P., Issa, R., & Haldar, J. (2019). Alternatives to conventional antibiotics in the era of antimicrobial resistance. *Trends in Microbiology*, *27*(4), 323–338. https://doi.org/10.1016/j.tim.2018.12.010.

Giamarellou, H., & Petrikkos, G. (1987). Ciprofloxacin interactions with imipenem and amikacin against multiresistant *Pseudomonas aeruginosa*. *Antimicrobial Agents and Chemotherapy*, *31*(6), 959–961. https://doi.org/10.1128/AAC.31.6.959.

Guner, R., Hasanoglu, I., Keske, S., Kalem, A. K., & Tasyaran, M. A. (2011). Outcomes in patients infected with carbapenem-resistant *Acinetobacter baumannii* and treated with tigecycline alone or in combination therapy. *Infection*, *39*(6), 515–518. https://doi.org/10.1007/s15010-011-0161-1.

Guo, B., Lin, J., Reynolds, D. L., & Zhang, Q. (2010). Contribution of the multidrug efflux transporter CmeABC to antibiotic resistance in different *Campylobacter* species. *Foodborne Pathogens and Disease*, *7*(1), 77–83. https://doi.org/10.1089/fpd.2009.0354.

Guo, C., Mandalapu, D., Ji, X., Gao, J., & Zhang, Q. (2018). Chemistry and biology of teixobactin. *Chemistry—A European Journal*, *24*(21), 5406–5422. https://doi.org/10.1002/chem.201704167.

Hugonnet, J.-E., Tremblay, L. W., Boshoff, H. I., Barry, C. E., & Blanchard, J. S. (2009). Meropenem-clavulanate is effective against extensively drug-resistant *Mycobacterium tuberculosis*. *Science*, *323*(5918), 1215–1218. https://doi.org/10.1126/science.1167498.

Jones, R. N., Ballow, C. H., Biedenbach, D. J., Deinhart, J. A., & Schentag, J. J. (1998). Antimicrobial activity of quinupristin-dalfopristin (RP 59500, Synercid®) tested against over 28,000 recent clinical isolates from 200 medical centers in the United States and Canada. *Diagnostic Microbiology and Infectious Disease*, *31*(3), 437–451. https://doi.org/10.1016/S0732-8893(98)80002-3.

Kalan, L., & Wright, G. D. (2011). Antibiotic adjuvants: Multicomponent anti-infective strategies. *Expert Reviews in Molecular Medicine*, https://doi.org/10.1017/S1462399410001766.

Karumathil, D. P., Yin, H.-B., Kollanoor-Johny, A., & Venkitanarayanan, K. (2016). Prevalence of multidrug-resistant bacteria on fresh vegetables collected from farmers' markets in Connecticut. *Journal of Food Protection*, *79*(8), 1446–1451. https://doi.org/10.4315/0362-028X.JFP-15-548.

King, A. M., Reid-Yu, S. A., Wang, W., King, D. T., De Pascale, G., Strynadka, N. C., ... Wright, G. D. (2014). Aspergillomarasmine A overcomes metallo-β-lactamase antibiotic resistance. *Nature*, *510*(7506), 503–506. https://doi.org/10.1038/nature13445.

Kohanski, M. A., Dwyer, D. J., & Collins, J. J. (2010). How antibiotics kill bacteria: From targets to networks. *Nature Reviews Microbiology*, *8*(6), 423–435. https://doi.org/10.1038/nrmicro2333.

Kortright, K. E., Chan, B. K., Koff, J. L., & Turner, P. E. (2019). Phage therapy: A renewed approach to combat antibiotic-resistant bacteria. *Cell Host & Microbe*, *25*(2), 219–232. https://doi.org/10.1016/j.chom.2019.01.014.

Koskiniemi, S., & Virtanen, P. (2019). Selective killing of antibiotic-resistant bacteria from within. *Nature*, *570*(7762), 449–450. https://doi.org/10.1038/d41586-019-01852-w.

Laws, M., Shaaban, A., & Rahman, K. M. (2019). Antibiotic resistance breakers: Current approaches and future directions. *FEMS Microbiology Reviews*, https://doi.org/10.1093/femsre/fuz014.

Laxminarayan, R., Duse, A., Wattal, C., Zaidi, A. K. M., Wertheim, H. F. L., Sumpradit, N., … Cars, O. (2013). Antibiotic resistance—The need for global solutions. *The Lancet Infectious Diseases*, *13*(12), 1057–1098. https://doi.org/10.1016/S1473-3099(13)70318-9.

Levy, M., Madinger, N. E., Nagpal, P., Courtney, C. M., Nagy, T. A., Detweiler, C. S., … Goodman, S. M. (2017). Potentiating antibiotics in drug-resistant clinical isolates via stimuli-activated superoxide generation. *Science Advances*, *3*(10), e1701776. https://doi.org/10.1126/sciadv.1701776.

Ling, L. L., Schneider, T., Peoples, A. J., Spoering, A. L., Engels, I., Conlon, B. P., … Lewis, K. (2015). A new antibiotic kills pathogens without detectable resistance. *Nature*, *517*(7535), 455–459. https://doi.org/10.1038/nature14098.

Lomovskaya, O., & Bostian, K. A. (2006). Practical applications and feasibility of efflux pump inhibitors in the clinic—A vision for applied use. *Biochemical Pharmacology*, https://doi.org/10.1016/j.bcp.2005.12.008.

López-Igual, R., Bernal-Bayard, J., Rodríguez-Patón, A., Ghigo, J.-M., & Mazel, D. (2019). Engineered toxin–intein antimicrobials can selectively target and kill antibiotic-resistant bacteria in mixed populations. *Nature Biotechnology*, *37*(7), 755–760. https://doi.org/10.1038/s41587-019-0105-3.

McKenna, M. (2017). *Big chicken: The incredible story of how antibiotics created modern agriculture and changed the way the world eats*. Washington, DC: National Geographic Partners. Retrieved from: https://books.google.co.in/books?id=T7e-DQAAQBAJ&pg=PA373&dq=Maryn+McKenna,+Big+Chicken+(Washington,+D.C.:+National+Geographic,+2017),+25&hl=en&sa=X&ved=0ahUKEwjUzumvm_XjAhVMpI8KHWKZB4gQ6AEINTAC#v=onepage&q&f=false.

Minato, Y., Dawadi, S., Kordus, S. L., Sivanandam, A., Aldrich, C. C., & Baughn, A. D. (2018). Mutual potentiation drives synergy between trimethoprim and sulfamethoxazole. *Nature Communications*, *9*(1), 1003. https://doi.org/10.1038/s41467-018-03447-x.

Mitchison, D. A. (2012). Prevention of drug resistance by combined drug treatment of tuberculosis. In A. R. M. Coates (Ed.), *Antibiotic resistance, handbook of experimental pharmacology* (pp. 87–98). Berlin, Heidelburg: Springer. https://doi.org/10.1007/978-3-642-28951-4_6.

Mizoguchi, J., Suginaka, H., & Kotani, S. (1979). Mechanism of synergistic action of a combination of ampicillin and dicloxacillin against a beta-lactamase-producing strain of *Citrobacter freundii*. *Antimicrobial Agents and Chemotherapy*, *16*(4), 439–443. https://doi.org/10.1128/AAC.16.4.439.

Mody, L., Washer, L. L., Kaye, K. S., Gibson, K., Saint, S., Reyes, K., … Zervos, M. (2019). Multidrug-resistant organisms in hospitals: What is on patient hands and in their rooms? *Clinical Infectious Diseases*. https://doi.org/10.1093/cid/ciz092.

Morones-Ramirez, J. R., Winkler, J. A., Spina, C. S., & Collins, J. J. (2013). Silver enhances antibiotic activity against Gram-negative bacteria. *Science Translational Medicine*, *5*(190), 190ra81. https://doi.org/10.1126/scitranslmed.3006276.

Nagarajan, D., Roy, N., Kulkarni, O., Nanajkar, N., Datey, A., Ravichandran, S., … Chandra, N. (2019). Ω76: A designed antimicrobial peptide to combat carbapenem- and tigecycline-resistant *Acinetobacter baumannii*. *Science Advances*, *5*(7), eaax1946, https://doi.org/10.1126/sciadv.aax1946.

Neu, H. C. (1992). The crisis in antibiotic resistance. *Science*, *257*(5073), 1064–1073. https://doi.org/10.1126/science.257.5073.1064.

Nguyen, T. V., Blackledge, M. S., Lindsey, E. A., Minrovic, B. M., Ackart, D. F., Jeon, A. B., … Melander, C. (2017). The discovery of 2-aminobenzimidazoles that sensitize *Mycobacterium smegmatis* and *M. tuberculosis* to β-lactam antibiotics in a pattern distinct from β-lactamase inhibitors. *Angewandte Chemie International Edition*, *56*(14), 3940–3944. https://doi.org/10.1002/anie.201612006.

Nicolas, I., Bordeau, V., Bondon, A., Baudy-Floc'h, M., & Felden, B. (2019). Novel antibiotics effective against gram-positive and -negative multi-resistant bacteria with limited resistance. *PLoS Biology*, *17*(7), e3000337, https://doi.org/10.1371/journal.pbio.3000337.

Nikaido, H. (2009). Multidrug resistance in bacteria. *Annual Review of Biochemistry*, *78*(1), 119–146. https://doi.org/10.1146/annurev.biochem.78.082907.145923.

O'Neill, J. (2014). Antimicrobial resistance: Tackling a crisis for the health and wealth of nations. *Review on Antimicrobial Resistance*, 1–16. Retrieved from: https://amr-review.org/sites/default/files/AMR Review Paper - Tackling a crisis for the health and wealth of nations_1.pdf.

Odonkor, S. T., & Addo, K. K. (2018). Prevalence of multidrug-resistant *Escherichia coli* isolated from drinking water sources. *International Journal of Microbiology*, *2018*, 1–7. https://doi.org/10.1155/2018/7204013.

Olajuyigbe, O. O. (2012). Synergistic influence of tetracycline on the antibacterial activities of amoxicillin against resistant bacteria. *Journal of Pharmacy and Allied Health Sciences*, *2*(1), 12–20. https://doi.org/10.3923/jpahs.2012.12.20.

Olajuyigbe, O. O., & Animashaun, T. (2012). Synergistic activities of amoxicillin and erythromycin against bacteria of medical importance. *Pharmacologia*, *3*(9), 450–455. https://doi.org/10.5567/pharmacologia.2012.450.455.

Parmar, A., Lakshminarayanan, R., Iyer, A., Mayandi, V., Leng Goh, E. T., Lloyd, D. G., … Singh, I. (2018). Design and syntheses of highly potent teixobactin analogues against *Staphylococcus aureus*, methicillin-resistant *Staphylococcus aureus* (MRSA), and vancomycin-resistant Enterococci (VRE) *in vitro* and *in vivo*. *Journal of Medicinal Chemistry*, *61*(5), 2009–2017. https://doi.org/10.1021/acs.jmedchem.7b01634.

Payen, M. C., Muylle, I., Vandenberg, O., Mathys, V., Delforge, M., Van den Wijngaert, S., … De Wit, S. (2018). Meropenem-clavulanate for drug-resistant tuberculosis: A follow-up of relapse-free cases. *The International Journal of Tuberculosis and Lung Disease*, *22*(1), 34–39. https://doi.org/10.5588/ijtld.17.0352.

Pena-Miller, R., Laehnemann, D., Jansen, G., Fuentes-Hernandez, A., Rosenstiel, P., Schulenburg, H., & Beardmore, R. (2013). When the most potent combination of antibiotics selects for the greatest bacterial load: The smile-frown transition. *PLoS Biology*, *11*(4), e1001540, https://doi.org/10.1371/journal.pbio.1001540.

Peterson, E., & Kaur, P. (2018). Antibiotic resistance mechanisms in bacteria: Relationships between resistance determinants of antibiotic producers, environmental bacteria, and clinical pathogens. *Frontiers in Microbiology*, *9*, https://doi.org/10.3389/fmicb.2018.02928.

Pfalzgraff, A., Brandenburg, K., & Weindl, G. (2018). Antimicrobial peptides and their therapeutic potential for bacterial skin infections and wounds. *Frontiers in Pharmacology*, *9*, https://doi.org/10.3389/fphar.2018.00281.

Phillips, I. (2003). Does the use of antibiotics in food animals pose a risk to human health? A critical review of published data. *Journal of Antimicrobial Chemotherapy*, *53*(1), 28–52. https://doi.org/10.1093/jac/dkg483.

Pletz, M. W., Hagel, S., & Forstner, C. (2017). Who benefits from antimicrobial combination therapy? *The Lancet Infectious Diseases*, *17*(7), 677–678. https://doi.org/10.1016/S1473-3099(17)30233-5.

Plotz, P. H., & Davis, B. D. (1962). Synergism between streptomycin and penicillin: A proposed mechanism. *Science*, *135*(3508), 1067–1068. https://doi.org/10.1126/science.135.3508.1067.

Quinn, J. P., Dudek, E. J., DiVincenzo, C. A., Lucks, D. A., & Lerner, S. A. (1986). Emergence of resistance to imipenem during therapy for *Pseudomonas aeruginosa* infections. *Journal of Infectious Diseases*, *154*(2), 289–294. https://doi.org/10.1093/infdis/154.2.289.

Read, A. F., & Woods, R. J. (2014). Antibiotic resistance management. *Evolution, Medicine, and Public Health*, *2014*(1), 147. https://doi.org/10.1093/emph/eou024.

Reardon, S. (2014). WHO warns against "post-antibiotic" era. *Nature*, https://doi.org/10.1038/nature.2014.15135.

Saud, B., Paudel, G., Khichaju, S., Bajracharya, D., Dhungana, G., Awasthi, M. S., & Shrestha, V. (2019). Multidrug-resistant bacteria from raw meat of Buffalo and chicken, Nepal. *Veterinary Medicine International, 2019*, 1–7. https://doi.org/10.1155/2019/7960268.

Seal, B. S., Drider, D., Oakley, B. B., Brüssow, H., Bikard, D., Rich, J. O., … Gay, C. G. (2018). Microbial-derived products as potential new antimicrobials. *Veterinary Research, 49*(1), 66. https://doi.org/10.1186/s13567-018-0563-5.

Shah, P., & Chen, C.-S. (2017). Combinational synergy of antibiotics and antimicrobial peptides. *Journal of Translational Proteomics Research, 3*, 6–11. Retrieved from: https://www.cosmosscholars.com/phms/index.php/jtpr/article/view/763.

Shakya, T., Stogios, P. J., Waglechner, N., Evdokimova, E., Ejim, L., Blanchard, J. E., … Wright, G. D. (2011). A small molecule discrimination map of the antibiotic resistance kinome. *Chemistry & Biology, 18*(12), 1591–1601. https://doi.org/10.1016/j.chembiol.2011.10.018.

Sharma, A., Gupta, V., & Pathania, R. (2019). Efflux pump inhibitors for bacterial pathogens: From bench to bedside. *Indian Journal of Medical Research, 149*(2), 129. https://doi.org/10.4103/ijmr.IJMR_2079_17.

Silver, L. L. (2011). Challenges of antibacterial discovery. *Clinical Microbiology Reviews, 24*(1), 71–109. https://doi.org/10.1128/CMR.00030-10.

Singh, N., & Yeh, P. J. (2017). Suppressive drug combinations and their potential to combat antibiotic resistance. *The Journal of Antibiotics, 70*(11), 1033–1042. https://doi.org/10.1038/ja.2017.102.

Smith, K. P., & Kirby, J. E. (2016). Validation of a high-throughput screening assay for identification of adjunctive and directly acting antimicrobials targeting carbapenem-resistant Enterobacteriaceae. *Assay and Drug Development Technologies, 14*(3), 194–206. https://doi.org/10.1089/adt.2016.701.

Stermitz, F. R., Lorenz, P., Tawara, J. N., Zenewicz, L. A., & Lewis, K. (2000). Synergy in a medicinal plant: Antimicrobial action of berberine potentiated by 5′-methoxyhydnocarpin, a multidrug pump inhibitor. *Proceedings of the National Academy of Sciences, 97*(4), 1433–1437. https://doi.org/10.1073/pnas.030540597.

Stokes, J. M., MacNair, C. R., Ilyas, B., French, S., Côté, J.-P., Bouwman, C., … Brown, E. D. (2017). Pentamidine sensitizes Gram-negative pathogens to antibiotics and overcomes acquired colistin resistance. *Nature Microbiology, 2*(5), 17028. https://doi.org/10.1038/nmicrobiol.2017.28.

Sun, J., Wei, Q., Zhou, Y., Wang, J., Liu, Q., & Xu, H. (2017). A systematic analysis of FDA-approved anticancer drugs. *BMC Systems Biology, 11*(S5), 87. https://doi.org/10.1186/s12918-017-0464-7.

Tamma, P. D., Cosgrove, S. E., & Maragakis, L. L. (2012). Combination therapy for treatment of infections with Gram-negative bacteria. *Clinical Microbiology Reviews, 25*(3), 450–470. https://doi.org/10.1128/CMR.05041-11.

Taylor, S. N., Morris, D. H., Avery, A. K., Workowski, K. A., Batteiger, B. E., Tiffany, C. A., … Dumont, E. F. (2018). Gepotidacin for the treatment of uncomplicated urogenital gonorrhea: A phase 2, randomized, dose-ranging, single-oral dose evaluation. *Clinical Infectious Diseases, 67*(4), 504–512. https://doi.org/10.1093/cid/ciy145.

Taylor, P. L., Rossi, L., De Pascale, G., & Wright, G. D. (2012). A forward chemical screen identifies antibiotic adjuvants in *Escherichia coli*. *ACS Chemical Biology, 7*(9), 1547–1555. https://doi.org/10.1021/cb300269g.

Tehrani, K.H.M.E., & Martin, N. I. (2018). β-Lactam/β-lactamase inhibitor combinations: An update. *MedChemComm, 9*(9), 1439–1456. https://doi.org/10.1039/C8MD00342D.

The antibiotic alarm. (2013). *Nature, 495*(7440), 141. https://doi.org/10.1038/495141a.

Theuretzbacher, U., & Piddock, L. J. V. (2019). Non-traditional antibacterial therapeutic options and challenges. *Cell Host & Microbe, 26*(1), 61–72. https://doi.org/10.1016/j.chom.2019.06.004.

Tran, T. B., Cheah, S.-E., Yu, H. H., Bergen, P. J., Nation, R. L., Creek, D. J., … Li, J. (2016). Anthelmintic closantel enhances bacterial killing of polymyxin B against multidrug-resistant *Acinetobacter baumannii*. *The Journal of Antibiotics, 69*(6), 415–421. https://doi.org/10.1038/ja.2015.127.

Tyers, M., & Wright, G. D. (2019). Drug combinations: A strategy to extend the life of antibiotics in the 21st century. *Nature Reviews Microbiology, 17*(3), 141–155. https://doi.org/10.1038/s41579-018-0141-x.

Ventola, C. L. (2015). The antibiotic resistance crisis: Part 1: Causes and threats. *Pharmacy and Therapeutics, 40*(4), 277–283. Retrieved from: http://www.ncbi.nlm.nih.gov/pubmed/5859123.

Viswanathan, R., Singh, A. K., Ghosh, C., Dasgupta, S., Mukherjee, S., & Basu, S. (2012). Profile of neonatal septicaemia at a district-level sick newborn care unit. *Journal of Health, Population'and Nutrition, 30*(1), 41–48. https://doi.org/10.3329/jhpn.v30i1.11274.

Wang, H., Gill, C. J., Lee, S. H., Mann, P., Zuck, P., Meredith, T. C., ... Roemer, T. (2013). Discovery of wall teichoic acid inhibitors as potential anti-MRSA β-lactam combination agents. *Chemistry & Biology, 20*(2), 272–284. https://doi.org/10.1016/j.chembiol.2012.11.013.

Webber, M. A., Ricci, V., Whitehead, R., Patel, M., Fookes, M., Ivens, A., & Piddock, L. J. V. (2013). Clinically relevant mutant DNA gyrase alters supercoiling, changes the transcriptome, and confers multidrug resistance. *MBio, 4*(4), https://doi.org/10.1128/mBio.00273-13.

Whitfield, J. (2007). Weak drug combos find new use. *Nature*, https://doi.org/10.1038/news070402-6.

WHO. (1999). *Fixed-dose combination tablets for the treatment of tuberculosis*. Geneva: World Health Organization. Retrieved from: https://www.who.int/tb/publications/fixed-dosecombination-report/en/.

WHO. (2009). In M. C. Stuart, M. Kouimtzi, & S. R. Hill (Eds.), *WHO model formulary 2008*. Switzerland: World Health Organization Press. Retrieved from: https://apps.who.int/iris/handle/10665/44053.

WHO. (2017a). *Antibacterial agents in clinical development, an analysis of the antibacterial clinical development pipeline, including tuberculosis*. Geneva: World Health Organization. Retrieved from: https://www.who.int/medicines/areas/rational_use/antibacterial_agents_clinical_development/en/.

WHO. (2017b). *Global Antimicrobial Resistance Surveillance System (GLASS) report early implementation 2016–17*. Retrieved from: https://apps.who.int/iris/bitstream/handle/10665/259744/9789241513449-eng.pdf.

WHO. (2018). *Global tuberculosis report 2018*. World Health Organization. Retrieved from: https://www.who.int/tb/publications/global_report/en/.

Winston, D. J., Emmanouilides, C., Kroeber, A., Hindler, J., Bruckner, D. A., Territo, M. C., & Busuttil, R. W. (2000). Quinupristin/dalfopristin therapy for infections due to vancomycin-resistant *Enterococcus faecium*. *Clinical Infectious Diseases, 30*(5), 790–797. https://doi.org/10.1086/313766.

Woerther, P.-L., Andremont, A., & Kantele, A. (2017). Travel-acquired ESBL-producing Enterobacteriaceae: Impact of colonization at individual and community level. *Journal of Travel Medicine, 24*(Suppl. 1), S29–S34. https://doi.org/10.1093/jtm/taw101.

World Economic Forum. (2013). *Global risks 2013 eighth edition an initiative of the risk response network insight report*. Retrieved from: http://reports.weforum.org/global-risks-2013/.

World Health Organization. (2015). *WHO library cataloguing-in-publication data global action, plan on antimicrobial resistance* Retrieved from: https://www.who.int/antimicrobial-resistance/publications/global-action-plan/en/.

Index

Note: Page numbers followed by *f* indicate figures and *t* indicate tables.

D

I

N

V

W

X

Z

FSC
www.fsc.org
MIX
Papier | Fördert
gute Waldnutzung
FSC® C083411

Kolibri 360 GmbH
Ferdinand-Jühlke-Straße 7
99095 Erfurt, Deutschland
produktsicherheit@kolibri360.de